ICU Resource Allocation in the New Millennium

T0234337

David W. Crippen
Editor

ICU Resource Allocation in the New Millennium

Will We Say "No"?

 Springer

Editor
David W. Crippen, M.D., F.C.C.M
Departments of Critical Care Medicine and Neurological Surgery
University of Pittsburgh School of Medicine
Pittsburgh, PA, USA

ISBN 978-1-4614-3865-6   ISBN 978-1-4614-3866-3 (eBook)
DOI 10.1007/978-1-4614-3866-3
Springer New York Heidelberg Dordrecht London

Library of Congress Control Number: 2012939429

Printed on acid-free paper

Springer is part of Springer Science+Business Media (www.springer.com)

This book is dedicated to my clinical colleagues in the Department of Critical Care Medicine at the University of Pittsburgh Medical Center, all of whom have dedicated their formidable talents to caring for the critically ill.

Foreword

Health care is expensive, and spending had increased to cope with growing demand and new, costly technology and treatments. In the USA, health expenditure increased from 15.7% of GDP in 2004 to 17.4% in 2009.[1] With the global recession over the last few years, the rate of increase in health care allocated funds has slowed worldwide, and in health care, as elsewhere, belts are tightening. But, demand for health care is still increasing, new treatments are increasingly expensive, and patient expectations are rising, so how can these aspects be balanced against the decreased funding? At some point, the seemingly endless pot of gold is going to run dry, and if we are to avoid a catastrophe, hard decisions regarding resource allocation need to be made now. Numerous approaches to curtailing costs or reducing expenditure have been proposed. Different countries have adopted various strategies, but health care remains a burden on financing authorities worldwide. More restrictions are likely to be necessary if we continue to be able to provide adequate health care for all.

Intensive care medicine is one of the fastest growing and arguably one of the most expensive areas of medicine. Aging populations, the ability to treat conditions that would previously have been considered fatal, more widespread use of immunosuppressive therapies, and improved organ support techniques all contribute to an ever-increasing demand for ICU services. The percentage of GDP allocated to intensive care is increasing in most developed countries as ICUs continue to expand. Moreover, it is not always a very cost-effective therapeutic option. Many patients treated in ICUs will have large amounts of expensive drugs, interventions, equipment, and nursing hours used on them only to die from multiple organ failure. Finances are limited and health care expenditure cannot continue to expand without boundaries, so how are we to tackle this problem, how best to reduce costs while maintaining the best possible service for all who need it?

---

[1] Data from OECD.

This is the topic of this book, the third in a series of volumes that confronts difficult and thought-provoking topics within the ICU environment from a global perspective. The first in the series[2] discussed how the so-called global village does or should deal with the clinical treatment of ICU patients at the end of life. The second volume[3] focused on how doctors communicate with patients and families about end-of-life issues across international, cultural, and religious boundaries. The present volume covers perhaps the most challenging topic of the three—how can we manage resource allocation, including end-of-life care, in the current era of financial constraints and resource rationing; and, equally, faced with increasing demands by patients and society that "all be done" and the refusal by many to accept that intensive care does not always (or even often) equate to successful care in terms of lives saved, but rather to expensive care. Is it acceptable to ration care in the ICU, refuse ICU admission, speed ICU discharge for patients in whom ongoing treatment is considered futile, in whom there is no longer any hope of recovery, or recovery with a meaningful quality of life? If so, who should make such decisions? Who should define "futile," "meaningful," and "no hope,"…: The patient (in advance?), the doctor, relatives, politicians, the courts, the payer, or some combination of several or all of these? Do we view rationing differently in different areas of the globe? Is it possible to develop a global approach to ICU resource allocation and spending, or is this simply not going to be achievable? What are the ethical implications of such decisions? Rationing is surely supported by the key ethical principles of beneficence, non-maleficence, and distributive justice, but what of the right to patient autonomy? Should individual patients be allowed to use or abuse the system to the potential detriment of others?

These are sensitive issues and many are reluctant to discuss them. But, while it is all too easy to put this debate on the back burner—after all, keeping patients alive is surely our priority as intensivists, and we are busy enough without taking time to debate who should not be treated or who should be left/"allowed" to die—this attitude is no longer acceptable. We need to reflect *now* on how we can best conserve resources, so that appropriate protocols can be put into place, and that in the future there will be sufficient funds for those patients who will actually benefit. We are, after all, the frontline, and we must be involved in decisions about resource allocation to our units, our patients. This is a highly emotive issue, but we cannot afford to be emotional—only with an objective eye can we truly make fair, equitable decisions.

In this book, leading health care experts take us through the current setup around the developed world in terms of resource allocation before discussing what health care reforms are underway or in the pipeline across the globe. A hypothetical health care act, the so-called Fair & Equitable Healthcare Act, is then presented in which

---

[2] Crippen D, Kilcullen J, Kelly D. Three patients: international perspectives on intensive care at the end of life. New York: Kluwer; 2002.

[3] Crippen D. End-of-life communication: a global perspective. New York: Springer; 2008.

what many would see as extreme proposals are put forward to improve ICU efficiency and reduce ICU spending. Ethicists and analysts are invited to explore the acceptability of these suggestions in today's climate and to draw their conclusions on whether we need or will ever be able to say "no" to certain aspects of therapy or, perhaps more controversially, to certain (types of) patients. This book is an imperative and impelling read for all involved in health care and particularly intensive care. By exploring differences in attitudes and approaches among countries and the reasons behind these differences, this volume helps us understand why *intensive* care is all too often *inefficient* care and challenges us to critically reflect on how we can face current and future resource restrictions in our ICU, our country, our continent, and our world in the context of ever-spiralling demand and costs.

Ixelles, Belgium   Jean-Louis Vincent

Introduction

"May you live in interesting times" notes the ancient Chinese curse—certainly relevant to today's progression of events in global health care. In this era of accumulation and management of information via the Internet, we are discovering many similarities within the global medical village, and different evolutionary directions as well [1].

Linked by the Internet, critical care physicians on virtually every continent examined the issue of end-of-life clinical intensive care, sharing global similarities and differences in the 2002 work *Three Patients: International Perspective on Intensive Care at the End of Life*. In 2008 [2], *End-of-Life Communication in the ICU: A Global Perspective* [3] further elaborated how members of the global critical care village approach end-of-life intensive care issues within their regions.

The sweeping new millenium financial crisis is expected to directly affect the delivery of health care worldwide [4]. Clinical patient care will evolve differently in the setting of economic instability and diminished resources.

In the past, some countries, most notably the USA, distributed health care on the basis of consumer demand and political expediency. The real cost of these services was either unknown or not revealed because it didn't matter. Limiting services to one degree or another was political anathema. As long as there was something left at the bottom of the barrel, or the barrel could be refilled in some way, health services were provided without restraint [5, 6].

Currently, there is a clear dichotomy between countries capable of health care service for most or all of their population through some politically acceptable form of rationing [7] and the USA, a country that serves only a portion of its population, at high cost and with manifest inefficiency [8, 9]. The decline in resources underwriting these services will directly affect the entire global village [10]. It will affect the USA more.

Many of those countries managing to provide health care to all their citizens through some variation of national health care are entering troubled times. Many have had troubled times for years before the financial crisis of 2007. Financial resources at the level of central government are now shaky throughout the global village and health care is not necessarily a high treasury priority. Many needy are standing longer in queue and seeing some services disappearing [11].

In the USA, health care facilities are largely owned and managed by the "private sector." However, 60–65% of health insurance is provided by the government in the public sector, including such programs as "Medicare, Medicaid, and the Veterans Health Administration" [12]. The US Census Bureau reports that 50.7 million American citizens, including 9.9 million noncitizens, or 16.7% of the population, were uninsured in 2009, not counting the underinsured (those with some insurance but not enough to cover catastrophic loss) [13].

In 2009, the US federal, state, and local governments, corporations, and individuals together spent $2.5 trillion, or $8,047 per person, on health care [14]. This amount represented 17.3% of the GDP, up from 16.2% in 2008 [15]. The Health and Human Services Department expects that the health share of GDP will continue its historical upward trend, reaching 19.5% of GDP by 2017 [16]. Growth in spending is projected to average 6.7% annually from 2007–2017 [17].

American health care is the most expensive in the world and the Congressional Budget Office reported that "about half of all growth in health care spending in the past several decades was associated with changes in medical care made possible by advances in technology" [18]. But the USA ranks poorly in preventable mortality, below Canada, Ireland, and Portugal [19]. A 2007 study found that 62.1% of all personal bankruptcies involved high medical expenses [20].

Those citizens not indemnified by some means are reduced to using emergency departments (EDs), where by law they must be seen regardless of their ability to pay. Their care is underwritten by the facility—frequently at a great financial loss [21]. These individuals tend to delay treatment, with the common result being escalation to emergencies and consequent admission to a hospital through an ED. This care is not fully reimbursed and the facility is free to institute legal measures through collection agencies to recoup expended resources. An American health care indemnifier for less advantaged individuals (Medicaid) requires reduction of a family to penury before a family member becomes eligible for care. An unexpected medical emergency has the potential for ruining a family's finances permanently [22].

Most, if not all, countries in the global medical village have accepted the political and social reality of saying no as a tool for health care resource allocation. For most of their history, countries with some form of national health care have, out of necessity, said no to goods and services with poor cost-effectiveness ratios. The citizens in these countries have become accustomed to this kind of allocation and have accepted it as a social reality. Americans have most assuredly not accepted being told no for health care demand. Some have become experts at gaming the system to ensure they get as much as they want for as long as they want [23]. This attitude has been buttressed by a governmental indemnity system (Medicare) that contains ambiguous language about cost and length of hospital stay.

This volume examines the mechanisms of how the global critical care village has said no in the past and how it will probably continue to do so. That said, American health care indemnifiers are famously reluctant to say no under any circumstances, no matter how compelling [24–26]. Of particular interest is how the USA, with the most expensive health care delivery plan in the world, will react to the current financial meltdown, especially to the inevitability as the rest of the global village has done. If for no other reason than progressive lack of resource affordability and reduction in credit to finance it, saying no will be inevitable in the near future.

Here is a representative (though fictitious) example of the challenges we face in the allocation of scarce medical resources in a system where saying no is not practiced:

A 56-year-old man develops a sudden headache and then becomes unconscious. He is evaluated in an ED and is found to have a massive subarachnoid hemorrhage. His physical examination is compatible with near brain death, but he spontaneously ventilates. He is evaluated by a neurosurgeon and is admitted to the neurology intensive care unit (ICU) after emergency endovascular coiling of the aneurysm. After 72 hours of aggressive care in the ICU, his condition is unchanged. His wife is told that nothing further can be done, the prognosis is hopeless, and comfort measures should be considered.

The patient's wife states that doctors are sometimes wrong and she is content to wait indefinitely for a miracle. She consults various Internet support sites, some offering more optimistic predictions not necessarily compatible with her situation. The palliative care team is consulted 5 days post admission and is told that their services aren't necessary, as she fully expects the patient to improve in time. The hospital ethics committee is consulted at 7 days, and their opinion is that this is not an ethical issue, because the patient's wife is competent to make any decision she desires.

Ten days post admission, the patient's condition remains unchanged. He has been treated for ventilator-associated pneumonia and pyelonephritis. At 14 days, the patient is still in the ICU and his condition is still unchanged. The subject of tracheostomy, enteral feeding, and transfer to a skilled nursing facility is broached.

The unit social worker informs the wife that the family's limited insurance policy will not pay for much more care and she will need to apply for state-funded medical assistance. That agency will require that she have no monetary resources other than a house and some other bare essentials. In essence, the family will be recipients of welfare to maintain the patient in his current state. The wife states that she will not consent to comfort measures because she anticipates the patient will start improving soon, and she continues to desire to have "everything done."

At day 18, the patient develops fulminating sepsis while receiving two antibiotics, and requires three vasopressors to maintain organ perfusion pressure. He then develops acute renal failure and ultimately goes into cardiac arrest and cannot be resuscitated.

One might wonder why caring surrogates would accept and act on such a meager potential for a good outcome. Several reasons have been offered:

- The Internet has become the new resource for families seeking information about a disease process. Looking for more optimistic data than that offered by their physician, they frequently find it, in the form of poorly authenticated opinions from pseudo-experts at seemingly authoritative sites [27].

- The popular media frequently presents stories of patients who have awakened after years of coma. The conditions of most of these individuals have been embellished to generate public interest, and subsequent investigators often cannot find these patients, or if they do, the patients are rarely as advertised [28].
- Once a patient goes on life support, decisions by his or her surrogate become more emotional and less rational. In some situations, surrogates make decisions embracing a long-shot cure rather than decisions that will result in inevitable death, believing that any chance for life is better than no chance [29].
- Cadavers on a morgue slab look convincingly dead. Moribund patients on life support look relatively comfortable, if poorly responsive or sedated. If the patient can be maintained comfortably for long enough, the surrogate thinks a cure may become possible [30].
- It is difficult to invoke the concept of futility in situations where vital signs can be interpreted as life. Any medical treatment capable of sustaining hemodynamics, ventilation, and metabolism is not technically futile. In those circumstances, virtually any treatment is fair game, even if it will do nothing to revitalize the patient [31].

It could be argued that this case example is an extreme case, not representative of mainstream intensive care. Most ICU patients' surrogates are much more reasonable. If such extreme cases are only a very small part of the whole, saying no might not be worth the friction and bad publicity for the hospital. However, many ICU physicians are reporting that situations like these are becoming more common.

The purpose of this volume is to pursue the volatile issue of doing everything versus saying no in situations similar to the one described above.

Given that doctors are sometimes wrong in their prognostications, what are the practical, clinical aspects of saying no? Does the individual have the right to ignore expert medical opinion in favor of the extreme long shot? Does a surrogate have an absolute right to speak for an incompetent patient if no wishes are definitely known? How many moribund patients similar the one described above are we willing to maintain in expensive units to ensure that one unexpected survivor doesn't surface? How will we weather the public's reaction to being told "no" in a world that rejects physician paternalism?

Tapping the potential of real-time communication via the Internet and drawing from the insights of critical care clinicians (both academic and proprietary), ethicists, legal experts, nurses, and others around the globe, this volume examines where we've been and where we're going in the relatively new world of saying no.

Pittsburgh, PA   David W. Crippen

References

1. Bleeding edge: the business of health care in the new century. New York: Aspen; 1998.
2. Crippen D, Kilcullen J, Kelly D. Three patients: international perspective on intensive care at the end of life. New York: Kluwer; 2002.
3. Crippen D. End-of-life communication in the ICU: a global perspective. New York: Springer; 2008.
4. Fronstin P. The impact of the 2007–2009 recession on workers' health coverage. EBRI Issue Brief. 2011;(356):1, 4–19.
5. Angell M: The case of helga. N Engl J Med. 1991;325:511–12.
6. Wilkinson DJ, Savulescu J. Knowing when to stop: futility in the ICU. Curr Opin Anaesthesiol. 2011;24(2):160–5.
7. Moreira T. Health care rationing in an age of uncertainty: a conceptual model. Soc Sci Med. 2011;72(8):1333–41.
8. The Oregon rationing plan: inspired or misguided? Healthweek. 21 May 1990.
9. Singer P: Why we must ration health care. New York Times. 15 July 2009.
10. Hurst SA, Forde R, Reiter-Theil S, Slowther AM, Perrier A, Pegoraro R, Danis M. Physicians' views on resource availability and equity in four European health care systems. BMC Health Serv Res. 2007;7:137.
11. Wade R. The first-world debt crisis of 2007–2010 in global perspective. Challenge 2008;51(4):23–54. http://www.challengemagazine.com/extra/023_054.pdf.
12. DeNavas C, Proctor B, Smith J. Income, poverty, and health insurance coverage in the United States: 2009. U.S. Department of Commerce Economics and Statistics Administration U.S. CENSUS BUREAU. Sept 2010. http://www.census.gov/prod/2010pubs/p60-238.pdf.
14. U.S. Department of Health and Human Services. CMS Financial Report. Fiscal Year 2007. The Centers for Medicare & Medicaid Services at a Glance. Washington, DC. http://www.cms.hhs.gov/NationalHealthExpendData/Downloads/proj2007.
15. Wolf R. Number of uninsured Americans rises to 50.7 million. USA Today, 17 Sept 2010 http://www.usatoday.com/news/nation/2010-09-17-uninsured17_ST_N.htm.
16. The World Health Organization. The World Health Report 2000. Annex Table 1 Health system attainment and performance in all Member States, ranked by eight measures, estimates for 1997. http://www.who.int/whr/2000/en/annex01_en.pdf.
17. http://en.wikipedia.org/wiki/Health_care_in_the_United_States.
18. Congress of the United States Congressional Budget Office. Technological Changes and the Growth of Health Care Spending. Jan 2008 http://www.cbo.gov/ftpdocs/89xx/doc8947/01-31-TechHealth.pdf.
19. Department of Economic and Social Affairs. Population Division. World population prospects the 2006 revision. New York, NY: The United Nations. http://www.un.org/esa/population/publications/wpp2006/WPP2006_Highlights_rev.pdf.
20. Parker-Pope T. Medical bills cause most bankruptcies. The New York Times. June 4, 2009. http://well.blogs.nytimes.com/2009/06/04/medical-bills-cause-most-bankruptcies/.
21. Smulowitz PB, Lipton R, Wharam JF, Adelman L, Weiner SG, Burke L, Baugh CW, Schuur JD, Liu SW, McGrath ME, Liu B, Sayeh A, Burke MC, Pope JH, Landon BE. Emergency department use by the uninsured after health care reform in Massachusetts. Intern Emerg Med. 2009;4(6):501–6. Epub 2009 Sep 24.
22. Himmelstein DU, Thorne D, Warren E, Woolhandler S. Medical bankruptcy in the United States, 2007: results of a national study. Am J Med. 2009;122:741–6.
23. Crippen D. Medical treatment for the terminally ill: the risk of unacceptable badness. Crit Care. 2005;9(4):317–8.
24. Why saying no to patients in the United States is so hard. N Engl J Med. 1986;314:1380–3.
25. Callahan D. Setting limits: medical goals in an aging society. New York: Simin & Schuster; 1987.
26. Learning to say no. N Engl J Med. 1984;311:1569–72.

27. Crippen D. Critical care and the internet. A clinician's perspective. Crit Care Clin. 1999;15(3):605–14, vii.
28. Golby A, McGuire D, Bayne L. Unexpected recovery from anoxic-ischemic coma. Neurology. 1995;45(8):1629–30.
29. Annas GJ. Asking the courts to set the standard of emergency care—the case of baby K. N Engl J Med. 1994;330:1542–45.
30. Crippen D. Brain failure and brain death. ACS Surgery: Principles and Practice. WebMD, 2005.
31. Crippen DW, Whetstine LM. Ethics review: dark angels—the problem of death in intensive care. Crit Care. 2007;11(1):202.

Contents

Contributors

Monica Viegas Andrade, Ph.D. Centro de Desenvolvimento e Planejamento Regional – CEDEPLAR, Faculdade de Ciências Econômicas da Universidade Federal de Minas Gerais, Belo Horizonte, Brazil

Derek C. Angus, M.D., M.P.H., F.R.C.P. Department of Critical Care Medicine, University of Pittsburgh School of Medicine, Pittsburgh, PA, USA

Anna M. Batchelor, M.B., Ch.B., F.R.C.A., F.F.I.C.M. Department of Peri-operative and Critical Care Medicine, Newcastle upon Tyne NHS Trust, Newcastle, UK

Frank H. Bosch, M.D., Ph.D. Department of Internal Medicine, Rijnstate Hospital, Arnhem, The Netherlands

Timothy G. Buchman, Ph.D., M.D. Department of Surgery, Emory University School of Medicine, Emory Center for Critical Care, Emory Healthcare, Atlanta, GA, USA

Lynn Barkley Burnett, M.D., Ed.D. Department of Medical Ethics, Community Medical Centers, Fresno, CA, USA.

Departments of Emergency Medicine & Forensic Pathology, Touro University, California College of Osteopathic Medicine, Vallejo, CA, USA

Richard Burrows, M.B., B.Ch., F.C.A. (SA) (Critical Care) (SA) Private Practice - Bon Secours Hospital, Galway, Ireland

Frederico Bruzzi de Carvalho, M.D. Intensive Care Unit, Hospital Eduardo de Menezes, Fundação Hospitalar do Estado de Minas Gerais–FHEMIG, Belo Horizonte, CEP, Brazil

Donald W. Chalfin, M.D., M.S., F.C.C.M. Ibis Biosciences, Carlsbad, CA, USA

Rubens Costa-Filho, M.D., F.C.C.P., M.B.A. Critical Care Centre, Hospital Pró Cardíaco, Rio De Janeiro, Brazil

David W. Crippen, M.D., F.C.C.M. Departments of Critical Care Medicine and Neurological Surgery, University of Pittsburgh School of Medicine, Pittsburgh, PA, USA

J.V. Divatia, M.D. Department of Anaesthesia, Critical Care and Pain, Tata Memorial Hospital, Mumbai, India

Christopher James Doig, M.D., M.Sc., F.R.C.P.C. Department of Community Health Sciences, University of Calgary, Calgary, AB, Canada

Multisystem Intensive Care Unit, Department of Critical Care Medicine, Foothills Medical Centre, Alberta Health Services, Calgary, Canada

Timothy C. Hardcastle, M.B., Ch.B. (Stell) M. Med. (Chir) (Stell), F.C.S. (SA) Department of Surgery, Nelson R Mandela School of Medicine, University of KwaZulu-Natal, Ethekwini-Durban, KwaZulu-Natal, South Africa

R. Eric Hodgson, F.S.C. (SA) (Crit Care) Department of Anaesthesia and Critical Care, Addington Hospital, EThekwini-Durban, KwaZulu-Natal, South Africa

Department of Anaesthesia, Critical Care, and Pain Management, Nelson R Mandela School of Medicine, University of KwaZulu-Natal, Ethekwini-Durban, KwaZulu-Natal, South Africa

Ross Hofmeyr, M.B.Ch.B. (Stell), Dip.P.E.C. (SA), D.A. (SA) Department of Anaesthesia and Critical Care, GF Jooste Trauma and Emergency Hospital, Cape Town, South Africa

Steven M. Hollenberg, M.D. Department of Medicine, Cooper Medical School of Rowan University, Camden, NJ, USA

John W. Hoyt, M.D. Department of Critical Care, Pittsburgh Critical Care Associates, Inc., Critical Care Medicine, University of Pittsburgh School of Medicine, Pittsburgh, PA, USA

Farhad Kapadia, M.D., F.R.C.P. Department of Medicine and Critical Care, P.D. Hinduja Hospital & MRC, Mumbai, India

Thomas Kerz, M.D. Klinik und Poliklinik für Neurochirurgie, Universitätsmedizin der Johannes Gutenberg Universität, Mainz, Germany

Jack K. Kilcullen, M.D., J.D., M.P.H. Department of Critical Care Medicine, Virginia Hospital Center, Arlington, VA, USA

W. Andrew Kofke, M.D., M.B.A., F.C.C.M. Department of Anesthesiology and Critical Care, University of Pennsylvania, Philadelphia, PA, USA

Michael A. Kuiper, M.D., Ph.D., F.C.C.M. Department of Intensive Care Medicine, Medical Center Leeuwarden, Leeuwarden, The Netherlands

Atul P. Kulkarni, M.D. Tata Memorial Cancer Hospital, Mumbai, India

Marco Luchetti, M.D. Department of Anaesthesia, Intensive Care, & Pain Management, A. Manzoni General Hospital, Lecco, Italy

Giuseppe A. Marraro, M.D. Department of Anesthesia and Intensive Care, Fatebenefratelli and Ophthalmiatric General Hospital, University of Milan, Milan, Italy

Mark Mazer, M.D. Department of Pulmonary, Critical Care and Sleep Medicine, Pitt County Memorial Hospital, East Carolina University, Greenville, NC, USA

Álvaro Réa Neto, M.D., F.C.C.P. Internal Medicine Department, Universidade Federal do Paraná–UFPR, Centro de Estudos e Pesquisas em Terapia Intensiva – CEPETI, Curitiba, Brazil

Michael A. Rie, M.D. Department of Anesthesiology, University of Kentucky College of Medicine, Lexington, KY, USA

Gilbert Ross, J.D. Sussman, Selig & Ross, Chicago, IL, USA

Leslie P. Scheunemann, M.D., M.P.H. Division of Pulmonary, Allergy, and Critical Care Medicine, University of Pittsburgh Medical Center, Pittsburgh, PA, USA

Eran Segal, M.D. Department of Anesthesiology, Intensive Care and Pain Medicine, Assuta Medical Centers, Israel

Israeli Society of Critical Care Medicine, Israel

Ian M. Seppelt, M.B.B.S., F.C.I.C.M. Discipline of Intensive Care, Sydney Medical School, University of Sydney, Sydney, NSW, Australia

Department of Intensive Care Medicine, Nepean Hospital, Penrith, NSW, Australia

Rodrigo Ferreira Simões, Ph.D. Centro de Desenvolvimento e Planejamento Regional – CEDEPLAR, Faculdade de Ciências Econômicas da Universidade Federal de Minas Gerais, Belo Horizonte, Brazil

Melanie S. Smith, R.N., M.S.N., C.C.R.N. Department of Neurovascular Intensive Care Unit, UPMC Presbyterian University Hospital, Pittsburgh, PA, USA

Charles L. Sprung, M.D. Department of Anesthesiology and Critical Care Medicine, General Intensive Care Unit, Hadassah Hebrew University Medical Center, Jerusalem, Israel

Stephen Streat, B.Sc., M.B., Ch.B., F.R.A.C.P. Department of Critical Care Medicine, Auckland City Hospital, Auckland, New Zealand

Andrew Thorniley, M.D. Department of Anaesthetics, The Hillingdon Hospital NHS Trust, Uxbridge, Greater London, UK

Jean-Louis Vincent, M.D., Ph.D., F.C.C.M. Department of Intensive Care, Université Libre de Bruxelles, Brussels, Belgium

Randy S. Wax, M.D., M.Ed., F.C.C.M. Department of Medicine, Queen's University, Kingston, ON, Canada

Department of Medicine, University of Toronto, Toronto, ON, Canada

Department of Emergency Medicine and Critical Care, Lakeridge Health, Oshawa, ON, Canada

Leslie M. Whetstine, Ph.D. Division of Humanities, Walsh University, North Canton, OH, USA

Douglas B. White, M.D., M.A.S. Program on Ethics and Decision Making in Critical Illness, Department of Critical Care Medicine, University of Pittsburgh Medical Center, Pittsburgh, PA, USA

Center for Bioethics and Health Law, University of Pittsburgh, Pittsburgh, PA, USA

Brian Wowk, Ph.D. 21st Century Medicine, Inc., Fontana, CA, USA

Part I
Contrasts in Global Health Care Resource Allocation

Chapter 1
Australia: Where Have We Been?

Ian M. Seppelt

Background

The Australian medical system in the twentieth century was to a large extent inherited from the British system prior to the advent of the National Health Service (NHS). Health Care had to be paid for, and about 75% of the population had private health insurance to help cover this. In the period since the First World War the system was such that the government paid about a third of health care costs, insurance for whoever had it paid another third and the individual paid the remainder. Those truly unable to pay were treated free, by both doctors and hospitals, and particularly by the charitable institutions and benevolent funds. Up to 17% of the population was uninsured but "elf funded", and a number of these were "caught in the gap" and ended up with the least opportunity for health care access [1].

Private doctors held "Honorary" appointments at public hospitals, and all the major hospitals were public, though many had private wings or other similar facilities. An "Honorary" [as in "Honorary Physician" or "Honorary Surgeon"] had the right to admit and treat private patients in the public hospital, in return for treating public patients for free. Most Honoraries took this responsibility seriously, and for the most part health care resources were managed fairly, in this era prior to technological medicine. But at the same time community expectations were often not high, and there were certainly individuals who missed out on health care as a consequence. In the 1930s seeing a specialist was almost unheard of, but with rapid advances in medicine and developing technology it became apparent that cost could dramatically blow out. World War II resulted in the reshaping of public policy in the direction of the "welfare state", with access to health care as an important

I.M. Seppelt, M.B.B.S., F.C.I.C.M. (✉)
Discipline of Intensive Care, Sydney Medical School,
University of Sydney, Sydney, NSW, Australia

Department of Intensive Care Medicine, Nepean Hospital, Penrith, NSW, Australia
e-mail: seppelt@med.usyd.edu.au

component, yet by the early 1970s private patients had risen from 20 up to 50% of bed occupancy in public hospitals.

In 1967 two health economists, Richard Scotton and John Deeble, first proposed a system of universal health insurance [2], which was taken up by the then Federal opposition party, as described by Prime Minister-to-be Gough Whitlam in November 1972:

> A federal Labor government will introduce a universal health insurance scheme. It will be administered by a single health fund. Contributions will be paid according to taxable income. An estimated 350,000 Australian families will pay nothing. Four out of five will pay less than their contributions to the existing scheme. Hospital care will be paid for completely by the fund in whatever ward the patient's doctor advises. The fund will pay the full cost of medical treatment if doctors choose to bill the fund directly, or refund 85 per cent of fees if the patient pays those fees himself.

The 1973 Health Insurance Act introduced this national mandatory health insurance scheme under the name Medibank. Over the ensuing decade, and two changes in government, the scheme was altered, and finally stabilized under the name Medicare in 1983 under the government of Prime Minister Bob Hawke. There was great concern about the changing nature of medical practice, from a relationship between doctor and patient to a relationship between patient and government. While initially Medibank payments to doctors were fair (and indeed generous) they have not kept up with the consumer price index since and Medicare rebates are now substantially lower than what many consider sufficient for a tolerable income, leading to the increasing prevalence of "Gap Payments" by patients, and also the advent of entrepreneurial rapid turnover medical centres which rely on volume to compensate for the inadequate rebates. With the availability of "free" (or heavily subsidized) health the number of Medicare services per patient doubled in a decade and a half, and cost blowouts in pathology and radiology services, as well as other technology and expensive therapies, have led to an appreciation that at current levels of expenditure [estimated 8.9% of Gross Domestic Product] the current health system is both under-resourced and unsustainable.

Medicare is funded by a mandatory levy (currently 1.5% of taxable income), plus an extra 1% surcharge for those with high income who do not have private health insurance in addition. The basic Medicare pays for primary care and public hospital treatment for the entire population. Medicare pays a rebate on specialist consultations, generally only a proportion of the total fee unless the specialist chooses to accept "rebate only". The are no restrictions on what services are paid for in the public hospital, but for elective admissions (mainly elective surgery) rationing is achieved by waiting, at times up to a year for non-urgent surgery, and patients are treated by a hospital-appointed doctor, rather than necessarily the doctor of their choice. Public hospital waiting lists are therefore a political football, as might be predicted.

A parallel private hospital system exists outside Medicare, for privately insured or self-funded patients. Private hospitals are predominantly for elective surgery, although an increasing amount of non-surgical and critical care services are being provided in the private sector. Private patients can also elect to be treated privately in public hospitals, and indeed it is in the hospital's interest to have such patients as services can then be billed to the insurance company rather than come directly from the hospital's operating budget. To further complicate the system, hospital specialists can be salaried as staff specialists, university-employed academics or private practitioners attending as "visiting medical officers", and remuneration and conditions for the three groups are quite different. A final layer of complexity is the demarcation between Federal Government and the individual states. The states run the public hospital system (using taxpayers' money transferred from the federal treasury) whereas outpatient services are funded directly from the Federal Government via Medicare. A large opportunity therefore exists for cost shifting, to transfer expenses to the federal government after hospital discharge rather than bearing the same expense on the hospital budget.

Over 30 years the system has been increasingly bureaucratized, and the rise of the health bureaucrat has paralleled the increasing complexity of medical services. The editor of the *Medical Journal of Australia* pointed out in December 2008 the following scenario:

Two patients limp into two different medical clinics with the same complaint. Both have trouble walking and appear to require hip replacement. The first patient is examined within the hour, x-rayed the same day, and has a time booked for surgery the following week. The second sees his family doctor after waiting 3 weeks for an appointment, then waits 8 weeks to see a specialist, has an x-ray which isn't reviewed for another week, and is finally scheduled for surgery a month later. Why the different treatment for the two patients?

The first is a golden retriever. The second is a senior citizen [3].

Having made that point about delays and bureaucracy, the delays primarily affect elective or non-urgent surgery, and overall the standard of care for these patients is still excellent, if delayed, and achieved at a total cost of 8.9% of GDP.

Most emergency care takes place in public hospitals. All major university teaching hospitals are public institutions [with one recent exception, when a totally private academic teaching hospital was opened by Macquarie University in Sydney in 2010], and the big public hospitals are where the major emergency departments and intensive care units are located. In this context the distinction between public and private becomes much less relevant as patients are triaged according to need, and at least at the most critically ill end of the spectrum insurance status becomes irrelevant as all are treated equally according to medical need.

The Beginnings of Intensive Care in Australia

The first intensive care units developed out of post-operative recovery rooms and infectious diseases wards, as they had elsewhere in the world. The Artificial Respiration Unit at Fairfield Infectious Diseases Hospital in Melbourne (established using Both tank ventilators in 1937) was using tracheostomy, negative pressure ventilation and a mechanical cough by 1957 and the first general intensive care units were at Prince Henry Hospital in Sydney and St Vincent's Hospital in Melbourne, both opened in 1961 [4]. From the beginning these units were run by dedicated and unusual individuals who were devoted to the concept of intensive care rather than any primary specialty, and the concept of intensive care and intensivists became accepted with demonstration that the intensivist, and intensive care nurse, could provide a higher level of care than a referring doctor could provide [5]. It was established early that intensivists had the right to care for patients by being present at the bedside, by their skills and by communicating treatment plans and treatment goals with both primary referring physicians and patients' families. As a consequence many of the "turf wars" seen elsewhere in the world have been avoided and from the beginning these ICUs have operated under what is sometimes now referred to as a "closed model" of intensive care, where the intensive care service coordinates management of all patients and exercise a duty of triage, to determine which patients were admitted to and discharged from the intensive care unit. A cynic would also point out that intensive care specialists were salaried hospital-employed doctors and did not alter the income of the primary referring physicians when they also became involved—this has caused problems elsewhere in the world.

The Professionalization of Intensive Care

The first formal intensive care training programs in the world were in Australia and New Zealand. The Faculty of Anaesthetists, Royal Australasian College of Surgeons, commenced a 4-year training and examination system in 1976, with the first final examination held in 1979 [6]. The Royal Australasian College of Physicians also established an intensive care training program in 1976. These two programs were merged in 2001 as the Joint Faculty of Intensive Care Medicine, and finally the independent College of Intensive Care Medicine of Australia and New Zealand (CICM) was established in 2010. The National Specialist Qualification Advisory Committee recognized intensive care medicine as a sectional specialty of both internal medicine and anaesthesia in 1980, and finally listed intensive care medicine as a recognized medical specialty in 2002.

The role of the intensive care nurse is crucial in any intensive care unit—no matter how clever the physicians are, poor-quality nursing will lead to poor outcomes in any intensive care unit. All unstable or mechanically ventilated patients are nursed 1:1 [one nurse to one patient] or more at all times, while more stable non-ventilated patients may be nursed 1:2. These ratios are specified in the document "Minimum Standards for Intensive Care Units" published by CICM [7], and while the document was written initially as a minimum standard for intensive care units to be accredited as training units, it has been tested in law and upheld as an enforceable standard by the Supreme Court of NSW [8].

Current Intensive Care Resources in Australia

The Australian and New Zealand Intensive Care Society (ANZICS) established an ICU registry in 1993 to describe the profession in ANZ, and this is now managed by the ANZICS Centre for Outcomes and Resource Evaluation (CORE), which also manages the adult and paediatric intensive care databases with data on over 900,000 discrete intensive care admissions. The Critical Care Resources section conducts an annual bi-national survey, with data from the 2005/2006 year published most recently, in 2010 [9].

There are 100 public sector and 51 private sector ICUs in Australia, with 1,990 physical beds (1,452 public, 538 private), 1,794 available beds and 1,314 ventilator beds (934 public and 380 private). Most ICUs were "general" (56.3%) with a further 36.4% combined ICU/CCU/HDU (generally smaller units), with 5.3% paediatric ICU (PICU) and only 1.9% specialty ICUs.

Based on most recent population data from the Australian Bureau of Statistics, there are 8.9 beds/100,000 population including 6.5 ventilator beds/100,000 population. There is some geographical variation, with 12.2 beds (8.4 ventilated beds)/100,000 in the state of South Australia down to 5.9 beds (4.6 ventilated beds)/100,000 in the geographically isolated state of Western Australia. Expressed as proportion of hospital beds, 1.4% of Australian hospital beds are designated ICU.

It is difficult to draw international comparisons and overseas data are not necessarily comparable. In 2004 there were 87,400 ICU beds in the USA [13.4% of hospital beds] with 29.2 ICU beds/100,000 population [10] while in England in 2006 there were 3,775 ICU beds (2.1% of hospital beds, 7.4 beds/100,000 population) and in Scotland in 2005 there were 147 beds (0.8% of hospital beds, 2.9 beds/100,000 population) [9]. European data are more difficult to ascertain, but in Germany and Austria it appears that there are as many as 23 beds/100,000 population [11].

In 2005/2006 there were 104,444 ICU admissions in Australia, including 21,365 elective and 43,444 emergency admissions from 79 public units with data and 20,006 elective and 6,394 emergency patients from 35 private units. As can be seen, the casemix between public and private ICU is quite different, with public ICUs taking predominantly emergency (surgical or non-surgical) admissions while private ICUs predominantly admitting elective surgery. Private cardiac surgery, in particular, is the reason for the growth in private intensive care units in the last 20 years and the ongoing economic rationale for their existence. Overall ICU bed activity was ~80% in the public sector and 20% private. Of available data 7,362 patients died from a total of 132,943 ICU admissions in Australia (6/8% mortality in public ICU and 2.8% mortality in private).

There were 306 full time equivalent (FTE) intensivists working in Australia in 2005/2006, and another 40 FTE specialists without a formal intensive care qualification [overall 1.13 FTE specialists/1,000 patient days in public, 1.39 FTE specialists/100 patient days in private]. A total of 9,718 registered nurses (RNs) are employed in Australian intensive care (7,435 FTEs, 5,517 public and 1,114 private) with an average 4.4 FTE RNs employed per available ICU beds or 18.4 FTE RN/1,000 patient days in the public ICUs.

Rationing

As can be seen, with ~80% of the intensive care bed activity in Australia occurring in the (chronically underfunded and under-resourced!) public sector, and bed occupancies often near 100%, some form of rationing is inevitable. Historically this has been a (deliberately) informal process, driven by a combination of primary physicians choosing not to refer patients to ICU, and the intensivist acting as gatekeeper. An increasing amount of time is now spent by the intensive care specialist assessing patients who clearly cannot benefit from intensive care as outside services become progressively more uncomfortable in dealing with end-of-life issues. The triage rule this intensivist teaches trainees is to ask a simple question about every referral: "Is there something that intensive care can offer, that the general ward cannot offer, that has a realistic chance of changing this patient's outcome?" If the answer is "Yes", then the patient should probably be admitted to ICU. If the answer is "No" then one of the two scenarios apply—either the patient is fine and likely to do well in a general ward or the patient is unsalvageable and should not be admitted to ICU for inappropriate ("futile") therapy that will not change the ultimate. In reality the situation is often not so clear-cut and it is often appropriate to admit a patient after much discussion and with clear limitations on what treatment will be offered.

A different group is patients who are likely to benefit from intensive care admission, but cannot be physically accommodated in a full ICU. Sometimes the rate-limiting step is availability of nurses and that can occasionally be corrected. Increasingly frequently the problem is the so-called exit block, the patient who no longer needs to remain in ICU but cannot be discharged as there is no suitable ward bed available. It is unethical to refuse admission to an appropriate patient who can benefit from intensive care, when the bed is blocked by a patient who can benefit no further from intensive care but cannot be discharged out.

Where there are genuinely no beds, or there is a need for advanced services not available at the primary hospital, patients are transferred to other hospitals. Precise mechanisms vary from state to state (and some super-specialty areas are coordinated nationally, such as burns) but all states offer one or sometimes multiple medical retrieval services staffed by senior physicians who can safely transfer critically ill patients between hospitals using a variety of vehicles (road, rotary and fixed wing aircraft). Every state has a version of a centralized "Medical Retrieval Unit" to coordinate such activity. One principle of medical retrieval is that a patient should only be exposed to the risks of interhospital transport in order to access services otherwise unavailable (a "higher level of care") [12] but in reality many transfers are arranged as part of a bed shuffling exercise, and frustratingly, some are merely "change of postal code of death".

2009/H1N1 Influenza Pandemic

The 2009/H1N1 Influenza Pandemic is a good example of the Australian intensive care system when stressed but not overwhelmed. Using the resources and good will of the ANZICS community and particularly the ANZICS Clinical Trials Group (CTG) it was possible to rapidly collect data on *all* patients admitted to intensive care in Australia and New Zealand with 2009/H1N1 influenza [13]. At the peak of the 12-week pandemic there were 7.4 patients per million population with proven infection occupying ICU beds, with the same number again with suspected but unproven influenza. While the pandemic seriously stretched resources it was manageable with a combination of cancelling elective surgery and "flexing up" ICU bed capacity. No patient who would otherwise have been admitted to ICU was refused ICU because of the pandemic. The rate-limiting step was not the availability of ventilators or other equipment, nor the availability of physical beds, but the availability of ICU nurses. The high standards of ICU nursing were not changed during the pandemic but highly skilled ICU nurses are not rapidly reproduced, and this constraint has not been adequately considered during pandemic planning by public health personnel without sufficient intensive care input. The cost of an ICU bed is estimated to be AU$5,135, and the total ICU cost of the 2009 pandemic was at least AU$65,000,000 [14].

While the Australian intensive care system coped under "usual rules" in 2009, it would not cope when faced with a more virulent pandemic. Under those circumstances either more rigorous triage rules will have to be applied, by a senior intensivist with full support of management, or nursing models will have to be changed, such that one qualified intensive care nurse can supervise the management of more than one critically ill patient, or both.

Conclusion

The Australian health system is publicly funded through a mandatory levy, with optional private health care, and overall the health system consumes 8.9% of GDP. The intensive care system in 2010 is a predominantly public but mature system, providing 6.5 ventilated beds per 100,000 population, staffed predominantly by qualified intensive care specialists and trained intensive care nurses rostered 1:1 per patient. Intensive care physicians are trained and expected to act as "gatekeeper" to the system, screening all referrals to the ICU and only admitting those patients who are likely to benefit from admission. The system is underfunded and at times stressed. There is a developing private intensive care system working in parallel, with a predominantly elective surgery casemix. Organizations such as ANZICS CORE and the ANZICS CTG allow comprehensive data collection and an accurate picture of "the state of intensive care" but caution is needed when drawing comparisons with other countries. Outcomes appear as good as those seen anywhere else in the world.

References

1. Khadra M. Terminal Decline: a surgeon's diagnosis of the Australian Health-Care System. Sydney, NSW: Random House; 2010.
2. Scotton RB. Medibank: from conception to delivery and beyond. Med J Aust. 2000;173: 9–11.
3. Van der Weyden MB. Doctor displacement: a political agenda or a health care imperative? Med J Aust. 2008;189:608–9.
4. Wiles V, Daffurn K. There's a Bird in my hand and a Bear by the bed – I must be in ICU. The pivotal years of Australian critical care nursing. Carlton: Australian College of Critical Care Nurses; 2002.
5. Dobb GJ. Intensive care in Australia and New Zealand. No nonsense "down under". Crit Care Clin. 1997;13:299–316.
6. Judson JA, Fisher MM. Intensive care in Australia and New Zealand. Crit Care Clin. 2006;22:407–24.
7. College of Intensive Care Medicine of Australia and New Zealand, IC-1 Minimum Standards for Intensive Care Units, revised 2011. http://www.cicm.org.au/policydocs.php.
8. NSW Supreme Court, Sherry v Australasian Conference Association (trading as Sydney Adventist Hospital) & 3 Ors [2006] NSWSC 75.
9. Martin JM, Hart GK, Hicks P. A unique snapshot of intensive care resources in Australia and New Zealand. Anaesth Intensive Care. 2010;38:149–58.
10. Halpern NA, Pastores SM, Greenstein RJ. Critical care medicine in the United States 1985-2000: an analysis of bed numbers, use and costs. Crit Care Med. 2004;32:1254–9.
11. Offenstadt G, Moreno R, Palomar M, Gullo A. Intensive care medicine in Europe. Crit Care Clin. 2006;22:425–32.
12. College of Intensive Care Medicine of Australia and New Zealand, IC-10 Minimum Standards for Transport of Critically Ill Patients, revised 2010. http://www.cicm.org.au/policydocs.php.
13. Webb SAR, Pettilä V, Seppelt IM, et al. The ANZIC influenza investigators, critical care services and 2009 H1N1 influenza in Australia and New Zealand. N Engl J Med. 2009;361: 1925–34.
14. Higgins AM, Pettilä V, Harris AH, Bailey M, Lipman J. Seppelt IM and Webb SAR on behalf of the ANZIC Influenza investigators, The Critical Care Costs of the Influenza A/H1N1 2009 Pandemic in Australia and New Zealand. Anaesth Intensive Care. 2011;39:384–91.

Chapter 2
Brazil: Where Have We Been?

**Frederico Bruzzi de Carvalho, Álvaro Réa Neto,
Rodrigo Ferreira Simões, and Monica Viegas Andrade**

Brazil is the world's fifth largest country, both by geographical area and by population. It is the only Portuguese-speaking country in the Americas and the largest lusophone country in the world. The Brazilian economy is the world's eighth largest economy by nominal gross domestic product (GDP) and the ninth largest by purchasing power parity. The population of Brazil is approximately 190 million and 83.75% of the population defined as urban. The population is heavily concentrated in the Southeastern (79.8 million inhabitants) and Northeastern (53.5 million inhabitants) regions [1].

Despite the political and economic stability achieved in recent years, Brazil is rated as 73th country in the Human Development Index, by the United Nations (UN), 75th in the Corruption Perception Index, by the Transparency International, and is one of the worst countries listed in income inequality metrics (and the very last considering medium to larger countries), including UN Gini coefficients [2–4].

The health care system mirrors these social and demographic characteristics.

F.B. de Carvalho, M.D. (✉)
Intensive Care Unit, Hospital Eduardo de Menezes, Fundação Hospitalar do Estado de Minas Gerais – FHEMIG, Rua Araguari 1670/302, Belo Horizonte, CEP 30190-111, Brazil
e-mail: fredbruzcarv@gmail.com

Á.R. Neto, M.D., F.C.C.P.
Internal Medicine Department, Universidade Federal do Paraná – UFPR,
Centro de Estudos e Pesquisas em Terapia Intensiva – CEPETI, Curitiba, Brazil

R.F. Simões, Ph.D. • M.V. Andrade, Ph.D.
Centro de Desenvolvimento e Planejamento Regional – CEDEPLAR, Faculdade de Ciências Econômicas da Universidade Federal de Minas Gerais, Belo Horizonte, Brazil

D.W. Crippen (ed.), *ICU Resource Allocation in the New Millennium: Will We Say "No"?*,
DOI 10.1007/978-1-4614-3866-3_2, © Springer Science+Business Media New York 2013

Historical Perspective

The Brazilian Health System has undergone important changes since the 1960s. But only after the 1988 Constitution this system was appropriately defined and the roles of each governmental level and private sector were established [5, 6].

The first major change happened in 1970, when the government determined the extension of the social coverage of the INPS (*Instituto Nacional de Previdência Social*) to autonomous workers and maids. The INPS was created in 1966 to unify the Institutes of Social Welfare of the various working classes. Until then each class of individual workers organized and formed their institute, aiming to provide working benefits (pension and retirement benefits) and the provision of medical care [5].

Thus, until 1970, only workers in the formal sector were guaranteed medical care. The extent of coverage for the other working classes and its growth has brought difficulties for the provision of medical services by INPS. Private institutions were thus recruited and much of the expansion of private institutions has been financed with public funds from the *Fundo de Assistência Social*, created in 1974.

In the early 1980s, the government began two key measures that mark a change in management of public health services in Brazil: the creation of the *Ações Integradas de Saúde* (AIS) and the change in the remuneration for services, which is now held by the *Autorização de Internação Hospitalar* (AIH).

The creation of the AIS was the first toward decentralization of health services from INPS. The AIS institutionalized a new relationship pattern among health sectors, including state and municipal government, it created mechanisms for transferring resources to these bodies that are now responsible for the medical care of the population. The AIH now replaces the system of payment per unit of service. The AIH pays the global historical cost of care, i.e., each payment is authorized by a specific diagnosis and not through the payment of each medical procedure.

From the mid-1980 the private sector begins to exert a complementary role to the public system. Until then, most private services were contracted through the public sector. The late 1980s and late 1990s were marked by great expansion of the private medical system. One factor behind this expansion was the tax incentives created by the federal government, i.e., individual health spending could be reduced from annual income taxes.

The *Sistema Único de Saúde*

The 1988 Brazilian Constitution, published after 21 years of military dictatorship, postulated that Heath Care is a right for all and a state's duty. Thus, the problem was not anymore universal care, but a financial and administrative one. Actually, except in rare excellence centers, public health services are generally of low quality and have a very disproportional balance between service and demand. Not surprisingly, despite said "universal coverage," 28% of the Brazilian population has a supplementary health care plan.

The principles delineated in 1988 were finally published in 1990 in the form of two federal laws: 8080 and 8142, regulating organizational and financial aspects of the public health care system, known as *Sistema Único de Saúde*, or SUS. Its main principles are:

- Universality—Everybody has the right to health care, independently of any social characteristics.
- Equity—Every citizen is equal to the SUS, but different communities may have different needs.
- Integrality—Actions must be done to protect, treat and rehabilitate individuals from disease.

From an administrative point of view the SUS is organized as follows:

- Regionalization and hierarchization—Care must be organized in regional systems and the entry on the health care system should begin from levels of less complexity.
- Resolutivity—systems should be designed to provide all levels of care.
- Decentralization—the main administrative unit of health care planning and execution is municipal.
- Direct citizen participation—through public councils, together with executive and legislative representatives.
- Private sector complementarity—should be organized in the same fashion of SUS, and may provide services to the public sector.

Albeit a little utopic, the principles of SUS were correct, but its implementation was not straightforward. Some of the political, economic, and cultural ambient in the 1990s considered its proposals not that appealing. Physicians and politicians, specially, had an important role on this. After 20 years of "Universal Care" in Brazil, we still have a somewhat fragmented health system, with many actors with very distinct interests, not a minority is divergent.

The major difficulty in the decentralization process of the health system has been the operationalization of the transfer system of resources between the three spheres of government. In terms of competence, the federal government should be responsible only for the mediation of the actions between the states in the field of public health and sanitary surveillance. States should conduct the mediation of activities among the cities in relation to public health and health surveillance, like establishing reference networks. In the SUS, cities are the basic units of management and health care delivery.

This process of decentralization of services has been quite complex and is ongoing. The economic and social heterogeneity between the different regions has certainly made this process even more complicated and diverse. One of the attempts by the federal government has been encouraging the formation of local consortia. Because there are no mechanisms to allow portability of funds to the city providing the service, the formation of consortia may not interest the city that has the installed capacity.

Private Care or Supplementary Health Sector

Currently the Brazilian Health System is characterized as a mixed health system. The private and public sectors coexist. Private medicine in Brazil is organized into different forms, including cooperatives, HMOs-like groups, self-management companies and health insurance plans that differ both in terms of access and payment system, as well as some of the benefits offered.

The Supplementary Health sector was also regulated in 1988 by the Law 9656, comprising all the private and corporate health care plans, including those associated directly or indirectly to any government agency, excluding SUS. This law brought many advances, those which were more interesting were:

- Previous chronic illness must be fully covered in new plans.
- An individual cannot be turned down unilaterally, unless in the absence of payment.
- All procedures and treatments listed by the National Agency must be fully covered.
- Limits to hospital or ICU length of stay, number of days in different admissions, number or types of exams needed or treatment is not allowed.
- Hospital transfer, when medically indicated, is also a responsibility of the health care plan.

The consequences to these new rules were that individual plans begun after 1988 became much more expensive, corporate, and collective plans being less expensive. Obstetric, ambulatory, and dental care could be optional.

The vast expansion of the medical market in Brazil occurred in the period 1987/94 when there was a growth of 73.4% of the population covered, from 24.4 million to 42.3 million policyholders. This enormous increase happened in an environment when SUS was not fully implemented. Estimates suggest today a contingent of about 28% of the total population in Brazil as purchasers of health insurance plans [6].

Health expenditures in Brazil was R$ 224.5 billion in 2008, meaning 8.4% of GDP. Hospital costs comprise around R$ 33.3 billion, or 15% of that money. Thus, more than half of heath care costs are paid directly by the individual or through his/her health care plan. Intensive Care costs are quite heterogeneous, but are responsible for 6–20% of hospital costs, being more important in the private or supplementary sector. Comparing to OECD's countries where government's heath care expenditures comprise up to 70%, leaving only 30% to individuals and families, in Brazil this proportion is, respectively, 41.6% and 57.4% [6].

Intensive Care in Brazil

Intensive Care Medicine appeared in Brazil in the late 1960s. Physicians that came from abroad were interested in caring for hospitalized patients opened the first ICUs in few excellence centers. Its development followed a slow but continuous growth in the 1970s. In 1980 the *Associação de Medicina Intensiva Brasileira* (AMIB) was created, and it was recognized as a distinct specialty by the *Associação Médica Brasileira* only in 2002.

The medical curriculum in Brazil has been strongly influenced by a focus in primary care, specially driven by successful experiences, especially from the Cuban health care system. Unfortunately some of these experiences were not adapted to an urban population, with health requirements different from other countries [7]. In contrast, the private sector has been strongly influenced by a hospital-based care, with all the consequences of this model, including very large ICUs, which are common.

After some time, we have seen changes in this scenario in the public sector, with the implementation of a national pre-hospital system (SAMU—*Serviço de Atendimento Móvel de Urgência*) in 2004 [8], and, more recently, with the creation of peripheral or micro-regional emergency units (UPA—*Unidade de Pronto Atendimento*) with national standardization [9]. Thus, the public health system now recognizes the need for resources in acute care, and this development has been changing practice, because more severely ill patients are arriving at hospitals, and intensive care units are an integral part of their treatment.

In the formal path to be an intensivist in Brazil, a physician, after 6 years in medical school, must have a 2- or 3-year prerequisite in anesthesiology or internal medicine or surgery followed by a 2-year fellowship in intensive care. Other possibilities are allowed, with no less time to be obtained and a board certification test is mandatory. During these 30 years AMIB certified 5,700 physicians in Brazil, but around 40% of them do not practice intensive care anymore (AMIB 2010, personal communication).

In 2009 AMIB did an ICU census and found 2,342 ICUs and 25,367 ICU beds in Brazil, with a relationship of 1.3 ICU beds per 10,000 habitants. The Brazilian Ministry of Health recommends a minimum of 1–3 ICU beds/10,000 Hab. The mean value described above is inside the recommended interval, albeit close to the inferior border. Almost half of the Brazilian states, most all of them in the North and Northeast regions, are below this lower limit [10].

The Agência Nacional de Vigilância Sanitária—ANVISA, Resolution #7/2010 states that every ICU has to have at least one intensivist in charge. Nevertheless, around 40% of the Brazilian ICUs do not have this physician working there. Most commonly, physician's work in Brazilian ICUs is structured in form of 12 h shifts, with intensivists covering or giving some support at daytime periods and weekends [10, 11].

Following ANVISA's Resolution #7/2010, the actual maximum number of patients per physician is 10. It is important to notice that the nurse to patient ratio is 1/8, the nurse–technician to patient rate is 1/2 and the physiotherapist to patient ratio is 1/10, covering 18 h per day [8]. We estimate that many small centers and those outside some regions cannot, or, are not, still complying with these legal definitions. Generally, physicians in public hospitals receive fixed salaries and those in the private sector work in a fee-for-service basis, receiving directly from health care plans. It is not uncommon to find physicians who work in two or three ICUs, normally trying to benefit from the best—and suffering from the worst—aspects of each type of employment or job.

Recent changes deserve consideration in Brazil and most probably will have an impact in the present and near future of intensive care in Brazil. It may be difficult to be documented but we have the impression that the technological gap between the public and the private services are shortening, with better equipped ICUs in public hospitals. Another recent change is the new revised medical ethics code, published in 2010, which emphasizes the role of patient autonomy and palliative care, two very important points when discussing resource allocation [12].

There is much to be done. But looking back there is a feeling that much has already been done and the path is ready for those who want to move forward and improve our practice. We hope that equilibrium in political interests finally will be achieved and maintained, and the allocation of resources will be directed to population needs, eventually benefiting most.

References

1. Instituto Brasileiro de Geografia e Estatística. http://www.ibge.gov.br/home/default.php.
2. United Nations Human Development Reports. http://hdr.undp.org/en/statistics/.
3. Central Intelligence Agency. https://www.cia.gov/library/publications/the-world-factbook/fields/2172.html.
4. Wikipedia. http://en.wikipedia.org/wiki/Gini_coefficient.
5. Constituição Federal do Brasil, Título VIII, Capítulo II, Artigo 196, 1988.
6. ANDRADE, Mônica Viegas. Ensaios em economia da saúde. Tese de doutorado. Rio de Janeiro: EPGE-FGV, 2000.
7. RESOLUÇÃO CNE/CES Nº 4, DE 7 DE NOVEMBRO DE 2001. Diretrizes Curriculares Nacionais do Curso de Graduação em Medicina.
8. Brasil, Ministério da Saúde. Política Nacional de Atenção às Urgências, 3ª edição, 2004. http://bvsms.saude.gov.br/bvs/publicacoes/politica_nacional_atencao_urgencias_3ed.pdf.
9. Brasil, Ministério da Saúde. PORTARIA Nº 1.601, DE 7 DE JULHO DE 2011.
10. Associação de Medicina Intensiva Brasileira. http://www.amib.org.br/pdf/CensoAMIB2010.pdf.
11. Agencia Nacional de Vigilância Sanitária. http://bvsms.saude.gov.br/bvs/saudelegis/anvisa/2010/res0007_24_02_2010.html.
12. Conselho Federal de Medicina. http://www.portalmedico.org.br/novocodigo/index.asp.

Chapter 3
Canada: Where Have We Been?

Christopher James Doig

Although touted as having a system of "National Medicare," Canada does not have one national system of publicly funded (government as payer) health care. Canada is a confederation of (at the present) ten provinces and three territories. The provision of health care is a responsibility of provincial or territorial governments. Therefore, it is more appropriate to state that all Canadians are covered, in the province or territory in which they reside, by a publicly funded universal health care insurance program. Publicly funded insured health services include, broadly defined, all medically necessary hospital services, medically required physician services, and surgical dental services performed in hospital. What distinguishes Canadian medicare is that insured services are similar and equitable between provinces. In particular, "catastrophic" costs that might be associated with serious acute illness or injury (those that might necessitate ICU care) are completely covered. However, not all Canadians' health care is covered by the government-funded single payer system. The share of health care funding that is "private" or not funded by the public payer system is ~30% [1]. The increasing proportion of provincial government expenditures that are directed towards health care may threaten the viability of the single payer publicly funded system. In the interim, as funding restrictions limit budget expenditures, programs will fight for the allocation of dollars within the system, and these threats will contribute to problems in managing ICU programs.

C.J. Doig, M.D., M.Sc., F.R.C.P.C. (✉)
Department of Community Health Sciences, University of Calgary, RM 3D39 TRW Building, 3280 Hospital Dr NW, Calgary, AB, Canada T2N 4Z6

Multisystem Intensive Care Unit, Department of Critical Care Medicine, Foothills Medical Centre, Alberta Health Services, Calgary, Canada
e-mail: cdoig@ucalgary.ca

D.W. Crippen (ed.), *ICU Resource Allocation in the New Millennium: Will We Say "No"?*,   17
DOI 10.1007/978-1-4614-3866-3_3, © Springer Science+Business Media New York 2013

The Development of Medicare in Canada

Canada was established in 1867 by terms contained within *The British North America Act*. This Act defined powers of the federal government, and those of the provincial governments. In simple terms, the federal government was given powers of all matters of the Nation, and all activities at the time thought likely to be costly (as the federal government had the broadest tax base). The provincial governments were given responsibility for local issues as these were simply thought unlikely to be costly. Part of the province's responsibilities was maintenance and management of hospitals. From this simple decision of hospital responsibility, health care, whether provided in hospitals or not, has subsequently been viewed constitutionally as under the authority of the provincial governments (with some minor exceptions, for example the military, the Royal Canadian Mounted Police, First Nations peoples, all which remain the responsibility of the federal government) [2].

Publicly funded universal health insurance programs were proposed in Canada, as in other countries, early in the twentieth century. Following the Second World War there was more serious deliberations for universal health insurance programs. In 1945, a Federal-Provincial Conference proposed a health care bill, in part modeled on the United Kingdom National Health System, with a proposed federal-provincial cost sharing. However, this plan failed as it was viewed as an incursion of federal power into an area of provincial authority. Provinces began to work towards universal medical insurance coverage, with the Province of Saskatchewan and its then Premier Tommy Douglas being the first to bring forward legislation. The Saskatchewan plan included coverage for hospital care, and for medically necessary physician services, the basis of the today's systems. Premier Douglas' plan created dissent in the both the medical profession and the public. At least one concern was the perception that if governments directly funded physicians, this would permit government (or its agents) to intrude into physician autonomy for clinical decisions for and on behalf of their patients. This in part resulted in the publicly supported "Doctors' strike" of 1962. The "Saskatoon Agreement" protected the professional autonomy of physicians, and the first universal medical care insurance program was enacted in Canada. Contemporaneously, Justice Emmett Hall of the Supreme Court of Canada was leading a Federal royal commission on medicare, and his 1964 report recommended the Saskatchewan plan as a template for universal health insurance coverage in all provinces. Hall also recommended the federal government share the cost with the provinces for these programs, and this was the basis for the *National Medical Care Act* of 1968.

Despite sharing half of the costs, the federal government did not have any direct authority over health care. As health care costs escalated in the 1970s, and with little federal control over how money was spent, in 1977 the federal government enacted Bill C-37. This legislation limited the federal government's contribution to 25% of the previous year's provincial health care spending, and tied subsequent increases in federal payments to growth of the gross national product (GNP). Concurrent with this plan were changes to taxation between the levels of government. The end result was to entrench not only the constitutional authority but also the fiscal responsibility for health care with the provinces.

With the changes in funding between levels of government, and the recognition of the constitutional authority of provinces for health care, the chief role of the federal government was to set broad principles expressing the nation's shared values. Today, the federal-provincial accord on health is defined by the *Canada Health Act* (1984), which is set to expire in 2014 [3]. The *Canada Health Act* continues the principles of previous legislation by requiring provinces to meet five criteria to receive federal transfer payments: (1) comprehensiveness, (2) accessibility (i.e., based on the need not ability to pay, therefore excluding user fees or extra-billing), (3) universality, (4) portability (coverage follows the person across provincial lines), and (5) public administration in the provision of insured health services.

Current Costs and Perceived Conflict in the Funding of Health Care

Canadians cherish their system of publicly funded health care insurance, and some consider it as part of the social identity of the country. However, recently there has been considerable public concern about timely access to non-emergent care, particularly but not limited to surgical services such as arthroplasty for degenerative joint disease, or cataract surgery (as examples).

When compared to other publicly funded health care systems in OECD countries, there are two unique characteristics of the Canadian system. The first is that insured services described above are almost entirely financed by general government revenue (personal, corporate, and sales tax), and federal legislation does not permit any additional charges by providers (user fees or extra-billing), or co-payments from patients for insured services. The second unique characteristic is that most OECD's universal public systems provide more comprehensive coverage such as pharmaceutical costs, or dental care. The mosaic of publicly funded health care insurance plans in Canada cover just over 70% of all health care costs experienced by Canadians. Examples, incomplete, of services which may not be covered include the following: (1) out-patient pharmaceutical costs (some provinces have variable coverage for seniors or individuals of lower income strata), (2) emergency medical (ambulance) services, (3) aids to daily living or medical devices such as crutches, eyeglasses, or hearing aids, (4) non-hospital services such as home care or physiotherapy, and (5) care provided by nonmedical health care professionals such as routine dental care. To cover these expenses, two-thirds of Canadians carry private insurance often through their employer, whereas 3–10% of Canadians have no supplemental insurance.

Health care in Canada, like the rest of the developed world, is expensive. Based on 2008 data [4], the estimated average health care expenditure in the ten Canadian provinces varied from \$4,654 to \$5,795 (CAD) per person, and as a percentage of GDP from 7.2 to 15.8% (a weighted national average of ~11.5%). These health expenditures are considerably less than those in the United States (weighted average of GDP of ~16%) but more than those in other OECD countries (weighted average of about 9%). Government funding for health care consumes between 39 and 52% of provincial budgets, and is in excess of 115 billion dollars per year. Care in the ICU is particularly expensive with old estimates calculating 8% of all hospital costs occurring within the ICU [5].

The annual government expenditures for health fall into two main categories, the funding of hospitals and the funding of physicians. Approximately 16% of the annual government expenditures on health are for physician services. These budgets pay not only physicians, but also cover their overhead including office staff. Most physicians, including ICU physicians, are not hospital or insurance plan employees. Rather, physicians are independent contractors who are either paid fee for service (payment for each individual patient service) or paid via an alternate payment plan scheme that provides a lump sum of dollars for the management of a group of patients. Little physician funding is tied to performance indicators or quality measures. Hospital-based funding receives the majority of the remaining dollars. In many provinces, hospitals within geographical areas are organized to coordinate care and respond to local needs. These regional boards or networks then distribute the dollars to the local hospitals primarily based on a block method rather than an activity-based model of funding (again with little consideration of quality indicator measures). In block methods of funding, reduction of services results in budget savings for hospitals, whereas in activity-based funding, dollars follow patients and flow to hospitals based only on services provided. As a result, hospitals have "solved" budgetary problems when block funded by restricting services such as closing beds, operating rooms, access to diagnostic services, and limiting staff (the major expense for hospitals). These restrictions have resulted in public concerns about access to services, and created problems for physicians.

Physicians paid fee for service and whose main activity is based on performing hospital-based procedures can suffer significant effects on their personal income from hospital responding to budget pressures by restricting access to services. Physicians serve as the gateway for most patients accessing non-emergent hospital services. To perform these services, they must work in collaboration or partnership with the hospital. Therefore physicians are often in a difficult position of explaining to patients why access might be limited (beyond their control) with the potential of being accused by the patient as being complicit, and at the same time advocating for patients within the hospital system for improved access, and being accused of being motivated by financial self-interest. There is a popular misconception that Canadian physicians are as a professional body opposed to publicly funded health care insurance plans. In fact, the Canadian Medical Association and its provincial chapters have endorsed the principles of publicly funded health insurance but called for a transformation including incentives to improve access and quality of care, and with the suggestion for more patient-centric care including a patient charter [6].

Public confidence in the health care system has been eroded mainly due to concerns about timely access to hospital-based therapeutic and diagnostic services. Given these concerns about access, and that many Canadians already carry supplemental health insurance, or pay out of pocket for services not covered by the publicly funded systems, it should not be a surprise that some Canadians expedite access to necessary health services outside of the publicly funded system [7].

First, patients may access out-patient nonhospital affiliated facilities, for example for diagnostic imaging procedures. Many of these facilities exist as they provide first and foremost services that are not defined as medically necessary and therefore are not paid for within the public system (for example, for medical imaging for insurance claims, or medicolegal cases). As these clinics have become more common, and waits for non-emergent imaging studies, particularly MRI, have lengthened, patients with the financial ability have "bought" images "for their own interest." Provincial and federal governments have often simply ignored that these practices occur.

Second, many citizens obtain surgical services at private clinics which function outside the publicly funded system. The most high profile of these clinics is the Cambie surgical centre in Vancouver where Dr. Brian Day, a former president of the Canadian Medical Association, operates. The clinic was originally opened to serve individuals not covered by the provincially administered health care insurance plan (for example RCMP officers). However, part of Dr. Day's rationale for moving to this clinic was that his local hospital in response to budget shortfalls had closed operating rooms, reduced surgical case volumes, had unacceptably high patient waiting times, and individual surgeon case volumes were falling below national recommended standards. Although the original reasons for starting the Cambie clinic are presented as laudable, the clinic now advertises its services to all Canadians: pay directly to avoid waiting. Now, against complaints from critics that private-for-profit surgical facilities offer a lower quality service, he has welcomed comparison of the surgical outcomes from his facility with those from publicly funded facilities. Furthermore, in response to requests from the BC Medical Services Commission to audit his practice on suspicion of unlawful billing practices in violation of the province's *Medicare Protection Act*, Dr. Day initiated a statement of claim with the British Columbia Supreme Court that the *Medicare Protection Act* is unconstitutional and therefore, the Commission lacks a valid legal authority to audit the Cambie clinic. Although the events have garnered national interest, private surgical clinics exist in other provinces, and some provincial governments purchase or contract publicly funded procedures through these facilities to improve access times. For example, in the province of Quebec, the provincial government has expanded the number of insured services that may be provided in private surgical facilities from 3 to 56 [8].

The final method of direct purchase is medical tourism whereby services are obtained from providers in other countries. Most of these services can be classified into one of the three groups: medically necessary insured services but with a restricted availability (for example organ transplantation), medically necessary insured services but with access delays (for example, joint replacement surgery), and medically necessary insured services but where expertise is obtained outside of Canada (although similar expertise is available in Canada).

One recent high profile example of the last type was the Premier of Newfoundland, an independently wealthy multimillionaire, who obtained cardiac surgery in the United States [9].

An important challenge to the current construct of federal and provincial legislations banning the purchase of medically necessary care outside of the publicly funded system is a recent decision from the Canadian Supreme Court, known as the Chaoulli decision [10]. In this case, Mr. Zeliotis, the co-claimant, claimed that he endured pain and suffering waiting for care in the Quebec public system, and given the long wait lists for services, the ban on obtaining services through private insurance was a violation of the right of personal security (within the province of Quebec's *Charter of Humans Rights and Freedom*). The Supreme Court ruled in his favor, but the result of this decision remains unclear [11, 12]. There are court challenges in other provinces that are currently before other levels of Courts, and some will ultimately reach the Supreme Court for a final decision. These cases, with the end of the Canada Health Act in 2014, will likely determine the future of the national system of health insurance plans and coverage.

The Impact on ICU Care

How do these challenges in the Canadian health care system impact ICU care? First, like other parts of hospitals, ICUs often struggle with budgetary restrictions that close beds either physically or functionally (by a reduction in staff). These closures have increased overall ICU capacity, and resulted in many patients being managed in nontraditional settings such as prolonged stays in the emergency department or recovery rooms and in earlier discharge of critically ill patients from ICU to non-ICU ward care. As ICU occupancies have increased, so has the complexity of individual patients. ICU staff and physicians are spread thinner (caring for more and sicker patients in the ICU, and critically ill patients elsewhere in the system), and anecdotally more time is spent coordinating and organizing beds for patients. This extra time is often at the expense of time spent with families. Second, the public is appropriately concerned about motivations within the health system to limit expenditures, particularly for very expensive care where outcomes are not guaranteed. Although anecdotal, it is not uncommon to hear ICU staff talk about how families now question the motivations of staff such as "are you suggesting that treatment be stopped because it is too expensive and grandma is too old" or "are you doing everything you can, or is the hospital restricting you because treatments are too expensive." These situations are very stressful on staff. Third, the public has heard many examples of emergency care gone awry. In my province, emergency physicians recently blamed the provincial government and Alberta Health Services (with operational responsibility for the delivery of health care) for problems in accessing emergency care and said that as a consequence there had been over 300 documented cases of death or near death, and warned of a "potential catastrophic collapse" [13]. The resulting media and political uproar resulted in the dismissal of the Alberta

Health Services CEO [14], and the expulsion from the government caucus of the junior minister of health [15], himself an emergency department physician who also blamed his own government for mismanaging health care. These high profile media cases have created or enhanced a climate of uncertainty within the public. Less time to talk to families, more patients, sicker patients, and staff stretched thinner, in an environment where the health care system is called in a crisis on the front page of newspapers, result in frequent difficulties dealing with ethical problems in the ICU. These exacerbate the frustration of staff, whom are simply doing their best in challenging times, and who feel disenfranchised as they bear the brunt of family concerns, but have little opportunity to influence change beyond the ICU. It is not surprising that a recent survey of ethicists identified waiting lists, access to care, and disagreement between patients/families and health care professionals as the preeminent ethical challenges in the Canadian health care system [16].

Conclusion

The Canadian health care system, as it is, has been described externally as achieving a reasonable balance of treatment quality, cost, and health outcomes, but with the recognition that reforms are needed to contain expenditures. Many of the challenges faced in the health care system have challenged care in the ICU, in particular, the volume and complexity of patients being managed, and the frequency of ethical issues in the ICU. Health care transformation to incent timely access and quality care is part of the needed change to maintain the viability of our publicly funded health care insurance programs.

References

1. Di Matteo L. Policy choice or economic fundamentals: What drives the public-private health expenditure balance in Canada? Health Econ Policy Law. 2009;4:29–53.
2. Vayda E, Deber RB. The Canadian health care system: an overview. Soc Sci Med. 1984;18: 191–7.
3. Canada Health Act (R.S., 1985, c.C-6).
4. OECD Economic surveys: Canada 2010. Retrieved from http://www.oecd.org/document/6/0,3 343,en_2649_34569_45925432_1_1_1_37443,00.html. Accessed 14 Dec 2010.
5. Jacobs P, Noseworthy TW. National estimates of intensive care utilization and costs: Canada and the United States. Crit Care Med. 1990;18:1282–6.
6. Health care transformation in Canada: change that works, care that lasts. August 2010. Retrieved from http://www.cma.ca/multimedia/CMA/Content_Images/Inside_cma/Advocacy/ HCT/HCT-2010report_en.pdf. Accessed 14 Dec 2010.
7. Flood CM, Thomas B. Blurring of the public/private divide: the Canadian chapter. Eur J Health Law. 2010;17:257–78.
8. Picard A. Private health care slips under the radar (2009, July 16). Retrieved from http://www. theglobeandmail.com/life/health/private-health-care-slips-under-radar/article1220145/. Accessed 14 Dec 2010.

9. Blackwell T. Danny Williams could have stayed in Canada for top cardiac care, doctors say. National Post (2010, February 2). Retrieved from http://www.nationalpost.com/news/Danny+Williams+could+have+stayed+Canada+cardiac+care+doctors/2514581/story.html. Accessed 14 Dec 2010.

10. Chaoulli v Quebec [2005] 1 S.C.R 791.

11. King J. Constitutional rights and social welfare: a comment on the Canadian Chaoulli health care decision. Mod Law Rev. 2006;69:631–43.

12. Flood CM. Chaoulli's legacy for the future of Canadian health care policy. Osgoode Hall Law J. 2006;44:273–310.

13. Braid D, D'Aliesio R. Alberta's health crisis enrages Tory MD Raj Sherman: says trust in Stelmach, cabinet "severely tarnished", Calgary Herald (2010, November 18). Retrieved from http://www.vancouversun.com/health/Alberta+health+crisis+enrages+Tory+Sherman/3845540/story.html. Accessed 14 Dec 2010.

14. D'Aliesio R, Komarnicki J. Duckett's post-cookie departure points to growing chaos in health care: critics: at least one board member resigns over decision. Calgary Herald (2010 December 9). Retrieved from http://www.calgaryherald.com/health/Duckett+fired+from+Alberta+Health+Services/3878118/story.html. Accessed 14 Dec 2010.

15. Dormer D. Outspoken MLA Dr. Raj Sherman kicked from Tory caucus. Edmonton Sun (2010, November 22). Retrieved from http://www.edmontonsun.com/news/alberta/2010/11/22/16265526.html. Accessed 14 Dec 2010.

16. Breslin JM, MacRae SK, Bell J, et al. Top 10 health care ethics challenges facing the public: views of Toronto bioethicists. BMC Med Ethics. 2005;6:5.

Chapter 4
Germany: Where Have We Been?

Thomas Kerz

Germany's health system organization is based on three major principles: compulsory insurance, compulsory contributions from both employers and employees and solidarity. A vast majority of German citizens follow the idea that health risks should be shared collectively by the community of all insured persons [8]. Irrespective of the amount each person had paid into the social system, all have access to comprehensive coverage. In order to understand Germany's peculiarities in health policy, this chapter provides an introduction into Germany's model of health care and then discusses which kinds of resource allocation strategies already have evolved. The second chapter deals with prospects for Germany's health care system in the near future and future. Both chapters do not discuss future developments in the area of medical technology.

Germany has two systems of health insurance: Firstly, there is a public sector which encompasses around 90% of the roughly 80 million inhabitants. Family members enjoy free coverage. One association of statutory health insurance physicians ("Kassenärztliche Vereinigung") in every one of the 16 federal states has the legal obligation to ensure "adequate, sufficient, and cost-effective treatment [2]" of all patients which also means to ensure an emergency service 24/7 h by office-based doctors. Access to health care is principally free, although, in order to cut health consumption and expenses, since 2003 there is a copayment of 10 Euro per quarter when a physician is consulted in his office. When sick, patients can go either to the office-based primary care physician or present directly to a hospital-based emergency department ("Notaufnahme"). When in hospital, there is a copayment of 10 Euro daily for a maximum period of 28 days per year, and the rule of "adequate, sufficient, and cost-effective treatment" also applies.

Premiums in the statutory system have risen from 8% of the gross salary in 1970 to 14% in 2007 and were, in the past, assessed by each health maintenance

T. Kerz, M.D. (✉)
Klinik und Poliklinik für Neurochirurgie, Universitätsmedizin der
Johannes Gutenberg Universität, Mainz, Germany
e-mail: kerz@uni-mainz.de

D.W. Crippen (ed.), *ICU Resource Allocation in the New Millennium: Will We Say "No"?*, 25
DOI 10.1007/978-1-4614-3866-3_4, © Springer Science+Business Media New York 2013

organization (HMO). For 2010, the premium was set by state authorities to 14.9% (2011: 15.5%) of the gross income (regardless of personal risk factors). The employee contributes 7.9% of the premium while the employer pays 7%. The maximum amount of salary from which premiums are deducted is 3,750 Euro/ month in 2010, so the employee's maximum share of premiums is near 300 Euro/ month.

The private sector accounts for around 10% of the population. Here, premiums are calculated according to patient's age and specific risks. Only persons earning more than 4,160 Euro/month before tax (2010) or self-employed persons are eligible to leave the compulsory system and join the private system.

Since 2007, otherwise uninsured persons are legally entitled to join a health insurance fund regardless of their risks, which solved the problem of almost 200,000 uninsured people that year. Pensioners have their own HMO. Similar to the rule for workers, the statutory pension fund contributes 7% to the premiums and the retired person 7.9%.

In 2009, the Health Fund came into effect for the public sector: All the contributions paid by employees and employers to the compulsory health insurance scheme flow into this fund which is supplemented by tax revenue. For each insured person, the health insurance companies receive a flat rate from the Health Fund. Companies which insure a particularly large number of old or sick people and low-income workers receive an additional subsidy. If a deficit should be encountered, up to 2% of the income can be charged by the companies from the insured. Although the fund will receive a tax subsidy of 16 billion Euro in 2010, there will be a deficit of around 4 billion Euro. This will likely be covered by charging an extra premium from the insured employees.

As of 2008, there were 421,700 physicians nationwide, and 120,000 of them were panel doctors [6]. In 2008, total health expenditures were 263.2 billion Euro, translating into a per capita total expenditure on health of 3,210 Euro (2007: 3,080 Euro) [13]. This represents 10.5% of Germany's gross domestic product (GDP) while this was 9.6% in 1992. The compulsory sector alone accounts for 6.5% of the GDP. In 2008 there were 2,083 hospitals left versus 2,400 in 1990. One-third of them are not owned by public institutions but by private investors. Total hospital costs amounted to 63.2 billion Euro, a 5% increase compared to 2007 [12]. Personnel costs account for 60%, and material costs for 38%. Mean length of hospital stay was 8.4 days. 17.5 million patients were admitted, and adjusted case costs were 3,610 Euro. Of 503,360 hospital beds, 23,890 were ICU beds with an occupancy rate of 80.5% [11], and 2 million patients were treated in an ICU. This translated into 29 ICU beds per 100,000 inhabitants, a ratio that has not much changed since 1991 when it was 25 beds/100,000 inhabitants. In 2003, costs for ICU care ranged from 672 to 1,090 Euro/day with a mean of 791 Euro [9].

Hospital billing for patients in the compulsory sector, and therefore for the vast majority of ICU beds, is based on the German adaption of the Australian-refined diagnosis-related groups (G-DRG): Since 2004, a patient's main diagnosis (leading to hospital admission) is coded into a DRG. Then, age, procedures, complications, minor diagnoses, and time of ventilatory support are incorporated at the end of the stay. Data are entered into a governmentally certified "grouper" software which translates them into a code and calculates the relative value of the DRG the patient falls into. The relative value then is multiplied by the base rate (issued every year in Euro by a joint commission of health insurance companies and the hospital association in each federal state) which, in turn, gives the amount of money paid to the hospital. So, the amount of money put into the system is coupled to the wages and can only increase when extra money is requested from the insured, but the distribution to the health care providers is self-governed by health insurance funds and hospitals. Although installation of the DRG system was effective to save costs by increasing efficiency and make costs comparable throughout Germany, several factors can invite hospitals to restrain costs and practice rationing:

- As one case translates into a more or less fixed amount of receipts, hospital administrators might want to urge doctors to prescribe cheaper therapies which could be less effective than more expensive therapies.
- As revenues do not increase infinitely as patients stay longer, hospitals might want to discharge patients earlier in order to save money (also called "bloodily discharge").
- Hospitals could aim for treating only the more simple cases, refuse treatment and send complex cases to specialized centres claiming limited resources.
- Patients whose DRG-related proceeds are low compared to patients with other, more financially rewarding diseases could see themselves beds refused or discharged early.
- Increasingly, procedures (and not diagnoses) determine revenues per case in the G-DRG system which could encourage doctors to perform operations/procedures because of financial considerations and not according to medical needs. For example, artificial ventilation from 250 to 499 h will result in only half the receipts compared to a case that is ventilated more than 500 h and more. It is conceivable that, prolongating artificial ventilation, a hospital can try to increase its revenues. Those economical factors could divert money from those who have a real need for a certain type of intervention away to those that assure higher hospital revenues.
- Cost savings might include reduction of medical personnel which in turn could mean less beds available for intense and attentive treatment.
- Hospital operators could aim to concentrate their beds at one site, thereby minimizing access to hospital care of those in more rural areas.

Which services will be provided and under which conditions they are reimbursed by the compulsory system are subject to decision by a federal joint committee ("Gemeinsamer Bundesausschuss", G-BA). Herein, national associations of doctors and dentists, the German Hospital Federation and the federal associations of health insurance funds discuss these issues. When new methods are to be discussed, the G-BA or the IQWiG (Institute for Quality and Efficiency in Health Care) investigate the benefits and possible harms, and the G-BA then evaluates necessity and degree of benefit and conducts a cost/benefit analysis. While, for outpatient care, methods are not reimbursable as long as the G-BA has not decided about a positive benefit of the treatment, the reverse is true for in-hospital care where methods are reimbursed by the statutory system as long as the G-BA has not decided about the negative benefit of the treatment. All decisions must be ultimately confirmed by the federal ministry of health. However, HMOs are free to reimburse other treatments such as homeopathic or acupuncture therapies.

The G-BA therefore represents the bottleneck through which new methods must pass before they can be reimbursed by the compulsory health insurance funds. Yet, assessment criteria such as benefit, morbidity, mortality, prevalence, costs of treatment and if alternative therapies are available have not been openly discussed; their members are named by the adhering associations and are not democratically elected. Representatives of patient's rights organizations have only advisory but no voting rights.

In general, rationing in medicine is discussed only occasionally in the public. In February 2010, Germany's federal health minister Philipp Roesler, himself a physician, when asked about how to finance the ever-raising costs of cancer therapies, uttered: "As long as I will stay in office and because of ethical reasons, I will never debate rationing or prioritization [10]". This statement is in-line with other statements from politicians who continue to promise unlimited benefits for every patient and incessantly claim rationalization efforts from the medical community. But, at the same time, by enacting health care reforms such as the health fund with its increasing out-of-pocket payments, more and more financial burden is put on patients directly. One year before, Joerg-Dietrich Hoppe, President of the German medical association, on the occasion of the yearly German Medical Assembly had given an inaugural speech entitled: "Equitable distribution through prioritization— patient welfare in times of shortage" [3]. He stated: " ... if we no longer get sufficient resources to provide patients with healthcare services—in other words, if our politicians consolidate the current shortage—then we simply must discuss the matter openly and honestly, and arrive at a fair distribution mechanism. ... we doctors do not want rationing ... ... but we also no longer want to be held responsible for the state-imposed shortage in surgeries and hospitals".

Given that, the medical community as well as researchers such as (health) economists, ethicists and others are well aware of an already ongoing practice of rationing, but little is discussed in public, and politicians fear to start the debate because of the anticipated loss of votes for those taking unpopular actions. As of now, there are no legal bodies or advisory committees entitled to take rationing decisions. Yet, from the words of Dr. Hoppe, it is clear that doctors would tend to treat their patients to the maximum, even at exploding marginal costs, as long as the statutory system assumes responsibility to cover the costs.

In the press, there have been occasional reports of patients waiting for hours in ambulances before hospitals agreed to accept their admission, be it because of a shortage of ICU beds or shortages in the emergency room. Hence, every state law specifies now that emergency cases must be accepted by any hospital, regardless of disease and insurance issues. More complicated cases are then redirected to appropriate facilities. Many patients have seen their operation date rescheduled many times because of shortage of ICU beds—around 12–15% of operated patients are estimated to need an ICU or monitored bed in an intermediate care unit perioperatively. Some hospitals work on an "operate-and-return" basis because of shortage of ICU beds, e.g. neurosurgical patients are operated in a specialized centre and then redirected to a general ICU. In a study by Jaeger et al., this was associated with a higher mortality in those who did not receive specialized care [5].

Already in 1999, 73% of physicians interviewed said that there was rationing in the German health system [7]. In a 2008 survey in German ICUs, Boldt and Schoellhorn [1] found 32% of the respondents to claim that therapeutic decisions were occasionally influenced by economic reasons while 9% found that this was often the case. Fifty-nine percent reported that this did not happen or happened only very rarely. When asked when and how patient admission to the ICU was refused, 35% of the respondents answered that occasionally or often patients were refused although there were other patients in the ICU whose treatment was considered futile. Thirty-five percent of the ICUs reported no policy of written contraindication to admission to the ICU. The same percentage of units did not accept moribund patients. Almost no ICU disposed of written standards when to freeze, limit or stop treatment. Eighty-eight percent said that age was no reason to withhold expensive treatments, and almost the same percentage saw no age limit for artificial renal support. In 30% of the units, the decision to administer expensive drugs was subjected to the head of the department.

In the same study, while 67% of the ICU physicians thought that there was rationing in German ICUs, 52% stated that there should not be any rationing, and 43% would prefer having a catalogue with clear indications when not to start ICU therapy.

Strech et al. [14] reported in 2009 that 68% of the responding physicians stated that they had already withheld medical services with a potential benefit for the patient because of cost considerations but only 17% of the screened cardiologists and 10% of the intensivists indicated that this happened often [14]. There was a wide variation regarding awareness and practising of rationing. Answers reached from "I am not responsible for the whole German health system, so I treat my patients maximally" to "As a doctor nowadays, it's my duty to find the balance between getting the maximum benefit for the patient and reducing costs for the health care system". No responder would dare to disclose to patients that the treatment decision was the result of a cost–benefit analysis but would always promote medical reasoning. Coping strategies most often included giving a less expensive therapy with a perceived similar benefit or administering a less expensive therapy at first hand, followed optionally by more expensive treatments. As a rule, rationing decisions were delineated as influenced by medical criteria (benefit, risk, general probability of patient's death) cost-effectiveness, age, compliance (e.g. to take aspirin/clopidogrel after stent implantation) and contribution of a person to society.

Compared to the results of a study done in the UK, Switzerland, Norway and Italy, rationing in Germany seems to occur in comparable fashion [4]. Yet, most interviewed German ICU doctors in Strech's study felt uneasy with bedside rationing (BSR) because they found themselves not competent in socio-economical questions, considered themselves without legitimation for BSR, and felt a negative impact of cost pressure on their work satisfaction and on the patient–doctor relationship. Furthermore, BSR was thought to result in inequity and arbitrariness [14]. Therefore, 74% of the respondents would like to see macro-level rationing policy instituted. Around one-half of the respondents refused to take economic effects into account, but the same percentage felt that rationing decisions should be taken by a doctor.

In the end-of-life phase in ICU patients, in Boldt's study, 14% of the ICUs continued all therapeutic attempts but did not administer CPR, 52% freezed therapy, 54% limited therapy to nutrition and hydration and 64% administered terminal analgosedation. Only 10% withdrew artificial ventilation, and 11% stopped all therapeutic actions. A 2005 study in ICUs of 17 European countries found in one-third of the cases that the end-of-life decision was not discussed with families. Generally, there was more family discussion in northern and central than in southern European countries (Greece, Israel, Italy, Portugal, Spain, Turkey) (Communication of end-of-life decisions in European intensive care units, 2005). From paediatric data, it seems that German doctors are amongst those in Europe who are more inclined to limit futile treatment because of poor neurological prognosis. Only physicians in Sweden and France reported more often to decide stopping treatment in such cases whereas doctors from Italy, Spain and some eastern European countries tended to press ahead with treatment (Neonatal EoL).

In conclusion, in the past, limiting the amount of money dispensed by health insurance funds through statutory regulations has shown to be effective in stabilizing the GDP share on health expenditures in Germany. Implementation of a DRG system has led to BSR but doctors feel uneasy because the burden of rationing is exclusively on them. A public debate on these issues has not yet taken place. Continuation of current policies and medical progress will result in rising health care expenditures which the public and politicians are probably unwilling to accept; a debate on these issues is on the horizon. Where this debate could lead to will be outlined in Chap. 2.

References

1. Boldt J, Schollhorn T. Ethics and monetary values. Influence of economical aspects on decision-making in intensive care. Anaesthesist. 2008;57:1075–82. quiz 1083.
2. Bundesministerium der Justiz. Sozialgesetzbuch V, § 72. In: Bundesministerium der Justiz, 2010. http://bundesrecht.juris.de/sgb_5/__72.html.
3. Hoppe Joerg-Dietrich. Equitable distribution through prioritization – patient welfare in times of shortage. Presented at German Medical Assembly, Mainz; 2009.
4. Hurst SA, Slowther AM, Forde R, Pegoraro R, Reiter-Theil S, Perrier A, Garrett-Mayer E, Danis M. Prevalence and determinants of physician bedside rationing: data from Europe. J Gen Intern Med. 2006;21:1138–43.
5. Jaeger M, Schuhmann MU, Samii M, Rickels E. Neurosurgical emergencies and missing neurosurgical intensive care unit capacity: is "operate-and-return" a sound policy? Eur J Emerg Med. 2002;9:334–8.

6. Kassenärztliche Bundesvereinigung. Grunddaten zur vertragsärztlichen Versorgung in Deutschland 2009. Berlin; 2009. Accessed 6 June 2012.
7. Kern A, Beske F, Lescow H. Leistungseinschränkung oder Rationierung im Gesundheitswesen? Deutsches Ärzteblatt. 1999;96:A-113–116.
8. Marstedt G. Solidarität und Wahlfreiheit in der GKV. In: Böcken J, Braun B, Schnee M, editors. Gesundheitsmonitor 2002: Die ambulante Versorgung aus Sicht von Bevölkerung und Ärzteschaft. Gütersloh: Verlag Bertelsmann Stiftung; 2002. p. 112–29.
9. Moerer O, Plock E, Mgbor U, Schmid A, Schneider H, Wischnewsky MB, Burchardi H. A German national prevalence study on the cost of intensive care: an evaluation from 51 intensive care units. Crit Care. 2007;11:R69.
10. Müller-Jung Joachim. Der Gesundheitsminister pfeift auf die Zukunft. In: FAZ.NET; 2010. http://www.faz.net/s/Rub7F74ED2FDF2B439794CC2D664921E7FF/Doc~EF009E5068D88 42739403500DE5FCC473~ATpl~Ecommon~Scontent.html. Accessed 6 June 2012.
11. Statistisches Bundesamt. Gesundheitsberichterstattung des Bundes. In: Bonn; 2009. http://www.gbe-bund.de/oowa921-install/servlet/oowa/aw92/WS0100/_XWD_FORMPROC? TARGET=&PAGE=_XWD_98&OPINDEX=4&HANDLER=_XWD_CUBE. SETPGS&DATACUBE=_XWD_124&D.000=3730&D.001=1000001&D.935=12147. Accessed 9 September 2010.
12. Statistisches Bundesamt. Pressemitteilung Nr.429 vom 12.11.2009. In: 2009.http://www. destatis.de/jetspeed/portal/cms/Sites/destatis/Internet/DE/Presse/pm/2009/11/PD09__429__2 31,templateId=renderPrint.psml. Acessed 9 september 2010.
13. Statistisches Bundesamt. Gesundheitsausgaben. In: 2010.http://www.destatis.de/jetspeed/portal/cms/Sites/destatis/Internet/DE/Navigation/Statistiken/Gesundheit/Gesundheitsausgaben/ Gesundheitsausgaben.psml. Accessed 9 September 2010.
14. Strech D, Danis M, Lob M, Marckmann G. Extent and impact of bedside rationing in German hospitals: results of a representative survey among physicians. Dtsch Med Wochenschr. 2009;134:1261–6.

Chapter 5
India: Where Have We Been?

Farhad Kapadia, Atul P. Kulkarni, and J.V. Divatia

Introduction

India has the world's second largest population with a majority residing in rural areas. It is a rapidly developing country but a significant proportion currently lives below the poverty line. Per capita income in the country is low. WHO data in 2007 put total expenditure on health per capita in India in the US $25–50 range, better than some South Asian and African nations (at <$25) but less than most of South America and North West Asia ($300–1,000) and much less than Europe and Australia ($1,000–5,000) or USA and Scandinavia (>$5,000).

India's annual healthcare spend continues to remain one of the lowest in the world, but the government is consistently increasing the allocation for the healthcare over the years. Healthcare spending for the year 2004–05 was 0.26% of the gross domestic product or 1.6% of the total budget expenditure, totalling around Rs. 8,086 crore (1 crore = 10 million or 10^7). This has increased to Rs. 25,154 crore for the financial year 2010–11 (0.36% of GDP or 2.3% of the total budget expenditure).

In global terms, 64% of health resources come from governments or social insurance, while private insurance and out-of-pocket or direct payments accounted for

Note to Reader: *Currency Indian Rupee (Rs.): US $1 = Rs. 45*
Conventionally stated in Lakhs and Crores (1 lakh = 100 thousand or 10^5 and 1 crore = 10 million or 10^7)

F. Kapadia, M.D., F.R.C.P. (✉)
Department of Medicine and Critical Care, P.D. Hinduja Hospital & MRC,
V.S. Marg, Mahim, Mumbai 400 016, India
e-mail: fnkapadia@gmail.com

A.P. Kulkarni, M.D.
Tata Memorial Cancer Hospital, Mumbai, India

J.V. Divatia, M.D.
Department of Anaesthesia, Critical Care and Pain, Tata Memorial Hospital,
Mumbai, India

18% each. As against this international pattern, in India it is estimated that 83% occurs in the private domain and only 17% is borne by the government. In the private sector, it is estimated that less than 7% of expenditure is covered by prepaid insurance plans. More than 92% of private healthcare financing is estimated to be "out of pocket", though it is likely that there will be greater contributions from insurance and other employer, corporate or "third-party" coverage in the future.

The Government of India funding in critical care is a very low priority. Appropriately, primary and secondary care and public health get priority. Critical Care services in the country are rapidly changing and evolving. There is minimal analysis of cost considerations in this relatively new field. It is estimated that approximately 10% of Critical Care services are from government hospitals and institutes. The rest is from the private sector. These private hospitals or institutes are run by individuals, trusts or companies and they charge patients for the treatment provided. They range from large tertiary care hospitals which are well equipped and form the bulk of the critical care facilities in the country to smaller hospitals and nursing homes which have moderate staffing and facilities, and are more affordable than tertiary care hospitals.

Indian ICU Data

There is little systematic data regarding cost implications in ICU. The published data which does exist pertains to the public sector hospitals in the 1990s. The cost expenses will vary greatly depending on the hospital, public, private or public–private mix. We review the data from the Public Sector and comment of the cost implication at the two institutes where the authors work. One a Public–Private Speciality (Cancer) Hospital (Divatia and Kulkarni) and the other is a Private Trust Tertiary Referral Hospital (Kapadia).

Public General Hospital

Rao et al. [1], from a public hospital in the city of Hyderabad, found the cost of intensive care in India to be about Rs. 5,000 per patient per day in 2000—however, details of the methodology used in these studies are not available. Parikh and Karnad [2] have done a more detailed analysis [2] from a large public hospital in Mumbai. In 1999, they conducted a prospective observational study of the quality, cost and benefits of intensive care in one of the largest public hospitals in Mumbai. They studied 993 patients admitted to the medical–neurology–neurosurgery critical care unit of the hospital. They found that the average cost of treatment per patient per day was Rs. 1,973 (US $56.36 or I $180.26). The average cost per survivor was Rs. 17,029 (US $486.54 or I $1,556.56). The cost per TISS point was Rs. 90 (US $2.25 or I $8.23). This low cost per TISS point was attributed to reuse of disposable equipment and the use of subsidized drugs. Staff salaries in India, especially those in government hospitals, are also considerably lower than corresponding ones in the Western world. Various factors could have contributed to the low cost found in this study. As this study was carried out in a public hospital ICU, it implied that a large proportion of costs were paid for by the Government. Hence, this data cannot be extrapolated to other critical care set-ups in India. It is also possible that patients may have spent on certain consumables from their pocket, and this data may have been missed out. It is interesting to note that the observed mortality in this study was 36%, with a standardized mortality ratio of 1.67. Also, Day-1 TISS scores were considerably lower than those reported from centres in Western countries. This could either represent variations in clinical practice, resulting in lower intensity of intervention, or could simply reflect the patient's inability to afford expensive interventions. The authors also attributed the high SMR to differences in case mix, with the diseases treated in the ICU not being represented in the APACHE database, and to lead time bias. The authors concluded that the cost of intensive care in India is lower than that in the USA, even after accounting for the low cost of living in India. However, it could also be argued that higher level of intervention and better nurse–patient ratios may have resulted in better outcomes, though at a higher cost. This brings us to the question: Would greater spending to attain higher levels of care in terms of equipment, drugs, interventions and staffing and organization have led to better outcomes? Would a higher cost have resulted in greater cost-effectiveness? The mean age of patients was 36 years. Approximately 40% of patients were admitted with tropical and other infections (malaria, leptospirosis, tuberculosis, tetanus, severe community-acquired pneumonia, others), organophosphorous poisoning or snakebite. All these conditions are treatable and are likely to have good survival with good quality of life after hospital discharge. Although a formal cost-effectiveness analysis looking at life years gained or quality-adjusted life years was not performed, one could speculate that intensive care would have been easily justified in this group of patients. Greater spending and costs to comply with acceptable international standards of intensive care would be more than offset if outcomes improved and intensive care proved to be more cost-effective.

Public–Private Hospital

Two of the authors (Divatia and Kulkarni), from Tata Memorial Cancer Hospital, conducted a study of costs in ICU in the year 2005 [3]. Costs were calculated for 101 patients, with a total of 311 patient days. The mean length of ICU stay was 74 (+62) h. The total cost of ICU care per patient day was Rs. 12,000 (US $273). Of this, the direct cost per patient day was approximately Rs. 8,000 and the indirect cost approximately Rs. 4,000. The cost per survivor was calculated to be approximately Rs. 47,000 (US $1,068). Even after accounting for inflation (an average of 7% per year), the costs were significantly higher than those calculated by Parikh et al. [2] in 1999. These higher costs could be attributed to sicker population of cancer post-operative and post-chemotherapy patients. A significant proportion of these patients were admitted to the ICU after a prolonged stay in the ward will possible have multidrug-resistant hospital-acquired infections requiring expensive antibiotic. Also, patients had received myelosuppressive chemotherapy and needed large quantities of blood and blood products, significantly increasing direct costs.

Arguably, with this case mix consisting of cancer patients, life years gained are likely to be significantly lower than with the case mix observed in the study by Parikh et al. [2]. However, the cost of intensive care should also be compared to the cost of therapy for the primary malignancy. While this data is not available, it must be remembered that the cost of chemotherapeutic agents is extremely high. For example, 1 cycle of rituximab, vincristine, adriamycin, cyclophosphamide and prednisolone given for treatment of non-Hodgkin's lymphoma in adults costs approximately Rs. 1,50,000 (US $3,750) for the medications alone, and 6 such cycles are given. Thus the cost per patient is Rs. 9,00,000 (US $22,500). Assuming a 70% 5-year survival, the cost per survivor is Rs. 12,86,000 (US $32,143). This does not include cost of hospitalization, professional charges and supportive care (growth factors, blood and blood products, antibiotics). Thus although intensive care is expensive, it is far less expensive than treatment of non-Hodgkin's lymphoma.

Private Trust Tertiary Referral Hospital

One of the authors (Kapadia) works at the Hinduja Hospital, which is a private trust hospital. All the revenue is from patient billing, and being a trust hospital, all profits mandatorily have to go back to the institute. This is then used to maintain, upgrade and strengthen the infrastructure. No individual or shareholder profit is feasible under the trust and charity laws. The following discussion concerns costs at the current time of writing (2010). As will be immediately apparent, this is significantly higher than that in the public and public–private systems discussed earlier, though it must be remembered that the public system analysis was from more than a decade ago, and the public–private hospital analysis was from 2005.

This is a tertiary referral hospital with a total of 350 patient beds of which 47 are in the adult ICUs. The critical care department is staffed with 4 consultants, 2 associate consultants, 21 medical residents, 12 respiratory technicians and 160 nurses and senior sisters. In a typical month, in the second half of 2010, the total income of the hospital would be approximately Rs. 2,500 lakhs and the outgoing Rs. 2,400 lakhs (1 lakh = 100 thousand or 10^5). This income is generated from bed charges (12%), operation theatre charges (5%), consultant doctors' fees (31%), diagnostic services (21%), materials and pharmacy (25%) and other miscellaneous income (6%). Approximately 45% of the outgoings are for staff salaries and medical

Table 5.1 Comparison of basic ICU charges versus total hospital costs during the ICU period of randomly chosen medical and surgical patients in a private institute in 2010

Diagnosis	Days in ICU	ICU (bed + service) charge (Rs.) (% of total bill)	Total hospital charge till ICU discharge or death (Rs.)
Hypoxic ischemic encephalopathy	13	60, 000 (31%)	1,88,000
Fulminant hepatic failure	3	14,000 (6.5%)	2,14,000
Convulsions (scar epilepsy)	2	14,000 (74%)	19,000
Alcoholic liver disease	4	19,000 (14.5%)	1,30,000
Transverse myelitis	11	51,000 (17%)	3,00,000
Uro-sepsis and chronic kidney disease	23	1,08,000 (22%)	5,00,000
Gastrointestinal bleed	3	14,000 (28%)	50,000
Coronary artery bypass graft	2	9,200 (5%)	1,95,000
Craniotomy	1	4,600 (4%)	1,05,000
Hip surgery (dynamic hip screw for # neck femur in 92-year male)	2	10,600 (5%)	2,10,500

Note: # indicates fracture

consultants' professional fees. Of a total income of Rs. 2,500 lakhs, only Rs. 70 lakhs are generated from direct ICU charges. The salaries of the ICU staff account for Rs. 45–50 lakhs and the rest is for the infrastructure costs of running the ICU. From this it can be seen that direct ICU income and outgoings form a relatively small component of the total hospital budget.

It is relatively difficult to get actual indirect costs of ICU care. The simplest method would be to calculate what percentage of the total bill is from direct ICU charges. The cost of an ICU bed depends on the class and varies from Rs. 4,600 to 10,400 per day. This covers infrastructure costs and payment of ICU staff. Additional daily charges exist for other consultant visits and for use of devices like mechanical ventilation, sequential compression devices for DVT prophylaxis and air mattress for bed sore prevention. Table 5.1 shows the charges of 7 randomly chosen medical patients and 3 surgical patients undergoing typical tertiary elective surgical procedures. As can be seen from the table, the above-mentioned ICU charge per se constitutes a small part of the bill, with the majority being due to drugs, other consumables and investigations. Depending on disease and duration of ICU stay, this direct ICU bill accounted from 7 to 74% of the total bill (mean 28% and median 22%) for medical patients and was only about 5% in elective surgical patients.

Conclusion

India is a vast country with a large population. Recently, development has been rapid, and correspondingly larger budgets are being allocated to health care. There is little systematic data regarding ICU expenditure, but the majority of this service is in the private sector and paid "out of pocket" by the patient. As medical expenses in general and ICU expenses in particular continue to soar, indirect payment by insurance or employers will necessarily increase in the near future.

There is a widespread acceptance that a critical illness is a very costly illness, but there is little systematically collected data in the peer-reviewed literature. Reviews from the 1990s in two large public hospitals in major cities estimated daily costs approximating between Rs. 2,000 and Rs. 5,000. A later analysis in a public–private super-specialized cancer hospital in the mid 2000s estimated the direct costs at Rs. 8,000 and an additional indirect cost of Rs. 4,000 per day totalling Rs. 12,000 per day. A random review of contemporary ICU billing in a private tertiary referral hospital suggests far higher expenses. The direct costs vary from Rs. 4,600 to Rs. 10,400 per day but this accounted for only a very small proportion of the total bill during the duration of ICU stay. This direct ICU cost ranged from 5 to 74% of the hospital bill (till ICU discharge or death). More systematic data from different regions of the country and different health services are needed to get a representative and holistic view of the cost of contemporary critical care.

References

1. Manimala Rao S, Suhasini T. Organization of intensive care unit and predicting outcome of critical illness. Indian J Anaesth. 2003;47:328–37.
2. Parikh CR, Karnad DR. Quality, cost, and outcome of intensive care in a public hospital in Bombay, India. Crit Care Med. 1999;27:1754–9.
3. Kulkarni AP, Divatia JV. Audit of intensive care unit cost for patients in Tata Memorial Hospital. A project for post graduate diploma in hospital & health care management. Symbiosis Institute of Health Care Management, Pune; 2006.

Chapter 6
Israel: Where Have We Been?

Eran Segal

Major Points

1. Israel's healthcare is very accessible. Access to critical care is thus available to all regardless of health insurance.
2. There is a severe shortage of ICU beds in acute care hospitals in Israel. Therefore, most patients requiring critical care are hospitalized in general ward beds.
3. Families tend to insist on maximal care in Israel even when care is deemed futile by the attending physicians.

Introduction

In 2004 a television reporter reported on the shortage of ICU beds in Israel. He found that there were many critically ill patients in Israeli hospitals who required critical care but were still hospitalized in regular floor beds. Even patients who were mechanically ventilated or requiring inotropic support were, more often than not, being cared for in general medical wards, with a nursing ratio that could be very low: Usually, 38–45 patients, of whom 3–5 are on mechanical ventilation, are cared for by 3–4 nurses in a medical floor at night. There is one on-call resident responsible for these patients, as well as to cover the 15–17 new admissions which come in every night in a busy medical department in Israel.

The impact of the television program was that an urgent meeting of the Health committee of the Knesset, the Israeli parliament was called. In this meeting,

E. Segal, M.D. (✉)
Department of Anesthesiology, Intensive Care and Pain Medicine,
Assuta Medical Centers, Israel

Israeli Society of Critical Care Medicine, Israel
e-mail: eranse@assuta.co.il

D.W. Crippen (ed.), *ICU Resource Allocation in the New Millennium: Will We Say "No"?*, 39
DOI 10.1007/978-1-4614-3866-3_6, © Springer Science+Business Media New York 2013

Dr. Ami Lev, the chairman of the Israeli Society of Critical Care Medicine, said: "There are dozens of people dying every year because of a severe shortage of ICU beds. The decision to admit a patient to the ICU may be a matter of life and death for these patients".

Structure and Finances of Healthcare in Israel

Israel has a socialized healthcare system. Most of the healthcare in hospitals is being delivered in either government hospitals or hospitals belonging to the largest HMO in the country, the Clalit Health services. A health insurance act that was passed in 1995 assures that all Israeli residents are eligible for a basic basket of health services including acute hospital care that in turn includes critical care. About 80% of the population purchases additional insurance, which provides a secondary layer of health services. But the government strictly controls ICU beds. The ministry of health licenses all ICU beds, and thus there is central regulation of this valuable resource. Hospitals can get special permission to open more ICU beds if they need them, usually at the expense of closing other beds in the hospital. At the same time, the government also dictates reimbursement for critical care. The reimbursement to hospitals for critical care delivered to patients is based on the principle that the less incentive there is for hospital directors to open ICU beds, the cheaper and more effective will hospital care be. Thus while the ratio of ICU beds in the USA for example is 8–15%, in Israel the number of ICU beds is about 3–5% of general acute care hospital beds. Simchen et al. showed that of patients requiring critical care who were screened in a number of large acute care hospitals in Israel, 49% were cared for in regular floor beds, 24% in specialized care units, and only 27% were admitted to the ICU [1].

The ministry of health has a dual function in Israel. On the one hand, it is the regulator of healthcare delivered, and is the main overseer of all health providers including all acute care hospitals. On the other hand, the ministry of health is the owner and administrator of 11 government-owned hospitals that account for about 50% of all acute care hospital beds in Israel. This dual function often leads to criticism of the ministry; since the ministry has a role of both regulator and owner, management of the regulatory aspects of the ministry may be in contrast to needs of the specific government hospitals. On the one hand, the ministry regulates the reimbursement of hospitals–and in this way controls the major income of hospitals, but on the other hand, the ministry directs the government hospitals with regard to providing and developing various services. The reimbursement for a day in the ICU was, until a few years ago, the same as for a day in a surgical floor. This of course meant that there was a great incentive for hospital administrators to direct to the maximum degree possible that even the most severely ill patients will be kept on regular floor beds rather than in proper ICU beds.

Pressure from the Israeli Society of Critical Care and hospital administrators led to change in the reimbursement for critical care use. The ministry of finance decided with the ministry of health that there should be a differential cost for ICU care. The price for an ICU day was significantly increased (at the price of reducing the reimbursement for medical admissions), but interestingly enough this was done only for the first 3 days of stay in the ICU. After these 3 days, a patient who requires more ICU days entitles the hospital for the same degree of reimbursement as stay in a general or surgical floor. The reason for this 3-day limit is interesting. The Simchen study [1] that compared the outcome of patients requiring critical care who were admitted to the ICU to that of patients who remained on regular floor beds showed that the difference in survival disappeared after 3 days. This was interpreted by the ministry as indication that there is no real value in extending critical care beyond the first 3 days. Therefore, the finances were designed in a manner which would pressure the ICU physicians to discharge patients earlier. In many general ICUs in Israel, average length of stay is in the range of 6–8 days with 15% mortality. Therefore, the option of discharging a patient on the 4th day is not a realistic option most of the time.

It has been shown that admission to the ICU confers a survival benefit for most patients although this benefit is not the same for all diagnosis and severity of illness. Shmueli and Sprung looked at a group of patients in their hospital who were cared for in the ICU compared to patients with the same diagnoses and severity of illness [2]. They found that the ICU led to increased survival but this advantage was greater in patients with central nervous system disorders, and in patients with APACHE II scores around 22.

Organization of Critical Care in Israel

Intensive care units in Israel developed either as offshoots of departments of anesthesia or as separate units independent of anesthesia. In many hospitals, the general or "respiratory" ICU is part of the department of anesthesia, while in other hospitals the general ICU is a separate department. Specific intensive care units have been established over the last years: medical intensive care units, surgical intensive care units, and there are also units which belong to specific services such as cardiothoracic surgery, and neurosurgery, which traditionally are managed by the surgeons in each specific department.

The general intensive care units, whether part of anesthesia or independent, are almost always closed units. Patients are admitted from a specific service, but are cared for with the intensivist as the primary physician. Surgeons and Internists are consultants for these patients. Although the approach to critical care delivery is a multidisciplinary one, primary responsibility for the patient is by the critical care physicians.

The shortage of ICU beds leads physicians in Israel to make different decisions regarding both admission and discharge of patients from the ICU. In a study of use of ICU resources, Einav et al. [3] found that Israeli physicians tended to admit patients to the ICU less than the US counterparts. At the same time, Israeli physicians were unlikely to discharge a patient from the ICU to make room for another patient.

Physicians also find it difficult to discharge patients from the ICU. In a study of discharge decisions in a tertiary hospital with an 11-bed general ICU, Levin et al. [4] found that in 18% of the patients which the attending intensivist considered ready for discharge to the ward, they could not transfer the patient. In 46% of the patients this was due to unavailability of a ward bed, or disagreement about the service, which should assume responsibility for the patient. In 21% of the patients this was due to disagreement by the ward team of the appropriateness of the transfer, and in 33% due to medical deterioration of the patient.

One of the solutions created for the shortage of ICU beds in Israel was the formation of high-dependency rooms in general floors in the hospital. These rooms that were established in almost all general medical floors are a place to admit many critically ill patients on the floor, including those requiring mechanical ventilation and hemodynamic support. The benefit for the hospital was that this created an environment, which is very cheap to maintain, which seems to have many of the attributes of true intensive care and thus alleviate the concerns of staff, family, and patients regarding the fact that there is no critical care bed to admit the patient. These rooms are usually equipped with monitors and mechanical ventilators, but do not have the capacity for advanced hemodynamic monitoring and most times even arterial lines are outside the scope of capabilities for them. The care of patients in these rooms is directed by the staff of the medical departments who are already overworked and overwhelmed by the workload in the general medical floors—Israel has one of the shortest general medical admission length of stay, number of general acute care beds, and therefore a relatively high rate of readmission following discharge. Medical floors are very busy and the ability of the physicians and nurses to care for critically ill patients with mechanical ventilation and hemodynamic instability is very limited.

Staffing of Critical Care

As in other parts of the world, there is a severe shortage of staff in critical care. Israel suffers from a shortage of physicians in general, and there are some specialties in which the shortage is very acute.

Specialization in critical care is relatively new, and was formed as a subspecialty. The program is a 2-year program into which can enter physicians with a primary specialty of Anesthesiology, Medicine, Surgery, or Pediatrics. During the training the resident must be exposed to general aspects of critical care including care of cardiac surgery patients or neurosurgical patients. For the most part, these patients in Israel are cared for in units that are part of the departments of surgery, and only in a minority of the cases are there dedicated intensivists responsible for these patients.

The shortage of physicians in intensive care led to a declaration of intensive care, together with anesthesiology, and neonatology as a specialty in an acute crisis. Because of this, at the end of a 10-year-long arbitration between the Israeli Medical Association and the government regarding physicians' salaries, physicians in these specialties were awarded a higher wage increase than the other specialties. Unfortunately, it seems that this salary increase is not enough to draw more physicians into the specialty, and the future of critical care with regard to physicians looks very bleak.

There is also a very severe shortage of nursing in ICUs and in many if not most ICUs patients will be cared for on a 1:2 nurse:patient ratio, even when caring for the most severely ill patients. It is thus not uncommon to see a nurse caring for two mechanically ventilated, unstable patients, one of whom is neutropenic with severe sepsis following chemotherapy and the other is after abdominal surgery with hemofiltration and invasive hemodynamic monitoring.

Some of the professions which are so important to providing well-organized and comprehensive critical care are not available in Israel, such as respiratory therapy, or are very limited in their presence—because of lack of funds—such as pharmacy.

The fact that there are no respiratory therapists means that the physicians and nurses manage ventilators. This leads to an added burden on the physicians and nurses and may lead to reduced compliance with protocols and guidelines. Again, since the majority of ventilated patients are cared for on general floors, the physicians caring for them and managing the ventilator are without special knowledge or training for ventilator management.

Impact of War on Critical Care

Although the vast majority of trauma cared for in Israeli hospitals is civilian, there is a significant impact of war on some of the hospitals, which care for patients during times of military conflict. Ziv hospital in Zefat, which is in the north of the country, received a significant number of war casualties during the second Lebanon war [5]. The authors describe that in a 1-month period, 1,500 military and civilian patients were treated in the hospital, and that the ICU, as well as the operating rooms, became a bottleneck for the ability of the hospital to care for these patients. This required transfers of patients to other hospital, in some cases to open up ICU beds and enable the hospital to continue to function.

During wartime or during terrorist attacks with many casualties, the allocation of resources is taken over by the medical corp. of the army which distributes patients according to the capacity of the different hospitals with regard to emergency room abilities, surgical capacity, and critical care availability. This is sometimes complex when trauma is inflicted in areas for which there is a single central medical center, such as Soroka Medical Center in the south of Israel, but the distances are relatively short, and the involvement of the military allows for Air-Medical evacuation to safely transport patients to further medical centers, with central direction.

Saying No

So what happens to the critically ill patient in an environment that has such a shortage of resources, and at the same time allows care to be delivered to the critically ill patient in a manner that is suboptimal?

The patient described in the introduction would very likely be admitted to the ICU after initial evaluation in the emergency department. Being 56 years old with an acute subarachnoid hemorrhage, most of the hospitals in Israel would manage to find an ICU bed, even if it meant discharging another patient somewhat earlier to the floor than was planned. If he required surgery or interventional radiological procedure, that would be performed and the patient then cared for in the ICU. Once it became clear over the next few days that the patient was not improving and the assessment of the physicians is that there is no reasonable likelihood of regaining meaningful recovery, preparations would be made to send the patient to the floor. Usually, an effort would be made to perform a tracheostomy and jejunostomy or gastrostomy to facilitate the more chronic care that the patient is to receive. However, at times when there is great pressure on the ICU to accept other patients, the patient would be sent to the floor even without these procedures. Although in theory the family could not demand that the patient remain in the ICU for as long as they feel that a miracle is still possible, in practice, many times patients are kept in the ICU because of family pressure. This leads to other patients, at least as deserving and often even with a greater chance of benefiting from the ICU admission, to be kept on a regular ward until an ICU bed becomes available, which may even take a few days. By this time the benefit of critical care for the patient waiting for an ICU bed may have diminished tremendously.

Despite the above, in a study of end-of-life care in 37 ICUs in 17 countries, lack of resources and need for an ICU bed were not a factor in decisions on limitation of care [6].

Once the patient is transferred to the floor, the ability to provide high-intensity critical care is dramatically reduced, and he or she will most likely succumb to an infectious complication such as a ventilator-associated pneumonia or CVL-related sepsis.

Diagnosing futility and providing end-of-life care are affected to a great degree by culture, and studies have shown differences between northern Europe, Eastern Europe, and Southern Europe. The approach by Israeli intensivists is similar to those of southern Europe. There is also a significant religious population in Israel, Jewish and Muslim, for whom acceptance of futility and allowing limitation of care when care is deemed futile are unacceptable. This may be the reason that Soudry found that although a similar percentage of physicians in Israel and the USA consider use of DNR orders, only 28% of Israeli physicians compared to 95% of north American physicians discuss this with the patient's family [7].

Summary

There is a significant shortage of resources for delivery of critical care in Israel. We are in great need of increasing the number of beds, as well as the number of clinicians from all disciplines who are involved in delivery of critical care. Still, an effort is being made to provide adequate care to all those who require it. In many cases, families demand maximal efforts to be maintained even when it seems futile, and the solution in Israel has been to provide care in other, less intense environments that enable treatment of even severely ill patients.

References

1. Simchen E, et al. Survival of critically ill patients hospitalized in and out of intensive care units under paucity of intensive care unit beds. Crit Care Med. 2004;32(8):1654–61.
2. Shmueli A, Sprung CL. Assessing the in-hospital survival benefits of intensive care. Int J Technol Assess Health Care. 2005;21(1):66–72.
3. Einav S, et al. Intensive care physicians' attitudes concerning distribution of intensive care resources. A comparison of Israeli, North American and European cohorts. Intensive Care Med. 2004;30(6):1140–3.
4. Levin PD, et al. Intensive care outflow limitation–frequency, etiology, and impact. J Crit Care. 2003;18(4):206–11.
5. Hadary A, et al. Impact of military conflict on a civilian receiving hospital in a war zone. Ann Surg. 2009;249(3):502–9.
6. Sprung CL, et al. Reasons, considerations, difficulties and documentation of end-of-life decisions in European intensive care units: the ETHICUS Study. Intensive Care Med. 2008;34(2): 271–7.
7. Soudry E, et al. Forgoing life-sustaining treatments: comparison of attitudes between Israeli and North American intensive care healthcare professionals. Isr Med Assoc J. 2003;5(11):770–4.

Chapter 7
Italy: Where Have We Been?

Marco Luchetti and Giuseppe A. Marraro

Introduction

Allocation is a frequently recurring concept from the economic literature which has sneaked into the field of medicine and calls on physicians, patients, their families and policy-makers to face up to complex challenges.

Resource allocation in medicine applies to two separate and complementary levels of care.

On the one hand, the stakes are the organisation of public health and the provision of general rules informing the management of the system (macro-allocation). The issues to be determined include: 1. which percentage of the gross domestic product a nation should earmark for public health expenses and its apportionment between the needs of research, prevention and care; 2. the means through which resources are identified for the purpose, through private insurance, mandatory contribution or taxation levied and administered directly by the state.

On the other hand, there is also the need to specify decision criteria for the daily practice of individual health care providers who are called on to decide on the utilisation of their allocated resources, while dealing with a demand side which often exceeds supply (micro-allocation).

M. Luchetti, M.D. (✉)
Department of Anaesthesia, Intensive Care, & Pain Management,
A. Manzoni General Hospital, Via dell'Eremo 9/11, Lecco 23900, Italy
e-mail: m.luchetti@ospedale.lecco.it; m.luchetti@fastwebnet.it

G.A. Marraro, M.D.
Department of Anaesthesia and Intensive Care, Fatebenefratelli and Ophthalmiatric
General Hospital, University of Milan, Milan, Italy

D.W. Crippen (ed.), *ICU Resource Allocation in the New Millennium: Will We Say "No"?*, 47
DOI 10.1007/978-1-4614-3866-3_7, © Springer Science+Business Media New York 2013

In today's society, the hospital as an institution is undermined by a now decade-old crisis of identity. Institutional structure is being remodelled along health care demand lines from a general population which is in a continuous state of flow.

In 2003, life expectancy in Italy was 2 years above the Organisation for Economic Co-operation and Development countries' average and infant mortality had improved dramatically (to 4.3 per thousand live births) since 1980, when it was three times higher [1]. The Italian population is now ageing: in 2004, the aged/working age population ratio was 50.6% and the ratio of Italians above 65 vs. those under 14 years of age was 137.7% [2]. Furthermore, it is estimated that the ageing index, i.e. the number of people under 20 against those 65 and over, as well as the population of working age, will more than double before 2050 and that future increases in public health care spending will be increasingly linked to the growth of long-term care expenditure [3].

In Italy, the number of hospital acute-care beds has collapsed from around 600 in 1990 to 400 per 100,000 inhabitants in 2002 [4]. A strategy to reduce hospital bed capacity should include adequate policies geared towards decreasing non-essential admissions while improving in-house care and facilitating early discharge. Alternative health care facilities and services should also be planned for and developed, even though this may mean that the overall costs of the health care system may not decrease at all, despite the diminishing number of beds [5].

In 2003, Italy spent (at US$ purchasing power parity) $2,258 per capita in health care expenses. The total health care expenditure amounted to 8.4% of GDP, which is similar to 8.0% in the UK but is less than what Greece (9.9%), France (10.1%) and Germany (11.1%) spent. Public spending (at 75.1%) made up more than three-quarters of the total health expenditure, above the OECD average by >3.4% points but below that of most Northern European countries [6]. Keeping public spending in check remains a major problem in Italy which has a high public deficit and is bound to bring under control such deficits as a member nation of the Euro zone.

The Italian National Health System
(Sistema Sanitario Nazionale)

As Article 32 of the Constitution of the Italian Republic states: "The Republic safeguards health as a fundamental right of the individual and as a collective interest, and guarantees free medical care to the indigent" [7].

Law 833 of 1978 sanctioned the institution of a universal health service under the governing principle of *soggettività collettiva* (collective liability), whereby no citizen should pay in relation to her or his state of health but only according to assessed taxable income. Individual income deductions thus accrue to a single tax-funded container from which the resources needed for health care are drawn [8].

Although the Italian Sistema Sanitario Nazionale (SSN) has scored highly in recent comparisons between the various national health care services [1], the system is currently having to face up to several challenges, the most important of which is to ensure geographical equity in access to care. One of the key features of Italian history is reflected in the persistence of marked North–South regional differences which tend to favour the northern and central regions in terms of wealth. The Italian system is thus characterised by strong inequalities at regional unit level. The growth of the economy has never been uniform and the gap between the central and northern regions and the south has never been bridged. Regional socio-economical disparity has created a wide-ranging dissimilarity in health service quality and efficiency between the country's regions, especially in the south, where a larger proportion of low-income families reside [9].

Over time, the evolution of the system has been such that, starting with cyclic attempts at reform to make ends meet with an ever-shrinking budget at the national as well as the European level, a powerful process of devolution has emerged during the last decade and is now in full swing [10]. Thus, one of today's main challenges of the SSN is the need to balance financial constraint with its universal and fostering ideals at a time of growing regional power autonomy [11].

During the last 10 years, numerous government emergency decrees and national laws have attempted to redress the balance between keeping the SSN efficient and the need to reduce public spending on health care or to take action on a more equitable distribution of resources. Eventually, Law 229 was passed in 1999 which established the following general principles [12]:

1. The dignity of the human person
2. Health care requirements
3. Savings in resource utilisation
4. The equality of access to care
5. The quality and adequacy of care

Law 229 establishes a novel concept into the SSN scheme, the essential levels of care (Livelli Essenziali di Assistenza, LEA), designed to ensure that all citizens may receive care of proven efficacy, with uniform requirements territorially. LEA are in substance the instruments of a political guarantee, with the aim to "determine essential levels of health care pertaining to civil and social rights which must be guaranteed on the national territory" [13].

If the current trend towards decentralisation continues, a scenario where as many different health systems as there are regions might develop. Of course, there could be room in theory for improved efficiency and more attributions within each local health care system, but to achieve this while preserving the basic principle of equal access to SSN care could be fraught with even more problems. There is a need to reflect on the fact that, however one may look at it, in spite of all the guarantees that we have regarding the health care system, strongly rooted egalitarian policies in this country have in actual fact produced a transfer of many of the issues regarding health care rationing to the level of the periphery. Indeed, the emerging picture in the public sector, whether within the remit of the territorial or of individual health care authorities, is that of ethical requirements conflicting with economics.

Ethics and Resource Allocation

The founding value which supports the constructs of traditional ethics in medicine is acting for the good of the patient ("beneficence"). The picture was altered when medicine crossed the pre- to modern period threshold to assume the features of a core value residing in the centrality of the human person and the ideal of self-determination. The patient has now become a "health care *user*" who consults a professional whose knowledge and expertise are made use of in order to arrive at options. The chief instrument which bioethicists have devised to ensure this qualitative improvement of choice is information, "informed consent" [14].

We have not yet described the characteristic model of modernity, and yet a novel mutation is under way towards a model, the symbol of which is the corporation and the task of rethinking health care within a framework of checks and fixed resources [15]. The basic question confronting anyone with responsibilities in decision-making will centre around quality components of a managerial nature: Which treatment will optimise resource utilisation and produce a satisfied patient–customer? Or, which patient can really benefit in proportion to received priority care rather than another, or others?

It would thus appear that good medical practice is the end product of "contract bargaining" which must take into account different criteria: clinical indication (the good of the patient), patient preferences and subjective values (informed consent) and *in fine*, appropriateness within the social context.

Controlling how these three elements interact with each other requires a constant commitment and synchronised interventions from the offer and supply sides as well• as from the perspective of available financial resources, implying the selection of essential levels of care to be universally and uniformly ensured, but also include the programming of supply and the regulation of individual attributions both at physician and collective levels.

The concept of the responsibility that now falls upon Italian physicians with respect to health care spending is quite novel and comes up against an opposition of a cultural nature which is, in part, justified. Physicians have become used to relate to their patients on a personal basis and, following the Hippocratic principle of "beneficence" which imposes it, to produce good for their patient's benefit. Thus, for cultural reasons, physicians are not even remotely associated with the costs of their interventions which they consider, quite rightly, to be incommensurable with the life and the restoration to health of the diseased person.

Physicians can therefore adduce their moral authority to be exempt from considerations of costs in their activities on behalf of health care. "Every time I treat a patient", as an illustrious clinician who had been personal physician to Chancellor Bismarck used to say at the turn of the last century, "I am alone with him on a desert island".

Today's public health is no longer island-bound but is in great hospitals and health trusts with intersecting and interacting activities. Physicians cannot delegate to managers and politicians their concern for the economical aspects of health care but should assume an active and responsible role.

All issues relative to the micro-allocation of resources and to the selective process that may develop in consequence arise only if one presupposes that there is a limit to the good to which more patients can aspire to access. The most difficult problem in the distribution of resources remains the finding of a convincing criterion to provide guidance, when often painful and dramatic choices have to be made, as health care providers have to do in the face of all too real inadequacies in the availability of resources [16].

"Who should have priority? What fairness criteria should be applied in this case? What type of legality are we appealing to?"

In order to carry out the choices they make, resource-allocators will have to establish a hierarchy between the very aims of fair allocation. In view of the intrinsic equal validity of all life, the aim to maximise the number of lives saved will be privileged, at least as a general principle [17].

The standard that appears fair and acceptable, even if exposed to a certain risk of subjective interpretation, is that of an adequate proportionality of criteria: health may not be for the exclusive good of socially useful individuals but at the same time it cannot be placed above all else, for the benefit of any individual, without consideration of the good of others. Perhaps there may be other people who objectively need that bed in intensive care or that organ more than one does. No lottery, no utilitarian privilege, but the serene consideration of the balance between the individual who is to recover his/her health and the entire community which must allocate scant resources.

The Situation in Italian Intensive Care Units

Intensive care is one of the most expensive specialities of medicine and accordingly demands the utmost effort in improving its efficiency. In Italy, the number of intensive care unit (ICU) beds is lower compared with other western nations but the Italian National Health Service comes up against ever-increasing financial difficulties.

Intensivists have come to realise that intensive care beds nowadays represent an expensive and limited resource [18, 19]. The lack of beds is a daily problem in many ICU [20, 21] and bed allocation has been considered one of the thorniest and stressful aspects of the intensivist's job [22].

Through a poll carried out in 2005, the Italian Association of Hospital Anaesthesiologists and Intensive Care Specialists (AAROI) surveyed the number of beds in all ICU across the country [23]. According to international standards, Italy would need 7,700 ICU beds to ensure an adequate level of care to intensive care patients. The survey reveals, however, that only 3,814 such beds are available on the whole Italian territory, which amounts to circa 50% of what would be necessary. From a detailed analysis of the data, it appears that, nationally, there is one hospital bed for 15,328 inhabitants, whereas the ideal average should be one for 7,592. In northern Italy, there is one bed for every 14,215 inhabitants, in Central Italy one for 13,601 and in the south of the peninsula and the main islands there is one hospital bed for 17,306 residents.

As already seen, the worst scenario is seen in the southern regions of the peninsula, with a number of hospital beds which represents barely a third of what is actually needed. The 1998 National Health Plan and most Regional Health Plans have indicated that 3% was the ideal ratio of intensive care beds to hospital beds. Not unexpectedly, this standard has not yet been reached and the current situation oscillates around 2.2–2.3% on average.

Human, technological and logistical resources in ICU are the determining factors for the level of care (LOC) each unit is able to deliver. The European Society of Intensive Care Medicine (ESICM) Task Force solves the problem of LOC classification of individual units by discriminating according to their nurse-to-patient ratio, since the nursing work force is the main determinant of care in intensive therapy as well as the main item of expenditure [24]. Studies have shown that resource use is often inefficient in European ICU. One of the main reasons for this inefficiency has been identified as nursing force "waste". In this respect, it is noteworthy that the organisation of labour is characterised by scant flexibility in this country, which impedes the workload to adapt to the complexity of care needed by every patient and induces proportionally much higher fixed costs [25].

The objectives of intensive care are the monitoring and support of deficient vital functions, the aim being to carry out the adequate diagnostic procedures and necessary medical and surgical treatments required to improve patient outcomes. In the past, when reimbursements were made on the basis of costs, the availability of ICU beds may have been facilitated the admission of patients who were unlikely to benefit from intensive care. In contrast, in the present day, other factors, such as the perception on the physician's part of his or her professional liability ("defensive medicine"), may play an important role in undermining the adequacy of admission to the ICU.

The definition of ethical, clinical and economical criteria for admission to ICU [26–29] and the drafting of the relevant guidelines [30–32] have been the object of a considerable international effort. The point fundamentally being that resources should be utilised appropriately and that the patient be of the right category, in the right place and at the right time. Furthermore, ethics dictates that resources be allocated where they are more likely to make an impact.

In 2003, the Italian Society of Anaesthesia, Analgesia, Resuscitation and Intensive Care (SIAARTI) issued its "SIAARTI Recommendations for admission and discharge and for the limitation of treatments in intensive care" [33]. According to this document, the clinical appropriateness of admission to, and discharge from, intensive care should be based on the following features:

1. The reversibility of the acute pathological state
2. The reasonable likelihood of expected benefits from intensive care, also considering treatment costs
3. Reasonable expectations regarding the resolution of the critical state

The rationale for ICU admission and discharge can be predicated on a priority scale which classifies patients on the basis of the expected benefit to accrue from intensive treatment. This decreasing priority scale runs from 1 (denoting the maximum expected benefit) to 4 (minimal or nil) [33]. However, it is easy enough to generate "on-paper" scenarios while upholding that patients who are too critical or not critical enough to benefit should not be admitted to intensive care, but in everyday practice to actually identify these patients is far from straightforward.

A recent study surveyed intensive care physicians' perceptions regarding inappropriate admissions and their attitudes on resource allocation in an Italian urban setting [34], also evaluating their assessment of "irrelevant" influences on the choice of allocation.

In the experience of the responding physicians, inappropriate ICU admissions were perceived as a common occurrence, though they were mainly attributed to difficulties in the evaluation of the appropriateness of the admission itself. Physicians were also aware that their decisions were not based merely on medical requirements but were often influenced by external factors such as pressure from their superiors, the referring physician and the patient's family or the threat of litigation.

Of note, the main perceived problems were all of a clinical nature and economical considerations, if any, were perceived as minor issues. The authors concluded that, when necessary, an apportionment of ICU resources should be carried out in an open fashion and adequately discussed with a full ethical review of all its aspects. A clear understanding of factors impinging on inappropriate admissions should be an important initial step to improve the decision process and equitably allocate the limited resources available to intensive care.

While a reasonable doubt, or uncertainty, may be entertained regarding the irreversibility of the clinical status, it is appropriate to initiate or continue intensive treatment. Conversely, if there is a reasonable certainty about the irreversibility of the clinical setting, it is appropriate not to initiate or to discontinue intensive measures to spare the patient the undue prolongation of the dying process [35].

Excessive treatment is ethically unsound and unanimously condemned, because it determines an inappropriate use of the means of treatment; it is uselessly painful for the patient, causes physical and mental harm and fails to respect the patient's dignity in death. Excessive treatment is also wrong from a moral point of view because it increases the suffering of family members, is frustrating for care providers and generates an inequitable distribution of resources by curtailing them for other patients [35].

The discontinuation of intensive treatment previously initiated because indicated and accepted, or because of the patient's clinical status and relevant prognosis, was insufficiently clear at the time and should be considered whenever the clinical picture counter-indicates treatment continuation, the patient withdraws consent or a previously defined therapeutic limit is reached [35].

In 2005, the Italian Group for the Evaluation of Interventions in Intensive Care Medicine (GiViTI), a network of Italian ICU researchers dedicated to the continuous improvement of quality of care, set up an investigation of end-of-life decisions in order to assess everyday practice in this sensitive field. By appending an ad hoc questionnaire to a validated electronic data-mining system [36], the researchers were able to correlate data on end-of-life decisions with many other already tabulated variables. This prospective, multi-centre, observational survey carried out in 84 ICU and enrolling 3,793 patients who had died while on intensive care or had been discharged in terminal conditions investigated the association between the mean propensity to limit treatments with overall survival [37]. The authors concluded that limiting treatment is a common occurrence in ICU and remains the overarching responsibility of the physician.

Units with a propensity to limit treatment less than the grand mean fared worse in terms of overall mortality, thus demonstrating that limitations per se are not against patients' interests: to limit treatment in terminal patients does not mean limiting their survival possibilities but only sparing them a useless and often harrowing protraction of the dying process. Thus, the propensity to limit end-of-life treatments can be viewed as a good departmental quality index. Obviously, it does not follow that intensive care physicians should increase treatment limitations in order to improve quality of care. To implement limitations simply means that intensivists act in their patients' best interests at the close of their lives, even if this implies difficult decisions from both emotional and professional points of view.

The capacity to accompany patients humanely towards their demise is characterised by several features: a dignified assisted death, controlling pain and suffering during the final moments of life, terminal sedation, communication between the people concerned, respect of the dying person's wishes and, whenever possible, seeking the patient's consent, giving up futile treatment and the introduction of palliative care measures [35].

Conclusions

Human beings have always been searching for immortality. Today's medicine appears to be a willing instrument in tackling this type of issues by making promises that will be hard not to break. The mythicisation of medicine and the endless expansion of its purview should be impugned with a critical mindset and a wide-ranging response capability in the cultural arena, so that the individual and society become aware of the actual and potential efficiency of health care services, without heeding the sirens of "mythical expectations" without corollary in scientific evidence.

The most urgent and useful form of action still to be undertaken regards these unwarranted expectations—confining with the miraculous—that society entertains about the efficacy of medicine and the message to put across as far as public opinion is concerned ought to be that death is inescapable and that the most severe diseases are incurable.

Once the inevitability of resorting to often dramatic measures in today's health care system is postulated, the problem arises of finding an ethical justification to subsequent decisions. On the basis of the choices made necessary by the scarcity of available resources, medical treatment would be "apportioned", i.e. distributed according to commitments and rules, with the inevitable exclusion—partial or total—of some from the utilization of the services themselves.

It should be evident that there is a need to make such choices in a transparent and consensual manner by eschewing spontaneity, the dynamics of which are less than controllable. In many countries, through the application of ethically correct methods, open decision criteria have been created. In Italy, institutional criteria are the "essential levels of care" (LEA) and the appropriateness of health care requirements.

Rationalisation, intended as best utilisation and fair limitation, is an economic necessity, juridically and ethically legitimate, albeit still patently inadequate, even if universally pursued. Increased efficiency is contrasted by actual limitations. To define an order of priorities may induce a targeted kind of apportionment, according to a hierarchy of emergencies which must be established at the diagnostic and treatment level. The objective must remain that of equitable apportionment. The basic concepts are those of rationalisation, rationing and priorities.

From a practical point of view, only the day-to-day gathering and periodical analysis of data on the various levels of care, on the severity of pathologies, on the respective workloads and the adoption of reliable parameters that are easy to obtain will provide the tools necessary to assess the adequateness of admissions to intensive care.

References

1. Organisation for Economic Cooperation and Development. OECD health data 2005. Paris: OECD; 2005.
2. Istat. Italia in cifre 2005. Rome: Istat; 2006. www.istat.it. Accessed 7 Oct 2010.
3. Italian Ministry of Economics and Finance, Department of General Accounts. Mid-long term Trends for the Pension and Health Care Systems Summary and Conclusions. The forecasts of the Department of General Accounts Updated to 2005. Rome: Ministry of Economics and Finance; 2005. Available in English at http://cerp.unito.it/datienormative/datiedocumenti/mid-long_term_trends1. Accessed 7 Oct 2010.
4. European health for all database [database online]. Copenhagen: WHO Regional Office for Europe; 2003. http://hfadb.who.dk/hfa. Accessed 10 Oct 2010.
5. McKee M. Reducing hospital beds: What are the lessons to be learned? European Observatory on Health Systems and Policies. Policy brief no. 6; 2004.
6. Giannoni M. Universality and decentralisation: the evolution of the Italian health care system. Eurohealth. 2006;12(2):10–3.
7. Constitution of the Italian Republic, Art. 32. Available at http://www.governo.it/Governo/Costituzione/1_titolo2.html. Available in English at http://www.senato.it/documenti/repository/istituzione/costituzione_inglese.pdf. Accessed 12 Oct 2010.
8. Legge 23 Dicembre 1978, n.833. Istituzione del servizio sanitario nazionale (G.U. n. 360 del 28 dicembre 1978). http://www.normativasanitaria.it/jsp/dettaglio.jsp?id=21035. Accessed 16 Oct 2010.
9. World Health Organization. World Health Report 2000. Geneva: WHO; 2000. www.who.int/whr/2000/en/statistics.htm. Accessed 18 Oct 2010.
10. Italian Ministry of Health. Piano Sanitario Nazionale 1998–2000. Un patto di solidarietà per la salute (Health Care Plan 1998–2000. A solidarity pact for health). Ministero della Sanità, Roma, 1998.
11. Giannoni M, Hitiris T. The regional impact of health care expenditure: the case of Italy. Appl Econ. 2002;14:1829–36.

12. Decreto Legislativo 19 Giugno 1999, n. 229. "Norme per la razionalizzazione del Servizio sanitario nazionale, a norma dell'articolo 1 della legge 30 novembre 1998, n. 419". Pubblicato nella Gazzetta Ufficiale n. 165 del 16 luglio 1999 – Suppl. Ord. n. 132. http://www.parlamento.it/parlam/leggi/deleghe/99229dl.htm. Accessed 18 Oct 2010.
13. Bindi R. Health care reform in Italy. Oxford: St Antony's College University of Oxford, European Studies Centre, Lecture series, Michaelmas Term 2002. Available in English at http://www.sant.ox.ac.uk/esc/esc-lectures/bindi.doc. Accessed 20 Oct 2010.
14. Hastings Center. Gli scopi della medicina:nuove priorità. Edizione italiana a cura di Maurizio Mori. Notizie di POLITEIA Anno 13 – N. 45, 1997.
15. Cantú E, Anessi Pessina E, editors. L'aziendalizzazione della sanità in Italia (Transformation of the Italian health care system into an 'enterprise'). Rapporto OASI, Milan: CERGAS-Bocconi; 2004.
16. La CI. medicina delle scelte. Torino: Boringhieri; 2000.
17. Berlinguer G. Bioethics, health, and inequality. Lancet. 2004;364:1086–91.
18. Szalados JE. Access to critical care: medical rationing of a public right or privilege? Crit Care Med. 2004;32:1623–4.
19. Cook D, Giacomini M. The sound of silence: rationing resources for critically ill patients. Crit Care. 1999;3:R1–3.
20. Vincent JL. European attitudes towards ethical problems in intensive care medicine: results of an ethical questionnaire. Intensive Care Med. 1990;16:256–64.
21. Metcalfe MA, Sloggett A, McPherson K. Mortality among appropriately referred patients refused admission to intensive-care units. Lancet. 1997;350:7–11.
22. Coomber S, Todd C, Park G, Baxter P, Firth-Cozens J, Shore S. Stress in UK intensive care unit doctors. Br J Anaesth. 2002;89:873–81.
23. Associazione Anestesisti Rianimatori Ospedalieri Italiani (AAROI). Censimento nazionale dei posti letto di rianimazione attivi al 30 giugno 2005. http://www.aaroi.it/Pagine/iniziative/iniziative_2005/censimento_nazionale_01.pdf. Accessed 25 Oct 2010.
24. Moreno R, Reis Miranda D. Nursing staff in intensive care in Europe: the mismatch between planning and practice. Chest. 1998;113:752–8.
25. Iapichino G, Radrizzani D, Rossi C, Pezzi A, Anghileri A, Boffelli S, Giardino M, Mistraletti G, Bertolini G, GiViTI Group. Proposal of a flexible structural-organizing model for the Intensive Care Units. Minerva Anestesiol. 2007;73(10):501–6.
26. Ron A, Aronne LJ, Kalb PE, Santini D, Charlson ME. The therapeutic efficacy of critical care units. Identifying subgroups of patients who benefit. Arch Int Med. 1989;149:338–41.
27. Sprung CL, Geber D, Eidelman LA, et al. Evaluation of triage decisions for intensive care admission. Crit Care Med. 1999;27:1073–9.
28. Bone RC, McElwee NE, Eubanks DH, Gluck EH. Analysis of indications for intensive care unit admission. Clinical efficacy assessment project: American College of Physicians. Chest. 1993;104:1806–11.
29. Rosenthal GE, Sirio CA, Shepardson LB, Harper DL, Rotondi AJ, Cooper GS. Use of intensive care units for patients with low severity of illness. Arch Intern Med. 1998;158:1144–51.
30. Society of Critical Care Medicine Ethics Committee. Consensus statement on the triage of critically ill patients. JAMA. 1994;271:1200–3.
31. American Thoracic Society. Fair allocation of intensive care unit resources. Am J Respir Crit Care Med. 1997;156:1282–301.
32. Task Force of the American College of Critical Care Medicine. Guidelines for ICU admission, discharge, and triage. Crit Care Med. 1999;27:633–8.
33. Gruppo di Studio ad Hoc della Commissione di Bioetica della SIAARTI. SIAARTI guidelines for admission to and discharge from Intensive Care Units and for limitation of treatment in intensive care. Minerva Anestesiol. 2003;69(3):101–18.
34. Giannini A, Consonni D. Physicians' perceptions and attitudes regarding inappropriate admissions and resource allocation in the intensive care setting. Br J Anaesth. 2006;96(1):57–62.
35. SIAARTI – Italian Society of Anaesthesia Analgesia Resuscitation and Intensive Care Bioethical Board. End-of-life care and the intensivist: SIAARTI recommendations on the management of the dying patient. Minerva Anestesiol. 2006;72(12):927–63.

36. Boffelli S, Rossi C, Anghileri A, Giardino M, Carnevale L, Messina M, Neri M, Langer M, Bertolini G. Continuous quality improvement in intensive care medicine. The GiViTI Margherita project—Report 2005. Minerva Anestesiol. 2006;72:419–32.
37. Bertolini G, Boffelli S, Malacarne P, Peta M, Marchesi M, Barbisan C, Tomelleri S, Spada S, Satolli R, Gridelli B, Lizzola I, Mazzon D. End-of-life decision-making and quality of ICU performance: an observational study in 84 Italian units. Intensive Care Med. 2010;36(9):1495–504.

Chapter 8
The Netherlands: Where Have We Been?

Frank H. Bosch

The Health Care System

Since 2006 the health care system has changed significantly in the Netherlands. It is based on the conviction that solidarity is a very important driving force in our society. Health care costs are financed by a dual system. All regular medical treatments (whether provided by a general practitioner or the hospital) are covered by a compulsory basic health insurance with one of the private health insurance companies. Medication costs are also covered by these insurance policies [1]. The insurance companies are obliged to provide a package with a defined set of insured treatments. The government decides every year which treatments are covered by the insurance. For instance in 2011 some treatments will no longer be covered (dental care for people older than 18 years; oral contraceptive agents for women over the age of 21) while others will be added (medication for smoking cessation). Some other costs, for instance the costs of a geriatric walker, are the subject of intensive political debate. On the one hand, the politicians try to cut costs, but on the other hand, they do not want to upset the public too much. So, for 2011, the walker can be reimbursed [2]. The insurance companies have to accept everybody for the same price for this basic package and they cannot refuse patients. The same premium is paid whether young or old, healthy or sick [3]. Affordability is guaranteed through a system of income-related allowances and individual and employer-paid premiums. In addition to this basic insurance package, it is possible to buy additional coverage for expenses that are not covered by this package. These additional packages differ markedly between the insurance companies and this is an area of intense competition. Furthermore, there is no obligatory acceptance policy with these packages and

F.H. Bosch, M.D., Ph.D. (✉)
Department of Internal Medicine, Rijnstate Hospital Internal Post 1241,
PB 9555, Arnhem 6800TA, The Netherlands
e-mail: fhbosch@rijnstate.nl

D.W. Crippen (ed.), *ICU Resource Allocation in the New Millennium: Will We Say "No"?*,   59
DOI 10.1007/978-1-4614-3866-3_8, © Springer Science+Business Media New York 2013

the insurance companies can refuse patients to buy these packages or demand a higher price.

Long-term treatments, especially those that involve semipermanent hospitalization and disability costs (wheelchairs), are covered by a state-controlled mandatory insurance. This is covered under the general law on exceptional health care costs (AWBZ), which came first into effect in 1968. This law states that there must be a medical reason as a reason why an individual cannot perform certain tasks. This AWBZ provides for instance for hospital costs after 1 year, nursing home costs, home care, a dog for the visually impaired, and abortion in a licensed clinic. The AWBZ also pays for the many vaccinations that are provided. The payments for services are usually done as a budget for which the benefitted must account for. The premium for the AWBZ is a percentage of the tax income. In 2008 the average premium per contributing person amounted to 4,000 Euros. The cost of the AWBZ is increasing due to the increase in the number of older persons in the Netherlands.

The population in the Netherlands

In 2009 there were 16,515,057 people living in the Netherlands; this number will increase to 17.8 billion in 2040 and decrease afterwards. We are facing an epidemic of older people. The number of people above the age of 65 will increase from 6% in 1900 to 15%. In some areas the percentage of older people will increase to 25% (see Fig. 8.1).

The age expectancy for a man is 75.55 years and for a woman 81 years (2001). This will probably increase due to the decreasing number of smokers and better health care. This increase in the number of older people will undoubtedly cause a rise in the health expenditures.

Health Care in Numbers

There are around 11,000 general practitioners in the Netherlands. It is customary to visit your general practitioner first for every complaint. This is called "The first line." In 2008 76% of patients visited a general practitioner. The average patient had 5,6 visits to a general practitioner in that year. Women obtained a prescription 7,8 times and men 5.5 times. The average general practitioner referred 212 women per 1,000 women to a specialist and 159 men per 1,000 men. Diabetes is commonly treated by the general practitioner (44.9/1,000 patients) as well as COPD (18.3/1,000 patients) and heart failure (8.8/1,000 patients) [4].

There are 145 hospitals in the Netherlands. Most hospitals are public general hospitals; there are eight university hospitals and a few dedicated hospitals (cancer, hemodialysis). The number of private clinics is increasing. Together these clinics represent a turnover of 15 billion Euros [5]. The number of hospital beds in the

Fig. 8.1 Expected percentage of older people in the Netherlands in 2025 (source: CBS)

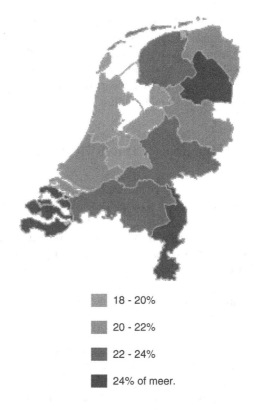

■ 18 - 20%

■ 20 - 22%

■ 22 - 24%

■ 24% of meer.

Netherlands is around 50,000. There is a steady decrease in the number of beds over the years. The aim of the government is to decrease the number of beds to 2 per 1,000 inhabitants in 2015. The average length of stay was 7.2 days in a university hospital, 5.6 days in a large teaching hospital, and 5.4 in a general hospital.

Quality Assurance

In addition to all the work that is being done by the medical scientific communities a new effort to make quality more visible and more uniform, a new initiative, was launched a couple of years ago: www.zichtbarezorg.nl. Its aim is to provide unbiased information about the quality of care of health care providers. "Care of good quality is patient directed, effective, safe, timely and customized to the needs of the individual patient." For general practitioners a set of quality indicators has been developed which will be evaluated in the near future. This set of indicators gives an impression about the care for patients with chronic diseases: for instance for diabetes mellitus the set investigates the presence of diabetes-oriented consultation hours, the number of blood pressure measurements, the percentage of patients with an HbA1c percentage <7 or >8.5, and the structured investigation for other risk factors for vascular damage [6]. For hospital care, many diagnoses have been analyzed and performance indicators have been made. The performance indicators have been divided into two groups: one based on the directives that have been made by the medical specialist and one based on the client preferences. The indicators are not yet used broadly by patients to choose between hospitals, but it is foreseeable that they will do so in the future.

Performance of Dutch Healthcare

The ministry of Health had commissioned the National Institute for Public Health and the Environment (RIVM) to produce the Dutch Health Care Performance Report every 2 years. In this report it is shown that health care is, generally speaking, in good health. Access to care is, mainly due to our many general practitioners, excellent. Most users are consistently positive about the care they receive. Dutch people are living longer, and the two additional life years gained are spent in good health. Disease prevention and health care are both critical factors in these health gains. Safety is firmly on the agenda with reports showing that the percentage of hospital patients with pressure ulcers or malnutrition is halved. The standardized mortality rate in hospitals declined somewhat between 2003 and 2008. Very important, it was shown that the growth in health care expenditures is mainly due to the increase in the volume of care. Of course, not all is well: waiting times and reach of services are not optimal in all cases [7]. And furthermore, a recent analysis by Nivel/Emgo demonstrated that the number of unnecessary deaths in Dutch hospitals is around 2,000 [8].

Medical Education

Medical studies can be done in eight universities: Amsterdam (2), Rotterdam, Utrecht, Groningen, Nijmegen, Maastricht, and Leiden. Admittance is restricted. In most cities there is a combined system. Roughly 50% of places are given on the basis of a weighted lottery (the higher your examination marks, the greater the chance that you are admitted). The other 50% are given on the basis of a motivational interview or other criteria. After the bachelor phase of 3 years, most students enroll into a master program. During these last 3 years they spend a large amount of time in the hospital in hospital-based training. After this graduation students can start training for a specialty. Admittance to these programs is based on motivation, expertise, and research interest. The number of trainees is determined by a government body which investigates the need for medical specialists in the future (capaciteitsorgaan). Many specializations know a subspecialty in the last 2 years of training. For instance, after 4 years of training in internal medicine candidates can apply for further training in intensive care medicine, hematology, oncology, infectious disease medicine, etc.

Intensive Care

Intensive care medicine in the Netherlands is not a separate specialty. Historically, the sickest patients were grouped together in the hospitals and they were cared for by a number of specialists. These days most hospitals have separate intensive care units, run by intensivists. The training programs for intensivists are in the process of revision. They will be adapted to the competence program of the European Society of Intensive Care Medicine (www.cobatrice.org). In the Netherlands almost all intensive care departments are closed units with the intensivist acting as the primary physician for the patients in that department. Most intensivists are internists or anesthesiologist by training. There are some cardiologist/intensivists, neurologist/intensivists, or pulmonologist/intensivists. Generally speaking, they have had at least 4 years of training in their primary specialty, followed by at least 2 years of additional training in intensive care medicine. The training of intensivists is governed by a regulatory body consisting of the presidents of several medical scientific communities.

In 2006 a report was produced by the Dutch Institute for Healthcare Improvement (CBO) regarding the organization of intensive care departments [9]. It gave detailed instructions about the number of physicians, nurses, and supportive personnel that should be available at the bedside. Furthermore it gave clear-cut definitions about levels of intensive care departments; for instance, a level 3 intensive care department is an intensive care that can deliver the most complicated intensive care, and has more than 3,000 ventilator days and most people at the bedside. A level 1 intensive care only provides basic-level intensive care. This report has had an enormous influence in the Netherlands. It was foreseen that it would take at least 5 years before all recommendations would be implemented, but it is fair to say that they are almost all implemented today. In 2011/2013 a new guideline about the treatment of intensive care patients will be published. This program is endorsed by the scientific communities of intensive care, internal medicine, and anesthesiology.

Costs of Medical Care

The costs of medical care in the Netherlands are regularly investigated by the RIVM. Their latest report regards the year 2005 and was produced in 2008. Total costs of healthcare amount to 68.5 billon Euros, around 4,200 Euros for every inhabitant. Most costs go to the treatment of psychological and psychiatric disturbances (14.2 billion) and cardiovascular diseases (5.5 billion). Hospital care is the most expense segment with 17.7 billion Euros. The costs of health care have increased yearly with 3.9% per year between 2003 and 2005. One percent can be attributed to an increase in prices, 0.8% to an increase in the number of inhabitants, and 2.1% to an increase in volume per patient. This is especially true for the increase in the productivity of the hospitals (4.7%). There is not much information about the costs of intensive care treatment. Historically, all the costs of intensive care treatment were hidden in the average cost of a specific diagnosis. For instance: If a patient was treated for pneumonia in the intensive care department with artificial ventilation, continuous veno-venous hemodiafiltration, etc, these costs would not be separately visible. In 2006 a new system was introduced [10]. This system aims to keep intensive care costs separate from the other costs, which are accounted for in a system of "Diagnose Behandeling Combinaties (DBC)," very much like the Diagnosis Related Groups in the USA. In this new system all intensive care treatment will be reimbursed on the basis of a few parameters: intensive care admission, intensive care treatment day, intensive care ventilation day, and intensive care hemofiltration. These parameters are reimbursed with a difference which is determined by the level attribution of the intensicare care department. For instance: A level 1 intensive care department is reimbursed less per ventilator day, in comparison to a level 3 intensive care department. Unfortunately, the system is only in place for a rather short period of time and there are up to now no structural analyses done.

References

1. Wikipedia. Healthcare in The Netherlands 2010: Available from: http://en.wikipedia.org/wiki/Healthcare_in_the_Netherlands.
2. VWS. Basisverzekering. 2010.
3. VWS. http://english.minvws.nl/en/themes/health-insurance-system/default.asp. 2010.
4. Nivel. http://www.nivel.nl/pdf/Rapport-kerncijfers-LINH-2008.pdf. 2010.
5. NVZ. http://www.nvz-ziekenhuizen.nl/Feiten_en_cijfers/FAQ. 2010.
6. huisartsen ZZ. https://zichtbarezorg.dmdelivery.com/mailings/FILES/htmlcontent/Eerstelijnsz org/eindrapport_definitief%20publieke%20indicatoren%20260609.pdf.
7. RIVM. http://www.gezondheidszorgbalans.nl/object_binary/o9931_Dutch_Health_Care_Performance_Report_2010_ES.pdf. 2010.
8. NIVEL/EMGO. http://www.nivel.nl/pdf/Rapport-zorggerelateerde-schade.pdf. 2010.
9. CBO. www.cbo.nl/Downloads/111/rl_ic_2006.pdf.
10. NZA. www.nza.nl/104107/139830/Advies-bekostiging-van-de-IC.pdf.

Chapter 9
New Zealand: Where Have We Been?

Stephen Streat

New Zealand

New Zealand is a small island nation (population ~4.4 million in 2010) in the south west Pacific, at least 2,000 km from any of its nearest neighbours—Australia and other South Pacific island nations. The population is increasingly ethnically diverse, including ~15% who identify as Maori (the indigenous people) and ~15% as Asian or Pacific people [1]. New Zealand ranks highly internationally on many measures of quality of life including educational attainment, economic, political and press freedom and lack of corruption. It was ranked third in the world in 2010 (behind Norway and Australia and just ahead of the USA) by the UN Development Programme [2] on the Human Development Index—a composite index of life expectancy, education and income. Per capita GDP at purchasing power parity (PPP) is $27,420 in 2010—or 58% of US per capita GDP [3]. Total health expenditure per capita at PPP was $2,497 in 2007 [2], around 75% of that in Australia, Canada and most countries in western Europe but only a third of US health expenditure at PPP in 2007 ($7,285).

The New Zealand Health System

New Zealand has a long tradition of progressive social policy—including universal suffrage (1893), an old-age pension (1898) and the establishment of a national department of health (1901), initially largely confined to policy matters. Following

S. Streat, B.Sc., M.B., Ch.B., F.R.A.C.P. (✉)
Department of Critical Care Medicine, Auckland City Hospital,
2 Park Road, Grafton, Auckland 1023, New Zealand
e-mail: stephens@adhb.govt.nz

D.W. Crippen (ed.), *ICU Resource Allocation in the New Millennium: Will We Say "No"?*, 65
DOI 10.1007/978-1-4614-3866-3_9, © Springer Science+Business Media New York 2013

the election of the first Labour (~socialist) Government in 1935, an extensive welfare state rapidly developed. This included compulsory union membership, a minimum wage and limits on working hours, publicly funded secondary education and a comprehensive taxpayer-funded social security system. There were benefits for the unemployed, sick or incapacitated and a universal superannuation benefit for all New Zealanders at age 65. A national healthcare system (including dental treatment for children), which had been subsidised from the 1920s, was increasingly expanded to include free hospital inpatient treatment for the whole population (1939), free hospital outpatient treatment, part-payment of general practitioners' bills and free prescription pharmaceuticals (1941), physiotherapy (1942), district nursing services and maternity care (1944), and laboratory diagnostics (1946). These components of a national health service were retained by the National (~conservative) Government which took office in 1949 and have largely been retained to the present day. A universal no-fault accident compensation scheme for all personal injury (now also including iatrogenic injury [4]) was introduced in 1974. The scheme covered (but not generously) the costs of medical expenses, rehabilitation and related transport. It included some earnings-related compensation, lump sum payments for permanent loss or impairment, pain and mental suffering and provision for funeral costs and lump sum payments to surviving spouses and children in case of accidental death. The scheme is fully funded from several sources including levies on employees, employers and direct and indirect taxation (on petroleum and motor vehicle registrations). It has been somewhat modified by legislation over the last 10 years, with an increasing focus on prevention and early rehabilitation, but remains an important feature of the New Zealand healthcare system (Accident Compensation Corporation, http://www.acc.co.nz/). In 1993 a national pharmaceutical management agency was established (PHARMAC, http://www.pharmac.govt.nz/) with the statutory purpose to "*secure for people in need of pharmaceuticals, the best health outcomes that are reasonably achievable from pharmaceutical treatment and from within the amount of funding provided.*" The activities of this agency include deciding which pharmaceuticals are subsidised and to what extent, negotiating (bulk, national) contracts with pharmaceutical suppliers and promoting the most effective use of pharmaceuticals to health professionals and the public. Public healthcare is provided by District Health Boards (DHBs) which are responsible for their regional health and disability services. This includes part-funding of primary healthcare organisations (which include general practitioners and other health professionals) on an annual payment per capita of persons enrolled (http://www.moh.govt.nz/primaryhealthcare). The Government estimated [5] that in 2008, 81% of total national healthcare expenditure was Government-funded and only 19% was paid for out-of-pocket or via personal health insurance. Although pharmaceuticals are heavily subsidised, there are small part-charges on prescription items (~US $2) for most people. Visits to a general practitioner are fully subsidised only for children under the age of six, but modest fees are charged for other children (typically ~US $20) and adults (typically ~US $40). These fees do act as barriers to accessing some healthcare services, particularly for low-income individuals [6]. Private (insurance-based and self-pay) healthcare has developed, but currently only 32% of New Zealanders have some form of private health insurance [7]. Most of the insured hold policies that cover elective surgical and specialist care but not day-to-day medical costs such as general practice consultation and prescription charges [7]. Recently, some publicly funded elective surgery has been contracted out to private providers.

Intensive Care Medicine

Intensive care medicine (as it is now called in New Zealand) began around 60 years ago. It has been suggested [8] that "Bjorn Ibsen's day" (26 August 1952) might, albeit arbitrarily, mark the beginning of modern intensive care medicine—combining as it did IPPV via tracheostomy for respiratory failure (due to poliomyelitis) and circulatory support for the cardiovascular collapse that ensued after thiopentone sedation. However the early Scandinavian pioneers knew of the previous large-scale successful use of IPPV via tracheostomy for poliomyelitis in Los Angeles [9] in 1949. Even before that, high-dose inotropic support and massive fluid infusion were used in the successful treatment of septic shock in 1946 [10], and 5 h of IPPV via tracheostomy had been used in 1850 to resuscitate an apparently lifeless child with airway burn [11].

Intensive care medicine began early in New Zealand—isolated patients with respiratory failure were ventilated in general wards during the mid-1950s and the first multi-disciplinary ICU probably began in Auckland Hospital in 1958 [12]—the earliest incarnation of the present Department of Critical Care Medicine. At that time, most patients were young and previously well, and were treated for airway diseases or respiratory failure accompanying poliomyelitis, tetanus, Guillain–Barré syndrome, barbiturate self-poisoning or surgery [12]. ICU mortality 1958–1961 was reported as 25% [12]. The work was arduous and intensive care stays were very long by comparison with today, but it was clear that many patients with what were previously thought to be lethal conditions could return to good health. Intensive care resource allocation was very limited. The necessity for specialised equipment and staffing was recognised and the model of 'closed units' (which is usual in New Zealand and Australia) was promulgated very early [12–16].

Intensive care medicine in New Zealand and Australia has always been closely linked, initially—in the late 1960s—by personal contact among a network of pioneers, and later by a strong professional organisation—the Australian and New Zealand Intensive Care Society—ANZICS, which they established in 1975 (http://www.anzics.com.au/). The considerable congruence between intensive care practice in Australia and New Zealand is maintained by strong and close personal and professional relationships within ANZICS, a long-standing common training pathway for intensivists and most recently the College of Intensive Care Medicine of Australia and New Zealand (http://www.cicm.org.au/). The recent considerable success (e.g [17–31]) of the ANZICS Clinical Trials Group (http://www.anzics.com.au/clinical-trials-group) is a further indication of the closeness of the Australia–New Zealand relationship.

Nevertheless, there are some important differences between intensive care practice in these two countries, which might well be diverging somewhat, particularly in respect of private intensive care services and remuneration practices, attitudes to ICU admission and treatment limitation [32], and responses to real or perceived threat of legal action or complaint. From this authors perspective, New Zealand remains relatively unchanged in these matters [33, 34] while Australia seems to be shifting in the direction of US practice [32, 35, 36].

During the 1970s, intensive care services were established in most of the large public hospitals and some of the smaller ones. Early pioneers suggested that 2–4% of acute hospital beds should be intensive care beds [16] in order to meet the perceived patient need and intensive care services were sized accordingly.

A close collegial network of the small number of intensive care specialists rapidly developed. Clinical practice variation within units was reduced by the establishment of small teams of specialists working closely together within a "closed-unit" model of care. Similarly clinical practice variation between units was (relatively) reduced by virtue of close cooperation and information-sharing—by mutual visitation of other units and by sharing clinical experiences at intensive care meetings within New Zealand and Australia.

During the second half of the 1970s in many New Zealand ICUs, including our own, there was a burgeoning of elderly surgical patients undergoing major elective surgical procedures and this, together with an increase in road trauma and admissions for asthma and (tricyclic) poisoning, led to an increase in the demand for intensive care services, including high-dependency admissions, but little if any increase in supply of such services.

ICU availability was largely limited (as it still is) by nursing staff, rather than by the number of physical beds or items of essential equipment. In both New Zealand and Australia intensive care nursing (for ventilated patients) remains an activity performed at 1:1 dependency. In other countries where one ICU nurse may look after 2 (or more) ventilated patients, other persons (e.g. nurse-assistants, respiratory therapists) might also be employed, but in New Zealand this has not occurred.

At a hospital level, the distribution of funds as budgets to services, including intensive care services, was made on the basis of perceived clinical need (rather than strictly supply and demand)—within a largely internal political process of continuous re-adjustment to demands, expectations, innovations, supply and achievements.

Even by the early 1980s, when we first compared intensive care service provision in New Zealand with overseas (US) intensive care [37], it was clear that there was substantially less resource allocated to intensive care in New Zealand. In that study [37] only 1.7% of hospital beds (30/1,781) in two large New Zealand hospitals were adult ICU beds compared with 5.5% (475/8,620) in 13 similar US hospitals.

Notably, in both New Zealand hospitals but only two of the 13 US hospitals did intensive care specialists ("ICU co-directors") "completely control admission, discharge and most treatment decisions" [37]. This "closed ICU model" was an early defining feature of intensive care practice in New Zealand and remains so to this day.

Table 9.1 Variation in adult ICU beds and admissions in ten countries

Country	Adult ICU beds/100,000	Adult ICU admissions/100,000	Adult admissions/bed
UK [40]	3.5	216	62
New Zealand [39]	4.6	389	85
Australia [39]	7.7	558	72
Spain [40]	8.2	N/A	N/A
Netherlands [40]	8.4	466	55
France [40]	9.3	426	46
Canada [40]	13.5	389	29
USA [40]	20	1,923	96
Belgium [40]	21.9	1,051	48
Germany [40]	24.6	2,353	96

Intensive care resources have increased somewhat in recent years, as has health funding from public sources and with it elective surgical services. This has been accompanied, as in Australia, by an ongoing increase in ICU admission of elderly co-morbid surgical patients [38].

The number of available (adult) intensive care beds per 100,000 population in New Zealand remains low by comparison with other developed countries—see Table 9.1. Nevertheless, adult ICU admission rates (per 100,000 population) in New Zealand are similar to those of many countries which have much greater ICU bed availability, reflecting the effects of higher ICU occupancy and faster "turnover".

This situation has been largely accepted as reasonable by the citizens of the country and by the health professionals involved in both intensive care medicine and other hospital-based specialities which refer patients for consideration of intensive care admission. The ways in which intensive care resources are, in effect, rationed by intensivists [33] include primary or secondary prevention of ICU admission, explicit rationing of access, increasing efficiency of throughput (and effectiveness of treatment) and reducing marginal costs. By way of example New Zealand intensivists have been involved in primary prevention initiatives in trauma [33], prescription drug poisoning and the misuse of alcohol and other drugs as well as secondary prevention of severe life threatening asthma [33]. Most ICUs have formalised their control of ICU admission within admission policies—of which these four excerpts are typical: *(1) The admission of any patient, emergency or elective, can only occur with the agreement of the Intensivist. (2) All admissions will ultimately be at the discretion of the attending Intensive Care Specialist. (3) All admissions to ICU and HDU must be approved by the ICU Registrar (senior resident) in consultation with the Duty Consultant, 24 h a day. (4) Decisions about suitability for ICU admission are made by the Duty Intensivist.* Currently only one of the 25 public hospital ICUs in New Zealand is run as an "open" unit—with admission available to other specialists—but that unit requires mandatory next-day-at-the-latest clinical review by the ICU director. Reducing marginal costs by the use of cheaper equivalent medications and supplies is an established aspect of the responsibility of ICU directors. Reduction in the ICU length of stay for non-survivors was possible (without increasing mortality) [33] by recognising and responding to unfavourable illness trajectory [41] or evidence of severe irreversible brain damage [34]. As an integral part of ICU admission, New Zealand intensivists often define a contract of "reasonable" treatment—i.e. some boundaries of what specific treatments might be offered or within which continued application of intensive treatments remains reasonable. Such "treatment limitations" are more likely in patients with pre-existing *severe functional impairment* due to co-morbidity. These "indicative limits" are made explicit to other involved professionals (e.g. surgeons) and the family. (Rarely these discussions are able to involve the patient directly, and usually only in the context of planning for major elective surgery where post-operative intensive therapy is thought necessary). Examples of such "indicative contracts" might include—*(1) "This elderly co-morbid patient has presented with an intra-abdominal surgical catastrophe. Assuming that remediable surgery is able to be completed successfully, we will admit the patient ventilated, with a thoracic epidural anaesthetic and accompanying inotropic support, provide 'good housekeeping' and expect a short ICU stay with progressive improvement. This seems reasonable to us. However, should oliguric renal failure develop we will not provide renal replacement therapy and should a major stroke or similar complication develop we will seek to withdraw intensive treatment and provide comfort care." (ii) "This patient has severe community-acquired pneumonia in the context of neutropenia after recent chemotherapy for haematologic malignancy. Some degree of extracellular fluid overexpansion has probably occurred secondary to administered blood products over the last few days. Oxygenation in particular is now threatened. We are prepared to add non-invasive ventilatory support, active diuresis and modest inotropic support for a few days, pending marrow recovery and control of infection. However, should oxygenation deteriorate further, progressive septic shock or other new organ failures develop, we will not escalate to include either invasive ventilation or renal replacement therapy and will discuss treatment withdrawal."*

Limiting (and withdrawing) intensive therapies are common practices in New Zealand [42, 43] and are usually well accepted by other health professionals, patients and their families [44–46]. It has been suggested that the "time" that is necessary to ensure that withdrawal of therapy is acceptable to all involved is a consequence of a salaried system of intensivist practice in New Zealand [45, 46]—where there is no financial incentive to focus on (usually procedure-based) billable items in fee-for-service payment systems.

Summary

Intensive care medicine began early in New Zealand, within a long and well-embedded cultural tradition of universal public healthcare. It developed in close partnership with Australia and remains very similar to Australian practice. However, in parallel with the relative wealth of the two countries, intensive care services are less available in New Zealand than in Australia and New Zealand has not developed intensive care services in private hospitals. Intensive care resources are rationed by intensivists in a closed-unit model of ICU management with a strong emphasis on effectiveness, efficiency and value for money. This includes well-developed practices of limiting and withdrawing intensive therapy when appropriate. This approach to national intensive care service provision remains culturally acceptable in New Zealand while small ongoing increases in resource availability continue to occur.

References

1. New Zealand Census Data 2006. Statistics New Zealand, Wellington. Available via http://www.stats.govt.nz/. Accessed 16 Nov 2010.
2. The Real Wealth of Nations: Pathways to Human Development. Human Development Report 2010. United Nations Development Programme, UN Plaza, New York, NY 10017, USA. Available via http://hdr.undp.org/en/. Accessed 16 Nov 2010.
3. World Economic Outlook database. October 2010. International Monetary Fund. Washington DC 20431, USA. Available at http://www.imf.org/external/pubs/ft/weo/2010/02/weodata/index.aspx. Accessed 16 Nov 2010.
4. Davis P, Lay-Yee R, Fitzjohn J, Hider P, Briant R, Schug S. Compensation for medical injury in New Zealand: does "no-fault" increase the level of claims making and reduce social and clinical selectivity? J Health Polit Policy Law. 2002;27(5):833–54.
5. The New Zealand Health and Disability System: Organisations and Responsibilities. Briefing to the Minister of Health. Ministry of Health, Wellington; 2008. Available at http://www.moh.govt.nz/moh.nsf/pagesmh/8704/$File/nz-health-disability-system-briefing2008.doc. Accessed 16 Nov 2010.
6. Schoen C, Doty MM. Inequities in access to medical care in five countries: findings from the 2001 Commonwealth Fund International Health Policy Survey. Health Policy. 2004;67(3):309–22.
7. Health Funds Association of New Zealand. Fact File - Health Insurance in New Zealand, October 2010. Available at http://www.healthfunds.org.nz/. Accessed 23 Nov 2010.
8. Trubuhovich RV. August 26th 1952 at Copenhagen: 'Bjørn Ibsen's Day'; a significant event for Anaesthesia. Acta Anaesthesiol Scand. 2004;48(3):272–7.
9. Trubuhovich RV. On the very first, successful, long-term, large-scale use of IPPV. Albert Bower and V Ray Bennett: Los Angeles, 1948–1949. Crit Care Resusc. 2007;9(1):91–100.
10. Christie C, Bowden TM, Mackenzie IR, Naismith J, Bryant J. Massive doses of adrenaline in acute toxic peripheral circulatory collapse. Lancet. 1947;250:206–7.
11. Poland A. Guy's Hospital. Scalded glottis; tracheotomy; artificial respiration for five hours and a half; recovery. Lancet. 1850;55:670–1.
12. Spence M. The emergency treatment of acute respiratory failure. Anesthesiology. 1962;23:524–37.
13. Spence M. A regime for the treatment of tetanus. N Z Med J. 1961; 325–34.
14. Irvine RO, Montgomerie MB, Spence M. Assisted respiration, noradrenaline infusion and the artificial kidney for glutethimide ("Doriden") poisoning. Med J Aust. 1963;2:277–9.
15. Campion DS, Spence M. Barbiturate intoxication occurring in Auckland. N Z Med J. 1964;63:206–10.
16. Spence M. An organization for intensive care. Med J Aust. 1967;1(16):795–801.
17. The SAFE Study Investigators. Saline or albumin for fluid resuscitation in patients with traumatic brain injury. N Engl J Med. 2007;357(9):874–84.
18. Myburgh JA, Cooper DJ, Finfer SR, Venkatesh B, Jones D, Higgins A, Bishop N, Higlett T, The Australasian Traumatic Brain Injury Study (ATBIS) Investigators for the ANZICS Clinical

Trials Group. Epidemiology and 12-month outcomes from traumatic brain injury in Australia and New Zealand. J Trauma. 2008;64(4):854–62.

19. Finfer S, Liu B, Taylor C, Bellomo R, Billot L, Cook D, Du B, McArthur C, Myburgh J, The SAFE TRIPS Investigators. Resuscitation fluid use in critically ill adults: an international cross-sectional study in 391 intensive care units. Crit Care. 2010;14:R185. Available at http://ccforum.com/content/14/5/R185.

20. The SAFE Study Investigators. Impact of albumin compared to saline on organ function and mortality of patients with severe sepsis. Intensive Care Med. 2011;37(1):86–96. Epub October 2010.

21. The Blood Observational Study Investigators on behalf of the ANZICS-Clinical Trials Group. Transfusion practice and guidelines in Australian and New Zealand intensive care units. Intensive Care Med. 2010;36(7):1138–46.

22. Robertson MS, Nichol AD, Higgins AM, Bailey MJ, Presneill JJ, Cooper DJ, Webb SA, McArthur C, MacIsaac CM, The VTE Point Prevalence Investigators for the Australian and New Zealand Intensive Care Research Centre (ANZIC-RC) and the Australian and New Zealand Intensive Care Society Clinical Trials Group (ANZICS CTG). Venous thromboembolism prophylaxis in the critically ill: a point prevalence survey of current practice in Australian and New Zealand intensive care units. Crit Care Resusc. 2010;12(1):9–15.

23. Flabouris A, Chen J, Hillman K, Bellomo R, Finfer S, The MERIT Study investigators from the Simpson Centre and the ANZICs Clinical Trials Group. Timing and interventions of emergency teams during the MERIT study. Resuscitation. 2010;81(1):25–30.

24. Bellomo R, Morimatsu H, Presneill J, French C, Cole L, Story D, Uchino S, Naka T, Finfer S, Cooper DJ, Myburgh J, on behalf of the SAFE Study Investigators and the Australian and New Zealand Intensive Care Society Clinical Trials Group. Effects of saline or albumin resuscitation on standard coagulation tests. Crit Care Resusc. 2009;11:250–6.

25. The ANZIC Influenza Investigators. Critical care services and 2009 H1N1 influenza in Australia and New Zealand. N Engl J Med. 2009;361(20):1925–34.

26. The RENAL Replacement Therapy Study Investigators. Intensity of continuous renal-replacement therapy in critically ill patients. N Engl J Med. 2009;361(17):1627–38.

27. NICE-SUGAR Study Investigators, Finfer S, Chittock DR, Su SY, Blair D, Foster D, Dhingra V, Bellomo R, Cook D, Dodek P, Henderson WR, Hébert PC, Heritier S, Heyland DK, McArthur C, McDonald E, Mitchell I, Myburgh JA, Norton R, Potter J, Robinson BG, Ronco JJ. Intensive versus conventional glucose control in critically ill patients. N Engl J Med. 2009;360(13):1283–97.

28. Chen J, Bellomo R, Flabouris A, Hillman K, Finfer S, MERIT Study Investigators for the Simpson Centre; ANZICS Clinical Trials Group. The relationship between early emergency team calls and serious adverse events. Crit Care Med. 2009;37(1):148–53.

29. Peake SL, Bailey M, Bellomo R, Cameron PA, Cross A, Delaney A, Finfer S, Higgins A, Jones DA, Myburgh JA, Syres GA, Webb SA, Williams P, ARISE Investigators for the Australian and New Zealand Intensive Care Society Clinical Trials Group. Australasian resuscitation of sepsis evaluation (ARISE): a multi-centre, prospective, inception cohort study. Resuscitation. 2009;80(7):811–8. Epub 2009 May 20.

30. Doig GS, Simpson F, Finfer S, Delaney A, Davies AR, Mitchell I, Dobb G. Nutrition Guidelines Investigators of the ANZICS Clinical Trials Group. Effect of evidence-based feeding guidelines on mortality of critically ill adults: a cluster randomized controlled trial. JAMA. 2008;300(23):2731–41.

31. Bellomo R, Chapman M, Finfer S, Hickling K, Myburgh J. Low-dose dopamine in patients with early renal dysfunction: a placebo-controlled randomised trial. Australian and New Zealand Intensive Care Society (ANZICS) Clinical Trials Group. Lancet. 2000;356(9248):2139–43.

32. Young PJ, Arnold R. Intensive care triage in Australia and New Zealand. N Z Med J. 2010;123(1316):33–46.

33. Streat SJ, Judson JA. Cost containment: the Pacific. New Zealand. New Horiz. 1994;2(3):392–403.

34. Streat S. When do we stop? Crit Care Resusc. 2005;7:227–32.
35. Faunce T. Emerging roles for law and human rights in ethical conflicts surrounding neuro critical care. Crit Care Resusc. 2005;7(3):221–7.
36. Faunce TA. Reference pricing for pharmaceuticals: is the Australia-United States Free Trade Agreement affecting Australia's Pharmaceutical Benefits Scheme? Med J Aust. 2007;187(4): 240–2. Epub 2007 Jun 13.
37. Zimmerman JE, Knaus WA, Judson JA, Havill JH, Trubuhovich RV, Draper EA, Wagner DP. Patient selection for intensive care: a comparison of New Zealand and United States hospitals. Crit Care Med. 1988;16(4):318–26.
38. Bagshaw SM, Webb SA, Delaney A, George C, Pilcher D, Hart GK, Bellomo R. Very old patients admitted to intensive care in Australia and New Zealand: a multi-centre cohort analysis. Crit Care. 2009;13(2):R45.
39. Drennan K, Hart GK, Hicks P. Intensive care resources and activity: Australia and New Zealand 2006/2007. ANZICS, Melbourne. 2008 Available at http://www.anzics.com.au/. Accessed 23 Nov 2010.
40. Wunsch H, Angus DC, Harrison DA, Collange O, Fowler R, Hoste EA, de Keizer NF, Kersten A, Linde-Zwirble WT, Sandiumenge A, Rowan KM. Variation in critical care services across North America and Western Europe. Crit Care Med. 2008;36(10):2787–93.
41. Streat SJ. Illness trajectories are also valuable in critical care (letter). BMJ. 2005;330(7502): 1272.
42. Ho KM, Liang J, Hughes T, O'Connor K, Faulke D. Withholding and withdrawal of therapy in patients with acute renal injury: a retrospective cohort study. Anaesth Intensive Care. 2003;31(5):509–13.
43. Ho KM, Liang J. Withholding and withdrawal of therapy in New Zealand intensive care units (ICUs): a survey of clinical directors. Anaesth Intensive Care. 2004;32(6):781–6.
44. Cuthbertson SJ, Margetts MA, Streat SJ. Bereavement follow-up after critical illness. Crit Care Med. 2000;28(4):1196–201.
45. Cassell J, Buchman TG, Streat S, Stewart RM. Surgeons, Intensivists and the covenant of care – administrative models and values affecting care at the end of life - Updated. Crit Care Med. 2003;31(5):1551–9.
46. Cassell J. Life and death in intensive care. Philadelphia, PA: Temple University Press; 2005.

Chapter 10
South Africa: Where Have We Been?

R. Eric Hodgson and Timothy C. Hardcastle

Introduction

The cultural and economic diversity of South Africa as a country has been overshadowed in 2010 by the shared goal of hosting a major world sporting event: the FIFA 2010 Football (Soccer) World Cup. However, in the aftermath of the Football World Cup these differences were brought starkly back into focus by a series of crippling strikes, including a strike by the majority of nurses and a number of doctors in the state health sector. The Football World Cup and the nursing strike contrast the tremendous progress with the long way still to go in the process of reversing the 30-year legacy of apartheid on the background of 300 years of European colonisation and destruction of indigenous African culture.

The process of reversing the legacy of colonialism is being significantly hampered by two epidemics that have a profound impact on healthcare and thus critical care service provision:

1. HIV/AIDS
2. Trauma—due to

    (a) Motor vehicle accidents
    (b) Interpersonal violence

R.E. Hodgson, F.S.C. (SA) (Crit Care) (✉)
Department of Anaesthesia and Critical Care, Addington Hospital,
Ethekwini-Durban, KwaZulu-Natal, South Africa

Department of Anaesthesia, Critical Care, and Pain Management,
Nelson R Mandela School of Medicine, University of KwaZulu-Natal,
Ethekwini-Durban, KwaZulu-Natal, South Africa
e-mail: iti20178@mweb.co.za

T.C. Hardcastle, M.B., Ch.B. (Stell) M. Med. (Chir) (Stell), F.C.S. (SA)
Department of Surgery, Nelson R Mandela School of Medicine, University
of KwaZulu-Natal, Ethekwini-Durban, KwaZulu-Natal, South Africa

D.W. Crippen (ed.), *ICU Resource Allocation in the New Millennium: Will We Say "No"?*,   75
DOI 10.1007/978-1-4614-3866-3_10, © Springer Science+Business Media New York 2013

The population of South Africa currently stands at 48 million people made up of 85% Black, 10% White and 5% Indian and Coloured South Africans [1]. Twenty percent of the population is covered by medical insurance (Medical Aid) and is cared for in private hospitals that consume 60% of the healthcare expenditure and employ 70% of medical specialists in South Africa [2]. The state healthcare system consumes 40% of South Africa's healthcare expenditure and employs 30% of the medical specialists to care for 80% of the population, half of whom are not formally employed. The state healthcare system also provides for the training of most healthcare professionals in South Africa [3], with the exception of some nursing and prehospital-care training.

Critical Care has been recognised as a medical subspecialty in South Africa for more than 10 years. Subspecialists from the primary specialties, including medicine, surgery, paediatrics, anaesthesia and emergency medicine, are registered with the Health Professionals Council of South Africa after 2 years of training in a unit accredited by the council [4]. For the last 10 years a certificate in critical care has been recognised by the Colleges of Medicine of South Africa, requiring not only 2 years of training but the completion of a written and oral examination. There are currently 60 registered intensivists and 115 pulmonologists in South Africa. There are 76 critical care training posts [critical care (anaesthesia and surgical primary specialty)—35, trauma surgery—10 and pulmonology (adult and paediatric)—31] [5]. Additionally there are 19 registered trauma surgeons who have completed at least 12 months of critical care training as part of their training for this new subspeciality of general surgery.

Healthcare Expenditure

South Africa spends ZAR21.5 Billion (US$3.034 Billion) or 8.7% of GDP on healthcare annually [3].

State Expenditure

Forty percent of funds spent on healthcare (3.7% of GDP) are spent by the state.

South Africa has a limited number of taxpayers (10% of the population = 5 million people) and the largest number of social grant recipients in Africa (25% of the population = 13 million people) with one individual taxpayer for every three grant recipients [3].

The state not only has to pay social grants but must address backlogs in housing, sanitation and education. Unfortunately these large demands on a limited tax base have been compounded by unsound financial management:

1. Inappropriate allocation of funds:

   (a) South Africa has spent US$40 billion on equipment for the Defence Force, including jet fighters, submarines and corvettes, despite there being no credible military threat to the country.

(b) Political patronage has resulted in the South African government expanding:

- The cabinet of the central government and resultant departmental bureaucracies have expanded from 27 ministers in 1994 to 32 in 2010 [6], compared with 22 in the UK and 30 in India.
- The four provinces in 1994 were increased to nine post independence, each with structures that duplicate not only one another but also central government departments.
- Health is a provincial responsibility. Using Kwa-Zulu Natal as an example the province now has eight health districts compared with three before 1994, each with a duplicated bureaucracy, resulting in the provincial heath department expanding from three floors of one building to full occupation of three buildings [7].
- In stark contrast to the massive increase in support staff, clinical posts have been reduced or remained static. Freezing of posts in response to budgetary restraints has a disproportionate effect on medical staffing due to the necessity of progression between posts e.g.: intern—community service medical officer—medical officer—registrar (resident)—consultant (attending). Staff who are denied posts at any stage after community service are likely to be lost to the state sector by being forced into private practice and, even worse, to the country by being forced to emigrate.

2. Inappropriate management of allocated funds:

(a) Staff may be deployed to state institutions on the basis of two main criteria other than competence:

- *Affirmative action.* Apartheid systems of racial classification have been maintained after independence. Members of the previously disadvantaged (Black and Coloured) populations are employed in preference to populations perceived as previously advantaged (White and Indian).
- *Political loyalty.* A major controversy in South Africa exists around the stated policy of the ruling African National Congress (ANC) of "cadre deployment" [8] where loyal members of the party are employed due to their membership of the party rather than demonstrated competence for the positioned to be filled.

The combination of these factors has resulted in senior management positions in the state sector being filled by managers with limited skills and qualifications. Under the circumstances, expected accounting practices and fund allocation are often sub-optimal, resulting in regular qualified audits of central and provincial government departments.

(b) *Inappropriate allocation of funds for goods and services.* Affirmative action principles are applied in the acquisition of goods and services by the state by the policy of Black Economic Empowerment (BEE) [9]. Companies supplying goods and services to the state are selected on the basis of BEE compliance rather than competence in providing the goods and/or services required. This has resulted in absent or defective delivery of goods and services and increased expense due to the requirement for prolonged and ineffective tender processes and repeated payment for goods and services by alternative providers after absent or defective delivery by the initial supplier selected.

3. Misappropriation of allocated funds

   While the weapon purchases by the South African government have generated the most debate and even criminal convictions regarding corruption, there are regular allegations of corruption in most government departments, including health departments. Corrupt practices that have been identified include:

   (a) *Nepotism.* Employment or allocation of contracts/tenders to family members.
   (b) *Accepting gifts in exchange for influence.* The successful prosecution of Tony Yengeni (accepted a motor vehicle in exchange for influence in arms purchasing) and Schabir Shaik (provided funds to government and ANC members in exchange for influence in arms purchasing) indicates the prevalence of corrupt relationships between influential members of government/ state employees and corporate entities [10].

The state has many higher priorities for spending than health and allocated funds are misspent for reasons outlined above. These factors have resulted in the decline of budgets for healthcare in real terms allocated to the state sector since independence in 1994.

By contrast, and despite the HIV epidemic, the population of South Africa is increasing by 2–3% a year. Progressive urbanisation and provision of primary healthcare services are bringing many more people within reach of secondary and tertiary care, where critical care is provided. The increasing demand for healthcare is being met with a reduced supply, resulting in an increasing deficit in healthcare provision to the majority of South Africans by the state healthcare service [3].

State Healthcare Spending on Critical Care

A decline in budgets with an increase in patient numbers has resulted in a decline in ICU beds to less than 1:100,000 population in 5 of the 9 provinces of South Africa [11] for patients in the state sector. Only three provinces approach the international recommendation per 100,000 [12].

Deficiency in critical care provision is most acute in rural areas of South Africa particularly central provinces such as the Free State, which meets the bed to population ratio, due to a sparse population, but most of the population live much more than 100 km from a hospital with an ICU [13]. Medical services in these rural areas are provided by junior medical staff [Community Service Medical Officers (CSOs) in their first year of practice post-internship]. These junior doctors have minimal senior supervision, the effect of which can be inferred by the steadily increasing rate of maternal death due to anaesthesia, resulting in anaesthetic death being the third most common cause of death identified in the most recent triennium of the confidential enquiry into maternal death in South Africa [14].

In urban areas the number of ICU beds is severely restricted, with only the Western Cape and Gauteng provinces approaching international norms.

The restriction on ICU resources requires state hospital intensivists to make decisions directly affecting patient management to maximise the utility of the scarce resource of critical care on the basis of:

1. *Triage*. Withholding therapy for patients with a low likelihood of survival with or without intensive therapy.
2. *Trial of therapy*. Withdrawal of therapy in patients who are failing to respond to intensive interventions.

South African intensivists typically perform triage decisions leading to withholding of therapy rather than deciding to admit patients for trials of therapy followed by withdrawal of therapy. However a decision to admit a patient for a trial of therapy tends to be followed by early withdrawal in the event of an inadequate response to therapy. Triage is difficult to assess by data that are routinely collected but can be inferred from the experience in Ethekwini-Durban.

State hospital ICUs in Ethekwini-Durban typically have occupancy close to 100%, resulting in up to half of all requests for admission being turned down on the basis of "no beds" [15].

Withdrawal may be assessed by means of duration of admission to ICU. The short duration of ICU stay in South African State ICUs is confounded by defects in healthcare throughout the state sector including:

(a)   Prehospital rescue and resuscitation
(b)   Emergency room management
(c)   Operative management

The resulting increase in disease severity of patients admitted to state ICUs will be combined with early withdrawal of therapy in patients who fail to respond to reduce the duration of ICU admission. Unfortunately, data to support the previous assertion are lacking due to the high clinical and educational workloads of South African state intensivists resulting in limited ability to collect and publish data.

The critical decisions required of South African state intensivists are whether a particular patient should be admitted and, for patients admitted for treatment, whether expensive treatment with limited availability is actually saving the patient's life, and is therefore justifiable, or prolonging his or her death, in which case treatment should be withdrawn [16]. These decisions are neither simple nor easy. To avoid the perception of bias in decision-making wide consensus is required, not only between the treating clinicians, as demonstrated by a joint study from Cape Town and London [17] but also between the treating clinicians, the patient and their family, where the current aim is to achieve informed assent of the family to the consensus decision made by the medical management team [18].

The legal validity of a medical decision to limit therapy in the face of family demands for continued treatment has not been tested in the South African ICU context, particularly in state ICUs. However, it is worth noting that the South African law does not place the burden of decision on the family, but rather on the clinician, as the decision whether a particular therapy is futile is deemed a medical decision, which the family cannot refuse [19].

Cases involving medical decisions in South Africa are heard by a magistrate with one or a judge with two appropriately qualified medical practitioners acting as assessors, making an adverse finding in the face of a reasonable decision made by consensus of a number of experienced clinicians extremely unlikely [19].

There is debate on the responsibility for adverse outcomes in the situation of resource constraint. American physicians working in Haiti provided a different standard of care to their Haitian patients compared with their patients in the USA, not due to differences in skills or professionalism but due to resource constraint. The standard of care thus relies on available resources and responsibility for adverse outcomes cannot be ascribed to clinicians practicing under these conditions, provided their professional standards are maintained [20].

In South Africa, legal opinion is that healthcare administrators may be required to take responsibility for adverse patient outcomes rather than clinicians where the adverse outcome can be ascribed to a resource deficiency that could have been reasonably foreseen and dealt with by the administrator. This point is particularly relevant if the resource deficiency has arisen due to maladministration or corruption on behalf of the administrator [21].

The attitude of the South African legal system may be inferred from a case in which renal dialysis was denied by a state hospital to a patient with chronic renal failure. The hospital argued that the patient did not meet the criteria for transplantation and thus had an incurable disease that should not be treated further as this would deny treatment to patients in whom cure was possible. The patient argued that his right to life guaranteed by the constitution was being infringed. The Constitutional Court found in favour of the hospital and the patient subsequently died [22].

Private Healthcare Spending

The private healthcare sector in South Africa spends 60% of the funds expended on healthcare (5.7%GDP) and employs 70% of specialists but only serves 20% of the population [4]. This percentage has not increased and may be decreasing in real terms as the cost of subscription to Healthcare Funding (Medical Aid) in South Africa has increased at a rate exceeding inflation. The number of lives covered by private medical aids has remained static despite a population increase [3].

The increase in costs to patients for Medical Aids has not been passed on to clinicians. Main cost drivers are hospital fees, pharmaceuticals and administration of the funds. South Africa has close to 300 medical funders so that even highly efficient administration of individual funds increases costs through duplication [4].

The focus of private healthcare is on curative services such as arthroplasty and coronary revascularisation. There is little spending on preventative health, rehabilitation or education of new medical professionals. Challenges similar those faced in the first world, such as increased medicolegal risk, have driven cost increases through increasing use of diagnostic testing, particularly imaging, as a surrogate for clinical assessment and judgement.

There is significant dissatisfaction with private healthcare among patients, due to increased costs, and clinicians, due to inadequate remuneration compared with non-medical professionals at similar levels of training (such as lawyers, accountants and engineers). Additionally, audit and peer review are compromised, due to the fact that practitioners are not employed by the private hospital groups, but rather are in solo practice with hospital access privileges and ICUs are largely run as open units.

Private Healthcare Spending on Critical Care

In the state sector the majority of units are closed and led by an intensivist as opposed to the private hospitals, where the majority of units are open with very limited if any intensivist input.

The situation of limited intensivist input to private ICUs is unlikely to change in the near future for the following reasons:

1. *Limited number of training posts.* Limited number of training posts is being addressed in two ways

    (a) There are currently 76 training posts in critical care in South Africa [4].
    (b) Introduction of an intermediate level subspecialty certificate in critical care that will not be recognised for subspecialty registration but may provide a means for funders to provide improved remuneration for ICU treatment by trained specialists.

2. *Poorly defined career path.* After 14 years of training from leaving school, specialist intensivists have a very limited earning capacity compared with potential referring clinicians (e.g. orthopaedic, cardiac and neurosurgeons, general or subspecialist physicians) [1].
3. *Limited referrals.* Non-intensivist specialists are unwilling to refer to intensivists, mainly due to the financial incentive of continued management of ICU patients.
4. *"Brain Drain".* Many South African trained intensivists leave the country to address the shortage of intensivists being experienced in the first world. Additional incentives for emigration include improved personal security, reduced clinical workload, greater research and teaching potential and a better work–life balance [3].

Who Provides the ICU Care in the Private Sector

Organ function can be supported in the ICU over prolonged periods. This support provides conditions that facilitate recovery of the patient's own organs but cannot regenerate, or may even damage further, the organs supported. Assessment of patient recovery is best done holistically, rather than by assessing organ function in isolation, as done by single organ specialists. Intensivists represent one of the last areas of medical practice where holistic assessment takes precedence over isolated organ function assessment [23].

Another skill that intensivists acquire is participation in end of life (EoL) care of patients and their families. These discussions are most appropriately initiated with patients with progressive, life-limiting diseases prior to the onset of critical illness but are often delayed while treatment is escalated by single organ specialists [24]. The majority of clinicians providing critical care in South African private ICUs, as in the USA, are not specialist intensivists but single organ specialists, most commonly pulmonologists [25].

In South African private practice, surgeons and anaesthesiologists tend to work in small or solo practices and are thus constrained from an intensivist role by their requirement to be present in the operating theatre while operating or providing anaesthesia, so they cannot cover emergencies in ICU. By contrast, physicians tend to be involved in consultative practice that provides the flexibility required to function as intensivists, despite having deficits in training and experience in ICU particularly in the perioperative setting.

Cost Implications of Non-intensivist Led Open ICUs

Initiation and continuation of critical care by non-intensivists increase the cost of critical care by a number of mechanisms:

1. *Inappropriate initiation of critical care.* Critical care is only of benefit to a limited group of patients who, according to the Goldilocks principle, should neither be too sick, or too well, but have reversible organ dysfunction that will reverse with appropriate critical care support [26]:

    (a) *Too sick.* Studies from the USA have shown that up to 50% of healthcare expenditure occurs in the last 6 months of a patient's life [27]. Admission of patients with terminal diseases including heart failure, chronic obstructive airway disease (O_2 dependent) and malignancy (metastatic and/or failed chemotherapy) substantially increase costs and patient discomfort with minimal improvement in duration of life that is of extremely poor quality. Admission to ICU for active intervention may be far less appropriate in these patients than referral for palliative care, which may take place in hospice or even at home.

    (b) *Too well.* Concerns over quality of nursing care in the general wards of private hospitals have led to the admission of patients to ICU/High Care after routine surgery such as arthroplasty or spinal decompression. Such admissions may increase morbidity as patients are exposed to monitoring procedures that are not justified by their disease severity [28]. This is similar to the challenges reported from the Developed World.

2. *Inappropriate maintenance of critical care.* ICUs have a strong ethos of life saving that may be inappropriately extended to death prolongation. Assessment by a single organ specialist may result in a focus on the organ of their specialty with limited appreciation of the overall progress of the patient. Continued active intervention limits stress both for the treating clinician and the family as the reality of impending death can be denied while concentrating on the minutiae of organ support [29].

There is also a financial incentive for both the treating clinician and the hospital to continue therapy as each day in the ICU generates income for the clinician and the hospital. This incentive has changed as private hospital ICU bed occupancy has increased from 50 to 60% in the 1990s to 80 to 90% in 2010. There have been days in Ethekwini-Durban in 2010 when every ICU bed in the city, both state and private, has been full, with other patients awaiting admission.

3. *HIV in ICU* [30]. HIV presents specific challenges that have substantially increased costs for Medical Aids, thus increasing subscription costs and reducing population cover. HIV testing requires informed consent with both pre- and post-test counselling. There are major societal disincentives to HIV testing following the AIDS denialism propagated by the government of Thabo Mbeki and his Health minister Manto Tshabalala-Msimang and the regular, well-documented episodes of unprotected sex by the current president, Jacob Zuma [31]. Patients presenting to GPs and physicians in private practice may have features consistent with HIV infection but, in the absence of consent, testing cannot be undertaken and initiation of antiretroviral (ARV) therapy is thus delayed until the onset of critical illness. AIDS defining conditions presenting as an episode of critical illness are difficult to manage for a number of reasons:

(a) *Profound immunosuppression.* The patient may present with a treatable disease such as Pneumocystis jerovecii pneumonia (PJP) or disseminate tuberculosis but are exposed to multi-resistant pathogens in the ICU to which they are unable to mount an adequate response [32].

(b) Initiation of ARVs. Only zidovudine (AZT) is available in a parenteral form. ARV monotherapy rapidly induces resistance in HIV and highly active antiretroviral therapy (HAART) now requires three drugs, one of which is a protease inhibitor. Administration of these drugs will need to be delayed until enteral nutrition is established which may not occur within the first week of admission [33].

(c) *Immune reconstitution inflammatory syndrome (IRIS)* [34]. ARVs are a two-edged sword in AIDS presenting as critical illness. Initiation of ARVs will result in IRIS, indistinguishable from the systemic inflammatory response syndrome (SIRS), in up to 30% of patients, particularly those with an active infection.

Mortality of ICU patients presenting with severe sepsis (infection with cardiac and/or respiratory dysfunction) is close to 70% with a prolonged length of stay exceeding 10 days [33]. This mortality and length of stay is not unusual for conditions managed in ICU, such as pancreatitis, polytrauma and burns, but the magnitude of the epidemic makes the approach of managing HIV infection only after the onset of critical illness completely unaffordable compared with early testing and initiation of HAART on an outpatient basis prior to catastrophic immunosuppression [35].

HIV is currently the commonest cause of death in South Africa. State intensivists have made a pragmatic decision to triage patients presenting with critical illness due to AIDS-related complications prior to initiation of HAART.

Clinicians admitting patients to private ICUs, who are largely not specialist intensivists, continue to admit patients with critical illness due to AIDS complications. Given the frequency of HIV in South Africa these admissions make up a substantial proportion of private ICU admissions, up to 70% in some private hospitals in Ethekwini-Durban. This large number of admissions generate correspondingly large costs, estimated at ZAR10,000/day (US$1,300/day) resulting in ICU admission for AIDS-related critical illness making up an estimated 20% of the total payments made by Medical Aids (personal communication) [3]. This level of payment is not sustainable and will make Medical Aid unaffordable to all but the richest 5–10% of the population in future.

4. *Delayed initiation of EoL management.* Initiation of EoL discussions is stressful for the clinician and family. Prolonged trials of therapy are thus generally well accepted by clinicians, families and hospital management. Conversely, an overly optimistic attitude to outcome is very prevalent in private ICUs despite clear signs of progressive deterioration despite escalating therapy [36].

A crisis point in the continuation of a trial of therapy arises at the point where therapy can no longer be funded by the Medical Aid due to exhaustion of benefits. At this point hospital managers become concerned by ongoing potential financial losses and families are concerned by generation of potentially crippling debt [37].

This crisis point seldom arises if realistic discussions with families, including warning of the risk of death, are initiated within 48 h of admission with ongoing communication regarding progress and prognosis. With such optimal communication, progression to EoL care is a seamless process [38]. In the absence of appropriate communication, particularly if the clinicians have been overly optimistic and have not warned of the risk of death, the response to exhaustion of funds will be unsatisfactory for both clinicians and families.

The first response is often to request a transfer of the patient to a state ICU. This is seldom a viable option:

(a) State ICU occupancy runs at 95–100% so the likelihood of a bed being available is small.
(b) Should a bed be available the patient from the private ICU seldom meets criteria for admission, given the necessity for triage in state ICUs. Application of different triage criteria for state and private patients could not be considered ethical.
(c) In the unlikely event of a bed being available and the patient being accepted the risk of transfer needs to be considered. Adverse events are far more likely

during transfer so the less stable the patient, the less likely that transfer can be safely undertaken.

In the absence of a bed to move the patient into, limitation of therapy may be seen as being applied due to coercion, even if such limitation is medically justifiable. Subsequent progression to transfer out of ICU and/or death will result in significant emotional trauma and guilt for the clinician and family. The clinician does not help the family by insisting on continued therapy while ignoring the financial burden faced by the family.

A useful intervention in these situations in South Africa is probably a request for consultation with a senior intensivist. The intensivist in this situation is able to take the role of an "honest broker" to reduce the guilt of both the treating clinician and family by an objective assessment and recommendation for continued therapy, including EoL care.

The Future

There is a proposal from the South African government that expenditure on health-care be consolidated to provide a national health service modelled on those in Canada, Sweden and the UK. This National Health Initiative (NHI) has generated massive controversy.

Proponents cite the necessity to narrow the gap in access to healthcare between the current state and private patients and to reduce the excess expense due to dupli-cation of administration and equipment in the private sector.

Opponents cite the increased expense of the NHI that will increase expenditure on healthcare by an additional 3–5% of GDP. There is no guarantee that patients who currently contribute to Medical Aids will continue to contribute at a similar or increased level to the NHI. There are also serious concerns regarding the manage-ment of the NHI, given the poor track record of the management of state enterprises, including national and provincial departments of health.

Conclusion

At present, South African intensive care faces a dichotomy of a state ICU service with severe resource constraints that require strict triage and limited duration of therapeutic trials in ICU. By contrast, the private system has adequate resources but inefficient utilisation and wasteful expenditure that may be addressed by greater involvement of specialist intensivists and addressing admission and discharge crite-ria with critical illness as a complication of AIDS. The diseases may be new, but the principles will keep intensivists "fielding on the boundary" for the foreseeable future [39].

References

1. http://www.southafrica.info/ess_info/sa_glance/health/health.htm.
2. http://www.medicalschemes.com.
3. http://www.hst.org.za/healthstats/108/data.
4. http://www.hpcsa.co.za.
5. http://www.collegemedsa.ac.za.
6. http://www.info.gov.za.
7. http://www.kznhealth.gov.za.
8. http://www.anc.org.za.
9. http://www.southafrica.info/business/trends/empowerment/bee.htm.
10. Feinstein A. After the party: a personal and political journey inside the ANC. 1st ed. Cape Town: Jonathan Ball; 2007. ISBN 978-1-86842-262-3.
11. Bhagwanee S, Scribante J. National audit of critical care resources in South Africa – unit and bed distribution. S Afr Med J. 2007;97(12):1311–4.
12. Lyons RA, Wareham K, Hutchings HA, Major E, Ferguson B. Population requirement for adult critical care beds: a prospective quantitative and qualitative study. Lancet. 2000;355: 595–8.
13. Bhagwanee S, Scribante J. National audit of critical care resources in South Africa – open versus closed intensive and high care units. S Afr Med J. 2007;97(12):1319–22.
14. National Committee on Confidential Enquiry into Maternal Death. 3rd confidential enquiry into maternal death in South Africa. National Department of Health of South Africa; June 2006. http://www.info.gov.za/view/DownloadFileAction?id=70073.
15. Burrows RC, Gopalan PD, Hodgson RE. DNR – the importance of the medical decision. Sydney, NSW: World Congress of Intensive Care; 2001. Free Paper 31.
16. Burrows RC, Hodgson RE. De facto gate-keeping and informed consent in intensive care. Med Law. 1997;16:17–27.
17. Turner JS, Michell WL, Morgan CJ, Benatar SR. Limitation of life support: frequency and practice in a London and a Cape Town intensive care unit. Intensive Care Med. 1996;22(10): 1020–5.
18. Azoulay E, Pochard F, Chevret S, et al. FAMIREA Study Group. Half the family members of intensive care unit patients do not want to share in the decision-making process: a study in 78 French intensive care units. Crit Care Med. 2004;32(9):1832–8.
19. Hardcastle TC. Ethical and Medico-legal aspects of Trauma. S Afr J Bioethics Law. 2010;3(1):25–7.
20. Annas GJ. Standard of care – in sickness and in health and in emergencies. New Engl J Med. 2010;262(22):2126–31.
21. McQuoid-Mason D. Establishing liability for harm caused to patients in a resource-deficient environment. S Afr Med J. 2010;100(9):573–5.
22. Thiagraj Soobramoney vs. Minister of Health, Province Of Kwazulu-Natal, South Africa. Durban Supreme Court Case Number 5846197, August 1997.
23. Moreno RP, Hochrieser H, Metnitz B, Bauer P, Metnitz PGH. Characterizing the risk profiles of intensive care units. Intensive Care Med. 2010;36:1207–12.
24. Murray SA, Kendall M, Boyd K, Sheikh A. Illness trajectories and palliative care. Br Med J. 2005;330:1007–11.
25. Gutsche JT, Kohl BA. Who should care for intensive care unit patients? Crit Care Med. 2007;35:S18–23.
26. Brilli RJ, Spevetz A, Branson RD, et al. Critical care delivery in the intensive care unit: defining clinical roles and the best practice model. Crit Care Med. 2001;29(10):2007–19.
27. Gill TM, Gahbauer EA, Han L, Allore THG. Trajectories of disability in the last year of life. New Engl J Med. 2010;362(13):1173–80.
28. Woolf SH. Potential health and economic consequences of misplaced priorities. Ann Intern Med. 2005;143:301–2.

29. Almoosa KF, Goldenhar LM, Panos RJ. Characteristics of discussions on cardiopulmonary resuscitation between physicians and surrogates of critically ill patients. J Crit Care. 2009;24:280–7.
30. Siegel MD. End-of-life decision making in the ICU. Clin Chest Med. 2009;30:181–94.
31. Huang L, Quartin A, Jones D, Havlir DV. Intensive care of patients with HIV infection. N Engl J Med. 2006;355:173–81.
32. Smith TC, Novella SP. HIV denial in the Internet era. PLoS Med. 2007;4(8):e256. doi:10.1371/journal.pmed.0040256.
33. Japiassú AM, Amâncio RT, Mesquita EC, et al. Sepsis is a major determinant of outcome in critically ill HIV/AIDS patients. Crit Care. 2010;14:R152. http://ccforum.com/content/14/4/R152.
34. Coquet I, Pavie J, Palmer P, et al. Survival trends in critically ill HIV-infected patients in the highly active antiretroviral therapy era. Crit Care. 2010;14:R107. http://ccforum.com/content/14/3/R107.
35. Beatty GW. Immune reconstitution inflammatory syndrome. Emerg Med Clin North Am. 2010;28:393–407.
36. Brinkhof MWG, Boulle A, Weigel R, Messou E, Mathers C, et al. Mortality of HIV-infected patients starting antiretroviral therapy in sub-Saharan Africa: comparison with HIV-unrelated mortality. PLoS Med. 2009;6(4):e1000066. doi:10.1371/journal.pmed.1000066.
37. Crippen D, Burrows RC, Stocchetti N, Mayer SA, et al. Ethics roundtable: 'Open-ended ICU care: Can we afford it?'. Crit Care. 2010;14:222. http://ccforum.com/content/14/3/222.
38. McMillan J, Hope J. The art of medicine. Justice-based obligations in intensive care. Lancet. 2010;375:1156–7.
39. Bateman ED. ICU care for tuberculosis patients-fielding on the boundary. S Afr Med J. 1997;87:635–7.

Chapter 11
United Kingdom: Where Have We Been?

Anna M. Batchelor

Background

To understand the NHS, it is necessary to understand a little of the history of England and her people.

The rule of law, parliament and the power of the state has deep roots for the English people and whilst on many occasions challenged have in the main led to a stable and at times affluent country. The forerunner of our modern Parliament (from the French parler i.e. a "talking shop") first met in 1265 instigated by Simon de Montfort, but the origins can be traced even further back to Anglo Saxon times from the eighth to eleventh centuries. Written records have been kept since 1278. The main reason for Parliament at the beginning was for the King to gain approval to raise taxes plus ça change, plus c'est la même chose ! Usually at this time taxes were raised to pay for wars, healthcare being a non-existent consideration.

Since 1327 when Parliament removed Edward II every Parliament has had representatives of the people; initially only freemen were allowed to vote but as a result of Reform Acts in 1832 small landowners, tenant farmers, and shopkeepers, followed in 1884 by householders and in 1884 all men were enfranchised. Women had to wait till 1918 before votes for over 30s and 1928 for equal rights to men.

During the late eighteenth and the nineteenth century the industrial revolution resulted in large population movements from the country into cities initially driven by cotton and textile manufacturing later by coal and iron. These industries and their owners had scant regard for the health or well-being of workers. Initial attempts to form workers organisations were suppressed by both employers and government, the latter fearful of revolutionary ideas spreading from mainland Europe, particularly France where the Revolution started in 1789. Trade unions started from craft

A.M. Batchelor, M.B., Ch.B., F.R.C.A., F.F.I.C.M. (✉)
Department of Peri-operative and Critical Care Medicine,
Newcastle upon Tyne NHS Trust, Newcastle, UK
e-mail: a.m.batchelor@newcastle.ac.uk

D.W. Crippen (ed.), *ICU Resource Allocation in the New Millennium: Will We Say "No"?*,   89
DOI 10.1007/978-1-4614-3866-3_11, © Springer Science+Business Media New York 2013

organisations, for example, the Amalgamated Society of Engineers dating from 1851; they levied high dues so that they could afford full-time officials and offered "friendly society" benefits.[1] This was also the time of the growth of cooperative societies sharing profits for the benefit of members. This concept of paying regularly into an organisation which in return helps when you are in need is an important precursor of the idea of a welfare state and the national health service. By 1874 trade unions had over a million members and 6.5 million by 1918.

> Out of this idea came the notion of the state…in the sense of the political and social expression of this collective will…the state as benefactor, as provider for those who could not provide for themselves[2]

The Labour party formed in 1900 by working people and trade unionists became a third political party with the existing Tory and Liberal parties. In 1906 with 29 Members of Parliament they entered into a pact with the Liberal party to form a government. They were able to institute some long-awaited social reforms including the institution in 1911 by the Liberal Prime Minister Lloyd George of National Health Insurance. National Health Insurance only paid benefits for the insured and none for his dependents. In case of sickness the insured had a right to free but limited care from a general practitioner on a list or panel. The term being "on the panel" for many years after still referred to someone who was seeing a doctor and was off work sick.

Community healthcare was provided by general practitioners who derived most of their income from patients who paid per visit and a smaller amount from panel patients. Hospitals where a hotchpotch of provision depending on location. Some were voluntary depending on charitable support or in the case of some ancient medical schools in London endowments. Local authorities provided beds for specific indications such as maternity hospitals, small pox and tuberculosis.

There were only two more short periods of Labour government before the Second World War, but the wartime coalition led by Winston Churchill had several Labour members in ministerial positions who thus were in charge of forming and implementing government policy. One of these ministers Ernest Bevin, Minister of Labour, asked William Beveridge, Director of the London School of Economics and Political Science and temporary civil servant, to chair a committee and produce a report on the existing social insurance and allied services. The Beveridge report was published in 1942 and recommended all working people pay national insurance contributions and would receive benefits in return if unemployed or ill. The report identified five Great Evils in society squalor, ignorance, want, idleness and disease. This report signalled the beginnings of the Welfare State and the National Health Service. The ideas were appealing to the working man, who up till then had limited access to healthcare; it also appealed to the Tory (right wing) party and employers because the costs of healthcare and pensions were transferred from the employer to the state, a particular benefit after a long and economically debilitating war.

---

[1] Professor Mary David, Centre for Trade Union studies, London Metropolitan University. http://www.Unionhistory.info.

[2] Tony Blair. A journey. Hutchinson; 2010.

In 1945 despite the success of Winston Churchill in leading the country to victory in the Second World War the general election saw a landslide win for the Labour party. The implementation of the Beveridge report was a high priority and one result was the National Health Act 1946 which stated " It shall be the duty of the Minister of Health…to promote the establishment…of a comprehensive health service designed to secure improvement in the physical and mental health of the people of England and Wales and the prevention diagnosis and treatment of illness and for that purpose to provide or secure the effective provision of services".

The NHS was officially launched on 5 July 1948 by the then Minister for Health Aneurin (Nye) Bevan and promised "everybody, irrespective of means, age, sex or occupation shall have equal opportunity to benefit from the best and most up-to-date medical and allied services available." The welfare state promised social benefits for all from the "cradle to the grave", sickness and unemployment benefits in return for national insurance contributions. The NHS promised free care for everyone and was not tied to the payment of contributions.

It is interesting to consider why the National Health Service was such a popular idea at the time of its institution and remains so today with every political party always promising whilst campaigning before elections to defend the NHS. The practices of wartime, command and control including for example moving workers to where they were needed including women to work on farms "land girls" to replace men called up to the armed forces, rationing, and the state taking a major role in organising and coordinating so many aspects of life meant people were prepared to accept the concept of state control including a welfare state and the NHS. Hospitals were in financial difficulties, a poor state of repair and there was no coordination about where or if at all a service was provided. An emergency medical service had been created in 1939 for the duration of the war and seen to work. A view that healthcare was a right was emerging. Matching this with the background of the slums left over from the industrial revolution and a society primed by craft associations, trade unions and cooperative societies, the idea fell on fertile ground.

After the war people were used to hardship and austerity measures, rationing (which on many things remained until 1953) was still in place and people expected little. A health service which provided for everyone, free at the point of care including prescriptions, dentistry and spectacles, must have seemed like manna from heaven.

The Labour government of 1945 instituted a policy of nationalisation of more than health, the Bank of England in 1946, the means of energy production including coal gas and electricity were nationalised in 1947 along with railways, ports and buses, and the iron and steel industry in 1950. The state in one of the world's oldest democracies now owned a substantial sector of the economy as well as the hospitals, and at the end of the Labour government in 1951, 20% of the economy was owned by the state and it employed two million people. With the exception of the Bank of England all of these industries were privatised in the 1980s and 1990s under the Conservative government led by Margaret Thatcher; however, not even she had the political weight to touch the NHS.

Reality

From the very beginning Bevan knew resources would struggle to meet expectations, he told a Nursing Conference "We shall never have all we need…expectations will always exceed capacity. The service must always be changing, growing and improving – it must always appear inadequate."

The basis of funding was taxation so everyone who was working contributed; those who earned more contributed more not only because they were paying a proportion of a larger salary but also by paying a higher percentage of earnings on amounts greater than preset tax thresholds. The national insurance contributions, paid by both employee and employer, whilst remaining to this day are not directly linked to the services provided by the NHS or any of the other aspects of the welfare state. All such taxation goes into the treasury and is allocated to various departments of state depending on the outcomes of the spending reviews and the budget.

Without a bottomless purse rationing by one means or another has been intrinsic to the NHS from the beginning. New developments were slow to be implemented and slow to spread out from teaching hospitals; in 1980 when I qualified, computerised tomography (CT scanning) despite having been commercially available since 1974 was not available in most District General Hospitals in the UK. The waiting list a feature of hospitals up until very recently was effectively a choke on activity and demand. In the early 1990s it was not unusual for patients to wait more than a year between referral by a general practitioner to an orthopaedic surgeon and even receiving a hospital appointment, and of course that was only the beginning; they then had to wait for their name to appear at the top of the operating list. This whole process could take a couple of years in someone whose mobility was limited by pain. Whilst no one liked these long waits they were accepted almost as the price one paid for a free health service, its just the way it was. This philosophy is almost certainly important in understanding how patients and relatives relate to the service in 2010.

Prescription charges were introduced in 1952 initially one shilling (12.5p) per item increasing to 20p in 1971 and 45p in 1979; today they are £7.20 per item although many people including children, pregnant women and people with chronic health conditions received free prescriptions. Ophthalmic care is no longer free and access to NHS dentistry is limited; however the general practice and hospital services remain free.

Intensive Care in the United Kingdom was rudimentary and driven by enthusiasts until 1962 when the Department of Health published "Progressive Patient Care".[3] This recommended "systematic grouping of patients according to their illness and dependence on the nurse, rather than by classification of disease or sex". As with all things state led a report such as this came with funding to establish

---

[3] Progressive Patient Care: Interim report of a departmental working group. Monthly Bulletin MOH and PHLS. 1962; 21: 218–26.

intensive care units, of course there was no skill base to draw upon for this new service and anaesthetists as people who understood ventilators became by default intensivists. The specialty was integral to the practice of anaesthesia in many if not all hospitals for the next three decades and still retains very close links today.

The Intensive Care Society was formed in 1970 by a group of enthusiasts; they organised meetings and whilst having no formal role in training did form a focus for the development of the specialty. Doctors undertaking training in the UK would on completion of a formal programme of training receive accreditation from their Royal College that this training fulfilled the requirements of the College and a trainee could then apply for a consultant post. There was no formal recognition of training in ICM and in 1988 a Joint Accreditation Committee in Intensive Therapy (JACIT) was formed by the Royal Colleges of Anaesthesia, Medicine and Surgery and an initial ten national posts created to train in ICM doctors after accreditation in a primary specialty. An Intercollegiate Committee was formed in 1993 followed in 1996 by the Intercollegiate Board for Training in Intensive Care Medicine (IBTICM); this body produced a training scheme whereby some ICM training could be added onto training in a base specialty such as anaesthesia, medicine, emergency medicine or surgery. Training consisted of 21 months in intensive care units and 6 months of complementary training; for an anaesthetist this was in medicine and for a physician in anaesthesia, and a surgeon would have to do both specialties. Specialty recognition for Intensive Care Medicine was achieved in 1999 and this can be recognised on the Specialist Register held by the General Medical Council (GMC). It is necessary for a doctor applying for a consultant post in the NHS to be on this specialist register, although it does not need to be for the specialty for which they are applying ! So an anaesthetist may apply for a post with input to ICM without being on the specialist register for ICM. The IBTICM training scheme is still in place today although it will be superceded by a single specialty training scheme in ICM from August 2012, which may be taken in conjunction with a training programme in another specialty, thus leading to dual Certificates of Completion of Training (CCTs). A Faculty of Intensive Care Medicine was formed in 2010 and Foundation Fellows will be admitted in 2011. ICM has become a mature specialty at last in the UK.

ICM has not escaped rationing; the availability of beds has always been used to limit expenditure. The UK uses a nurse to patient ratio of 1:1 for level 3 patients, i.e. those requiring support for more than one organ system and intensive nursing input; this is obviously expensive with 70% of the cost of a bed bay made up of staff costs. The population of England is about 52 million; there are about 115,000 general and acute hospital beds, of which less than 4,000 are critical care and only 2,000 are level 3 intensive care.

A step change in bed numbers can be seen between January 2000 and January 2001 (see Fig. 11.1). This coincided with a period of intense pressure on intensive care beds and the usual system of shunting patients between hospitals to manage limited capacity was overloaded. At the end of the 1990s, bed pressures led to several negative events both for the patients and for the government politically. These episodes involved in some cases children with head injuries being moved between hospitals more than once in an attempt to find a vacant bed, the inevitable

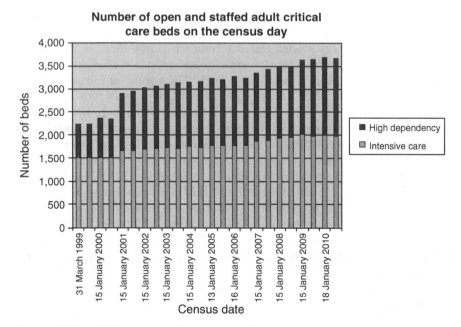

Fig. 11.1 Number of open and staffed adult critical care beds on the census day (Data from Department of Health kH03a data collections)

poor outcomes for the patients resulted in an outcry and £140 million pounds was infused into the critical care service. This coincided with the Production of a Department of Health document Comprehensive Critical Care in May 2000 which made several recommendations:

- Integration of intensive and high dependency units into the whole hospital, and intensivists taking an interest in acutely ill patients on wards before they became sick enough to need ICU, i.e. by preventing some patients needing critical care reducing the need for beds. Accomplished with the aid of outreach teams of nurses.
- The development of networks of ICUs across populations of two to four million, thus ensuring the development of common standards and protocols and assisting each other in bed crises.
- Workforce development including recognition of need for more trained intensivists.
- A data collecting culture to promote an evidence base and enable comparative audit.

The infusion of money led to a small increase in the number of level 3 ICU beds from 1,500 to 2,200, but a much larger increase in the number of high dependency level 2 beds for patients requiring either just single organ support or a higher level of nursing care than can be provided on general wards. High

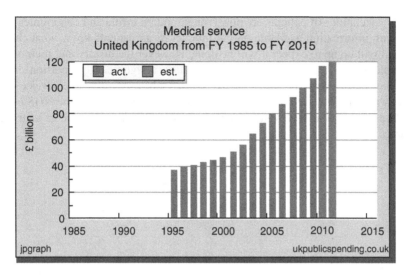

Fig. 11.2 Medical service: United Kingdom from FY 1985 to FY 2015 (Source: ukpublicspending.co.uk)

dependency units are now established in almost every hospital in most cases run by intensivists but in the case of specialist HDUs by the parent specialty, e.g. hepatology or vascular surgery.

Rationing is not only through a limitation of beds but also through limitations on access to drugs through the National Institute for Clinical Excellence (NICE). New drugs will be examined for cost effectiveness before being approved for use. Hospitals and GPs can prescribe them before approval but in a cash-limited environment the absence of NICE approval makes it unlikely that a drug will be widely prescribed. Whilst this most commonly affects new cancer treatments activated protein C (Xigris) received a UK product licence in 2001 but was not reviewed and supported by NICE until the end of 2003. Most health authorities and hospitals had supported either no or very limited use prior to its approval by NICE.

Despite the claim of the NHS to be free, since the 1980s there are many procedures particularly cosmetic which are no longer available on the NHS, for example, breast reduction or augmentation; however, where psychological issues can be demonstrated even these are permissible. Varicose veins followed cosmetic surgery and exclusion of others is likely.

The drive to improve the quality of healthcare saw the last Labour government (1997–2010) increase the funding into healthcare dramatically both in real terms £40 billion to £112 billion and as a percentage of GDP which rose from 6.6% in 1997 to 8.7% in 2008 (see Fig. 11.2).

The quality and efficiency of the NHS have undoubtedly improved considerably over the last 10 years and intensive care has paralleled this with improvements in facilities, staffing, training and data collection through the Intensive Care National Audit and Research Centre (ICNARC).

The expectations of patients are now rising; modern media and communications along with information available on the Internet mean patients know what can be done and will come to expect it will be done. However the history and memory of the development of our health care system mean a significant proportion of the population particularly the more elderly are acutely aware that the service they receive was not always available to their parents and grandparents. The NHS is still cherished and popular with voters; only time will tell if the tolerance of restricted resources persists.

Chapter 12
United States—Private Practice: Where Have We Been?

John W. Hoyt

It is important in this chapter not to draw any bright lines between Academic Medicine and Private Practice when discussing the delivery of critical services in the healthcare system of the United States of America (USA). Lawyers frequently speak of "eating what you kill." In some situations in the practice of law the attorney's income is completely based on their own billing. They are not an employee of the government or some large corporation that provides a salary and benefits. In the practice of medicine there is a diminishing number of situations where the "eat what you kill" model applies to medical practice. The payments for physician services, known as evaluation and management billing codes for cognitive work and procedure billing codes for technical or procedure work, do not support competitive incomes for physicians. This is particularly true for critical care if it is practiced in the intensivist model. Thus critical care physicians enter into some sort of contracted or employee model with a hospital or health system to support their income and provide a competitive salary in a market where there is a large demand for intensivists and a short supply. The hospitals that contract for intensivists might be quite academic with many resident training programs or not be academic at all with no resident training programs. The hospitals might be affiliated with a medical school or owned by a medical school or completely independent of any medical school and still have many resident training programs and be quite academic. The hospitals might be for profit or not for profit so they do not have to pay taxes. The biggest issue for the critical care physician is the intensivist practice model since the "eat what you kill" model or true unadulterated private practice will never apply to intensivists.

The roots of critical care in the USA can be traced back to the middle 1960s in a few medical school hospitals such as Presbyterian University Hospital in Pittsburgh under the leadership of Peter Safar, M.D., and Ake Grenvik, M.D., and

J.W. Hoyt, M.D. (✉)
Department of Critical Care, Pittsburgh Critical Care Associates, Inc.,
Critical Care Medicine, University of Pittsburgh, Pittsburgh, PA, USA
e-mail: HoytJ@pccaintensivist.com

D.W. Crippen (ed.), *ICU Resource Allocation in the New Millennium: Will We Say "No"?*,   97
DOI 10.1007/978-1-4614-3866-3_12, © Springer Science+Business Media New York 2013

the Massachusetts General Hospital under the leadership of Henning Pontoppidan, M.D. Anesthesia departments in these hospitals developed intensive care units from post anesthesia recovery units to care for surgical patients who needed a longer stay than the usual recovery room could offer. For the next 15–20 years the growth of critical care largely remained in this medical school owned/managed hospitals. In 1986 the American Board of Medical Specialists allowed the American Boards of Pediatrics, Internal Medicine, Anesthesiology, and Surgery to establish training programs in critical care and offer certificates of special competence in critical care. Shortly thereafter special billing codes were established through Medicare and private health insurance companies to pay physicians for delivering critical care services. Physicians who never did critical care fellowships were grandfathered into sitting for the critical care exam based on a practice history of taking care of ICU patients. Suddenly there were several thousand physicians declared competent to take care of ICU patients. Some of them were in private practice where they were called upon to consult on treatment decisions in patients with life threatening illnesses. Some of the critical care certified physicians were in university hospitals where they managed the ICU for the hospital and supervised the residents in their ICU care. Critical Care fellowships popped up in the base medical specialties and created a stream of new critical care physicians to sit for the certificate of special competence in Critical Care. This stream of freshly minted critical care physicians falls short of the growing workforce manpower needs as more and more hospitals decide to have dedicated critical care physicians in their intensive care units.

Somewhere in the 1990s the word intensivist was coined for a physician who dedicates his or her entire medical practice to taking care of ICU patients. Many think the word intensivist came from Leapfrog, a healthcare advocacy group developed out of a west coast healthcare buyers group that wanted to leap from a current level of healthcare quality to a much higher level of quality. In order to accomplish this, Leapfrog advocated for three things. They wanted computerized physician order entry, complex surgery done in hospitals with a high volume of this surgery, and finally and most important for this chapter that all intensive care units be staffed by intensivisits. An intensivist is a critical care fellowship trained physician with full critical care life support skills who dedicates his or her entire clinical time to working in the ICU seeing all the ICU patients without competing clinical responsibilities outside the intensive care unit. Over the past 10 years the intensivist model has taken shape. The intensivist usually works seven days in a row to preserve continuity of care. The intensivist manages 12–16 patients depending on the severity of illness. The intensivist partners with the admitting physician as an automatic co-attending and works with other consultants when additional expertise is needed. All orders go through the intensivist who functions as the ICU "captain of the ship." The intensivist leads multidisciplinary rounds each morning with the ICU team to include bedside nurses, respiratory therapists, pharmacists, etc. The intensivist team develops a plan of the day for each patient and carries out that plan communicating with the admitting physician and consultants as needed. The intensivist works no more than 18–20 shifts per month.

Leapfrog suggests that an intensivist should be present in the ICU at least 8 hours per day to be compliant with their guidelines. That recommendation routinely falls short of the needs of the critically ill patient who is sick 24 hours per day. A compromise position is for intensivists to staff the ICU 16 hours per day with one physician working from 7 a.m. to 5 p.m. and then a second physician working evenings from 5 p.m. to 11 p.m. and taking report from the two daytime physicians. In ICU situations of high severity of illness the intensivist team should be present 24 hours per day. Working 16 hours per day allows the intensivist to see and stabilize the majority of new admissions from the emergency department who are commonly admitted to the ICU from 5 p.m. to 10 p.m. Attending coverage 24 hours per day allows for a high level of intensivist experience in the ICU around the clock. In the 16 hour or 24 hour model the daytime intensivist should be present 7 days in a row. This daytime intensivist sets the plan of the day, meets with the admitting physician and consultants, and talks frequently with the families of the ICU patients. The evening intensivist does not need to work seven days in a row. Two or three intensivists could divide up the evening rotations in any given week. The evening intensivist works up and stabilizes all new admissions, does procedures, and treats emergencies in existing ICU patients. At least 75% of the physician component billing is done on days and 20% is done on evenings. Very little billing is possible on nights. In academic medical centers residents and fellows contribute little to this coverage model if billing and collections are important to the organization that employs the intensivist. The CPT codes are quite clear that billing an hour of critical care, CPT code 99291, requires the attending physician to actually do this work and make the note in the chart. In the private practice the model intensivists must capture all this evening billing to support the intensivist practice.

A common private practice model would be for a hospital to contract with a group of intensivists to cover a 15-bed medical ICU and a 15-bed surgical ICU 16 hours per day 365 days per years. This would require one intensivist in the MICU on days and another intensivist in the SICU on days (7 a.m. to 5 p.m.) A third intensivist would cover both intensive care units on evenings from 5 p.m. to 11 p.m. and take pager call for both intensive care units from 11 p.m. to 7 a.m. with the ability to return to the hospital in 30 minutes for emergencies. Five full-time intensivists would be required to work this schedule that has 90 shifts per month. They would work 18–20 shifts per month. The annual salary and benefits for these intensivists would be in excess of $2,250,000. An additional $250,000 would be needed to pay for billing and collections, transcription, legal and accounting fees, taxes, and a support staff person to do the clerical work for the group. This brings the total cost of the intensivist group to $2,500,000. Average revenue in such a private practice situation would be about $180 per patient day in the ICU. A 15-bed MICU and a 15-bed SICU are likely to have a combined 9,000 patient days in a year. With a good payer mix, little in the way of self pay patients, the predicted collections on ICU physician component billing would be about $1,700,000. This leaves a $800,000 gap between expenses and revenue. This gap must be made up by the hospital if it wants to have full-time intensivists in a private practice model. In exchange for this contribution on the part of the hospital, patients have a shorter length of stay, less complications, and a decreased death rate. Most hospitals view this subsidy cost for intensivists as a reasonable investment. Hospitals commonly subsidize emergency medicine groups, hospitalists, and anesthesiologists. These hospital-based practices make a significant contribution to the running of the hospital. In the case of critical care the hospital is likely to benefit at the rate of $1,000,000 per ICU from this intensivist model. They have a return on investment of $2,000,000 for a cost of $800,000.

Right now private hospitals in the USA with few to no training programs are embracing the intensivist model as described above. They do it by employing the physicians or by contracting with a physician group. They do not have the residents to take care of the minute-to-minute decision making in the ICU and the residents do not have the education or experience to manage critically ill ICU patients. Fellowship trained intensivists in academic practice or private practice save hospitals money and save patient lives. Numerous well-designed prospective studies have demonstrated the value of intensivists in reducing complications and saving lives in the ICU. Private hospitals, particularly in smaller and more rural communities, find that they get a more efficient use of ICU beds, better adherence to national standards in the treatment of critically ill patients, and they can keep patients in the community rather than shipping them out to larger metropolitan areas. Over the next 10–15 years in the USA it is anticipated that we will go from a 20% penetration of the private hospital/private practice model by intensivists to a 90% penetration. The improvement in the quality of care will be worth every subsidy dollar paid by the hospital for the intensivist group.

Chapter 13
United States—Academic Medicine: Where Have We Been?

David W. Crippen

To understand the rise of physicians' interest in academic medicine, it is necessary to outline the forces that drove them there.

Traditionally, academic medicine at major medical centers was the home of researchers with little if any patient contact on the clinical services [1]. Researchers enjoyed a protected environment and university credentials. There was little financial incentive for private practice providers to become hospital based. Community physicians made the best living and maintained the most professional autonomy in a fee-for-service system [2]. The concept of fee-for-service prospered in smaller hospitals dedicated wholly to direct patient care, many formally or informally associated with religious groups [3].

Since the rise of third-party insurers, however, fee-for-service has developed cost-efficiency problems. The bill for a typical 6-day hospital stay for childbirth in 1951 was $85—well within the range of most families. A 6-day hospitalization for cardiac workup at a large urban hospital in 2010 has recently been calculated to be $19,254; the facility lost $2,695 of that amount after reimbursement [4].

This cost situation arose in part because physicians in such a system have little motivation to reduce costs, given that the care is paid for by a third party relatively unable to process the value of need versus desire. Similarly, consumers of health care are not the purchasers thereof and so have little motivation to assess cost versus value. More is always better, especially when it is free. Medicare law actually prohibits consideration of cost in reimbursing for medical need [5]. Invoices are paid on the basis of correct paperwork.

D.W. Crippen, M.D., F.C.C.M. (✉)
Departments of Critical Care Medicine and Neurological Surgery,
University of Pittsburgh School of Medicine, Pittsburgh, PA, USA
e-mail: Crippen@pitt.edu

D.W. Crippen (ed.), *ICU Resource Allocation in the New Millennium: Will We Say "No"?*,   101
DOI 10.1007/978-1-4614-3866-3_13, © Springer Science+Business Media New York 2013

Medicine is no longer just an art and a science. It is clearly a business now.

In earlier days, graduates from medical college were congratulated as "professionals" and hung out a shingle proclaiming their services for those needing them. Now, medical school graduates are congratulated as small businessmen/women and sent out to market and create demand for their services. In fee-for-service, producers of medical services can create demand and are reimbursed incrementally more for providing more.

In the business of medicine, hospitals and physicians practice independently from each other. They bill differently for services and are reimbursed through different fee schedules [6]. Fee-for-service rewards physicians financially for overtreatment or redundant treatment, treatment that uses hospital resources without charge to the physician. Hospitals are penalized for waste, real or imagined. It is quite possible for a physician to be reimbursed indefinitely for treating a patient in a prolonged hospital stay, every day of which the hospital loses money. This quirk of medical economics makes it improbable that providers not aligned with the logistics of provision will make rational cost-effectiveness decisions.

Beginning in the 1970s as the cost of technology increased, many smaller community hospitals became unable to afford expensive duplication of services. Consolidation would be necessary for survival. Given inflation, expensive advances in technology and decreasing reimbursement, reimbursers and indemnifiers are moved to limit increasing expenditures. However, because limiting services for entitled consumers is political anathema, a different tack was taken to reduce costs—namely, controlling demand for services by creatively rationing payment at the provider level after services were rendered. The result? The consumer gets a full plate, and the provider makes up the shortfall any way possible.

In the 1980s, the era of rationing by inconvenience [7], the approach used for limiting expenses was to create blitzes of complex paperwork for reimbursement. Uncrossed t's and undotted i's delayed or stopped payment for services. Providers became expert form-fillers. Efficiency decreased, and administrative costs of providing health care increased.

In the 1990s, managed care operational algorithms were created for doing what was cheapest for the entire population [8]. This ploy failed because health care recipients are individuals, not a population. Patients lined up at emergency departments for care. Off-site gatekeepers with an incentive to deny fielded the patients' reasons for requesting treatment. The result was angry patients, most of them voters.

Capitation, which came next, was an attempt to decrease costs by putting providers in a position in which the more they used resources, the more it cost them personally [9]. Least busy and emptiest hospitals were the richest. Excessive treatment meant cost for providers, not patients. Providers were put in a situation where their economic lifestyle became a consideration in their professional obligations to patients. Even though capitation was somewhat effective, at least in California, it was never popular with providers [10].

As consolidation of health care institutions grew, and as overhead and legal liability increased for private practice, becoming a "company person" became more attractive. Medical malpractice insurance, office space, equipment hardware, computers, heat, air conditioning, lighting, and personnel (nurses, administrative assistants, attorneys, accountants) are usually provided as part of the contract for a medical center [11]. Institutional salaries, though not as high as in private practice, offer stability, security, and resistance to the vicissitudes of a fluctuating economy.

As an academic medical center employee, a physician does not need to do more to get paid more. A typical critical care physician is paid the same salary every month no matter how many patients seen or procedures done [12]. His or her duties are clearly outlined in a contractual agreement. If the physician chooses to do more, it is sometimes reimbursed, with some limits to ensure that the institution benefits as much from the extra work as the individual. The hospital also retains authority to maintain professional physician standards as it sees fit.

In smaller community hospital-based private practice, more autonomy sometimes means less oversight simply because there are fewer authoritative overseers. In academic medical centers, employee physicians are expected to practice "good medicine," and the institution rigorously enforces this according to objective standards. Physician managers in charge of quality assurance are empowered to seek out and investigate irregularities in clinical patient care. If a physician engages in some troublesome activity (sentinel event), he or she may be in serious trouble, all the way up to the level of the state medical licensing board.

As health care facilities grew larger, their ability to negotiate reimbursement from both governmental and private health care indemnifiers grew with them [13]. Reimbursement for medical services can be facilitated by an arrangement in which the provider of medical services is monitored by the economic interests of the center. Because physicians are necessary for admissions and discharges, hospital facilities have an incentive to own (and control) these individuals. Growing facilities thus endeavored to bring outlier providers into the fold so that these providers would work for the benefit of the institution rather than themselves, effectively controlling their incentives.

The following is an illustration of how different species of providers might deal with a "problem" ICU patient:

A 92-year-old patient with pneumonia, sepsis, and new renal failure is transferred from a skilled nursing facility to a medical center. She is emaciated and has multiple contractures, decubiti, a tracheostomy, and a feeding tube. She has advanced dementia and does not respond, other than by grimacing when in pain. She is receiving vasopressors to maintain adequate blood pressure. The last time she was admitted (the previous month, for the same complaint), her kidney function was marginal. Now she is in acute renal failure and requires dialysis. Her family comes in and explicitly states she would want to be kept alive by every means possible. The patient's attending physician, an internist, fills out the appropriate form stating the patient is to have everything done, commonly referred to as "full code."

In this circumstance, private practice physicians have a strong financial incentive to provide open-ended care, including consultations to other specialists and subspecialists with similar incentives. Many physicians feel it is not their job to question the benefit of care, instead using a consumer satisfaction standard: "They want everything done, and it's not for me to question their motivations." Because reimbursement is based not on rational need but on correctness of paperwork, no one is looking too closely at how much this treatment will actually benefit the patient.

On the other hand, in academic practice, there are many more onlookers willing to become involved in the question of benefit to the patient. Open-ended care of such a patient has little if any financial incentive for individual providers and is mostly a financial detriment to the institution. Institutional care is more likely to push for "consensus without consent" in caring for this patient. Palliative care services can quickly get involved and start questioning the goals of surrogates in maintaining this patient in an ICU, getting the surrogates' expectations defined and then critically examining these expectations. If goals are not found to be realistic, the institution ethics committee is more likely to become involved in saying no—setting limits on care.

Those are the benefits of an institutional medical practice. There are liabilities as well.

As institutions accumulate power, they develop the incentive to act more like a business and less like a profession [14]. As major medical centers grow, they gain more power in dealing with assorted enemies, including other medical centers and most reimbursers for medical services. The business plan of a large medical institution could shift to accumulating both monetary and political power through collective bargaining, like a union. These medical centers develop an incentive to kill off competing facilities and to hire nonmedical CEOs to interfere in clinical practice for the purpose of accumulating more political power.

Departments bringing in more money can find themselves in favored positions within the institutional hierarchy, not sensitive to the rules and regulations of others. Employees who resist corporate power mongering can find themselves without a job. De facto rationing can occur not to conserve scarce resources for patient care but to maintain corporate war chests.

Health care reform is no longer a promise for the future. It is more or less here, but its form and substance are still evolving [15]. The current health care reform effort (Affordable Care Act of 2010) is said to be budget neutral, but the legislation has several provisions that will directly affect distribution of care [16]. It effectively ends insurance payment discrimination and adds an estimated 31 million potential patients to an already overloaded system. As medical consumers increase, reimbursement will drop and the workload will increase.

How will this evolution affect academic medical centers? Politicians have promised universal health care for all Americans for years, but they have characteristically been vague as to details. All of them know that no economy in the world can meet an unlimited demand for expensive, tax-based services, but they also know that the public sees universal health care as all the care desirable for as long as desired. Politicians are also aware that their constituents are prone to use political coercion and lawsuits to fulfill their desire. Providers know that there are no hard legal precedents to protect them in saying no to medically inappropriate care, unless that care is entirely ineffective in sustaining vital signs.

It is difficult to imagine a health care plan that will achieve all the desired goals without some form of rationing, creative prioritization, and tough collective bargaining on both sides of the reimbursement table. Private practice doctors will have difficulty competing for reimbursement with facilities that offer the same services for less and can endure more financial hardship.

The only way businesses stay competitive is by becoming more efficient: doing more with fewer resources. Critical care medicine is most expensive when the act of doing more continues past the point of diminishing returns. Expenses mount when patients are kept a little longer just to be observed for a very unlikely unexpected decompensation potential, when daily blood draws and radiographs become a habit, and when patients and their families dictate bed utilization.

We can afford to take this approach to care now, but in the near political future, things will change dramatically. If providers cannot or will not say no when cost exceeds benefit, the consequences are likely to be real and exigent. In the current medical-industrial milieu, there is an incentive to accumulate patients, because today's business strategy of extracting reimbursement for patients is successful in most circumstances. It will be interesting to see how this business strategy evolves when marginal increases in patients entering the system generate an incremental financial loss.

References

1. http://health.uc.edu/health/aboutus/history.html.
2. Bleeding edge: the business of health care in the new century. New York: Aspen Publishers; 1998.
3. http://www.cpsonline.info/capcmanual/content/elements/financingcommunityhospitals.html.
4. Levine M. St. Vincent's is the Lehman Brothers of Hospitals. 2010. http://nymag.com/news/features/68991/.
5. Evans RG. Illusions of necessity: evading responsibility for choice in health care. J Health Polit Policy Law. 1985;10(3):439–67.
6. Stoline AM, Weiner JP. The new medical marketplace: a physicians guide to the health care system in the 90s. Baltimore: The Johns Hopkins University Press; 1993.
7. (No Authors) Health care rationing through inconvenience. N Engl J Med. 1989;321(9):607–11.
8. Iglehart JK. The struggle between managed care and fee-for-service practice. N Engl J Med. 1994;331:63–7.
9. Town R, Feldman R, Kralewski J. Market power and contract form: evidence from physician group practices. Int J Health Care Finance Econ. 2011;11(2):115–32.
10. Robinson JC, Casalino LP. The growth of medical groups paid through capitation in California. N Engl J Med. 1995;333(25):1684–7.
11. Robinson JC. Consolidation of medical groups into physician practice management organizations. JAMA. 1998;279(2):144–9.
12. Robinson JC, Shortell SM, Rittenhouse DR, Fernandes-Taylor S, Gillies RR, Casalino LP. Quality-based payment for medical groups and individual physicians. Inquiry. 2009;46(2): 172–81.
13. http://www.upmc.com/MediaRelations/NewsReleases/2010/Pages/FY2010-financials.aspx.
14. Matthews Jr M. Medicine as a business. Mt Sinai J Med. 2004;71(4):225–30.
15. http://www.pnhp.org/.
16. http://docs.house.gov/energycommerce/SUMMARY.pdf.

Chapter 14
Australia: Where Are We Going?

Ian M. Seppelt

Introduction

In 2011 Australian intensive care is evolving from the "closed shop" of a small and dedicated specialty to a much more diffuse form of practice, with many new stresses as the expectations of referring doctors change, the expectations of the general community become at times unrealistic, as well as changes in practice forced by the different expectations of a new generation of doctors. The bureaucratization of health care and growth of managerialism adds a greater burden. The scientific background of intensive care is developing and there is both the privilege and challenge of conducting good quality research in this difficult patient population, as old-fashioned approaches such as "it works for me" or "in desperate times do desperate things" are no longer considered appropriate.

The Growth of Mega-ICUs

Most major hospitals are seeing a growth in and increased utilization of their critical care services, while at the same time the individual patient acuity is decreasing. That in and of itself is not a problem—there is much more "value for money" spending a few days getting simple things right on a patient who can clearly benefit, rather than investing a huge number of resources in a deteriorating patient in multiorgan failure who is clearly going to die. As each major hospital is redeveloped, the service

I.M. Seppelt, M.B.B.S., F.C.I.C.M. (✉)
Discipline of Intensive Care, Sydney Medical School,
University of Sydney, Sydney, NSW, Australia

Department of Intensive Care Medicine, Nepean Hospital,
Penrith, NSW, Australia
e-mail: seppelt@med.usyd.edu.au

D.W. Crippen (ed.), *ICU Resource Allocation in the New Millennium: Will We Say "No"?*,   107
DOI 10.1007/978-1-4614-3866-3_14, © Springer Science+Business Media New York 2013

grows, thus two major hospitals in Sydney have gone from, in one case a 12 bed general ICU including liver transplant service and surgeon run cardiac and neurosurgical units to a 48 bed intensive care "hot floor," and in another case an 8 bed district hospital ICU has over 20 years become a 60 bed ICU in a major trauma and tertiary referral centre.

The implications of a mega-ICU are many. Staffing is much more challenging, both at medical and nursing levels. Five specialists can be a comfortable and collegiate group running a 12 bed ICU, but a department of 20 specialists functions with a very different dynamic. There is evidence that a patient load of 10–12 is optimal for an intensive care physician—fewer leads to inefficiency, while more leads to exhaustion and mistakes, as well as quantifiable differences such as prolonged length of stay [1]. Practice models such as "1 week in 4 clinical" with teaching, administration, and research during the remaining time become impossible with such numbers. It is impossible to justify the nonclinical time for so many staff specialists so instead specialists are employed specifically for clinical service delivery, which fosters a "shift-work" mentality without the commitment to the department, including teaching and development. In many places this leads to two tiers of specialists, a small core of those who are employed full time to run and develop the department, and a larger number of "workers" who come, treat the patients and leave. In this fragmentation, sometimes patient continuity can be lost.

The clean "closed model of care" is also potentially lost in a mega-ICU. Postoperative patients are booked in by surgeons, bypassing the "triage" process, and the increasing number of outreach services, including rapid response or medical emergency teams (MET) increase the responsibility of the ICU to "solve" the problems of the rest of the hospital. The large number of people involved means it is very difficult for an outside specialist to identify the "intensivist in charge" and frequent handovers lead to lack of continuity. It is impossible to be on call after hours for such a large number of patients and various models are being tried to provide the safety of specialist cover after hours which maintaining a satisfactory lifestyle and some degree of continuity of care. While few Australian intensivists want to do this, it is recognized as inevitable that eventually specialists will be sleeping in-house while on call.

The Bureaucratization of Health Care

The number of hospital administrators and bureaucrats has increased fivefold over the last 30 years and as a consequence it is increasingly difficult to get anything done. Simple procedures, such as hiring new junior staff, now become increasingly complicated when controlled centrally, and delay is often used as a means to save money. Reorganization at will, and the rapid turnover of senior administrators, leads to a large lack of corporate memory [2]. Bureaucracy is unfortunately a fact of life in any public health care system, but this is one of the factors that is driving senior skilled clinicians out into the private sector.

Public Expectations: "You Must Do Everything"

The public understanding of intensive care medicine remains extremely poor, and informed more by the popular media than by science or fact. When asked directly the vast majority of chronically ill patients do NOT want extreme measures taken when approaching the end of their lives, but merely want some reassurance that they will be cared for and not abandoned. It is time consuming but a good use of time to sit and talk to patients or particularly families who want "everything done" as that can usually be readily easily turned into wanting "everything reasonable done with good comfort care and without invasive therapies or intensive care admission." Palliative care is a growing part, and often a very satisfying part, of the job of the intensivist [3]. The College of Intensive Care Medicine of Australia and New Zealand (CICM) has a strong emphasis on communication skills in trainees, and these are formally taught and assessed as part of the examination process [4]. In addition, formal teaching on death and dying is required as part of the CICM training program.

Unrealistic expectations are often as much the problem of the referring doctor, as the patient or family. With the advent of super-specialization, and particularly the "single organologist," the intensivist remains the one true "generalist" in the modern hospital, with an overview of the whole of the patient's condition, not just a single organ or pathology. The danger of course is that the intensivist merely stands back and manipulates the ventilator, which is an abrogation of the privileged role and responsibility of the intensive care physician.

The Importance of Research in Critically Ill Patients

Australia is blessed with the intensive care culture and is rapidly developing the infrastructure to perform high quality research in critical illness. The Adult Patient Database with over 900,000 records is a unique resource, and the ANZICS Clinical Trials Group (CTG) has in a decade demonstrated to the world that it is possible to rapidly recruit to high quality trials and answer fundamental questions of relevance to intensive care practitioners around the world [5, 6].

A recurring paradox is that outcomes in Australia are frequently as good as or better than those seen in the intervention groups of randomized controlled trials, without applying the intervention. Thus, outcomes from Acute Lung Injury and ARDS in three Australian states [7] are equivalent to those seen in the low tidal volume arm of the ARDSNet trial [8] [control group 39.8% mortality, intervention 30.0% mortality, Australian practice 32% 30 day mortality] while an observational cohort study of sepsis management in emergency departments [9] showed better outcomes than seen in the original trial of Early Goal-Directed Therapy [EGDT] [10] despite patients being matched to the inclusion criteria of the EGDT trial, and EGDT not being used [EGDT control group 46.5% in hospital mortality, 30.5% intervention group mortality, 28.0% Australian matched cohort mortality].

There are many possible explanations for this paradox. In general, mortality in sepsis and ARDS has been improving over time, so it is reasonable to expect a more recent cohort study to show improved outcomes, and it is likely that there are significant unmeasured differences between cohorts which affect outcome, but there are some more provocative explanations which were explored in a recent paper by Professor Rinaldo Bellomo and colleagues [11].

Bellomo emphasizes the importance of national collaborative data collection (and there are only three national intensive care databases in the world) and collaborative involvement in investigator initiated research. Looking at multicenter randomized non-pharmaceutical investigator led intensive care trials from 1992 to 2007, Australia and New Zealand randomized 7,500 patients into published trials and another 9,500 in ongoing trials, Canada randomized 5,500 (published) and 5,800 (ongoing) while the USA with 10 times the population and 40 times the number of ICU beds randomized 2,000 (published) and 2,900 (ongoing). Expressed per million population, Australian and New Zealand randomized 380 patients per million, Canada randomized 156 while the USA only randomized 8 patients per million population. It is well established that patients recruited into clinical trials, or even just admitted to units which perform clinical trials, do better on average than those which do not [12], so it is not unreasonable to argue that an intensive care culture which includes an active collaborative research program is likely to have better outcomes.

The rise of research bureaucracy has the potential to strangle a lot of this, of course. In the last 10 years, research ethics and research governance procedures in Australia have become significantly more complicated despite efforts by bodies such as the National Health and Medical Research Council to harmonize research ethics processes around the country and it has reached the point that in a risk-averse culture it is almost impossible to do basic pathophysiological research on critically ill patients, and even simple clinical trials are so tied up in process that many researchers are starting to find the processes all together too hard.

Summary

Bellomo [11] lists a number of features that make up the "Australian way" in intensive care medicine, most of which are developing but some are under threat. The most important is probably the 1:1 nurse to patient ratio. In many parts of the world that will be considered an expensive luxury, but the highly trained ICU nurses fulfill a great many other responsibilities, including operating ventilators, managing respiratory therapy, managing dialysis machines and titrating vasoactive drugs, in the context of a collaborative and nonhierarchical "team" environment. If anything defines the essence of Australian intensive care it is the strong presence of ICU nurses at the bedside, and in adequate numbers, and objective outcome data support this assertion. Allied health, particularly physiotherapists, and pharmacists are likewise crucial members of the ICU team. Intensive care training through the College of Intensive Care Medicine is rigorous and highly regarded, even as the number of trainees seeking the qualification escalates. The "closed model" of intensive care, while under threat in the mega-units, is still considered the ideal, allowing the patient's management to be supervised by one senior individual who has an overview of the whole of the patient. Finally, participation in a nationwide database and many investigator initiated clinical trials provides robust and objection information as a basis for quality assurance and comparison.

Conclusion

As we enter the second decade of the twenty-first century, there are a number of challenges and threats to the current standard of intensive care medicine in Australia. As the specialty developed it has almost become a victim of its own success, with increasing calls for outreach and the development of mega-ICUs while other specialties contract into their specialized and often single organ niches. This requires increased numbers of personnel with the consequent fragmentation of practice. This is exacerbated by all the problems seen with the increasing bureaucratization of health care, and unrealistic expectations by members of the public and often more importantly their doctors. The special features which define Australian intensive care remain, including 1:1 nursing, a rigorous and well-regarded intensive care training program and a strong focus on quality assurance and collaborative clinical research.

References

1. Dara SI, Afessa B. Intensivist-to-bed ratio – association with outcomes in the medical ICU. Chest. 2005;128:567–72.
2. Oxman A, Sackett DL, Chalmers I, Prescott TE. A surrealistic mega-analysis of redisorganization theories. J R Soc Med. 2005;98:563–8.
3. Cosgrove JF, Nesbitt ID, Bartlet C. Palliative care on the intensive care unit. Curr Anaesth Crit Care. 2006;17:283–8.
4. Corke C, Milnes S. Communication for the Intensive Care Specialist: planning for effective, efficient and compassionate interactions. York: Erudite Medical Books; 2007.
5. The SAFE Study Investigators. A comparison of saline and albumin for fluid resuscitation in the intensive care unit. N Engl J Med. 2004;350:2247–56.
6. The NICE-SUGAR Study Investigators. Intensive versus conventional glucose control in critically ill patients. N Engl J Med. 2009;360:1283–97.
7. Bersten AD, Edibam C, Hunt T, et al. Incidence and mortality of acute lung injury and the acute respiratory distress syndrome in three Australian states. Am J Respir Crit Care Med. 2002;165:443–8.
8. The Acute Respiratory Distress Syndrome Network. Ventilation with lower tidal volumes as compared with traditional tidal volumes for acute lung injury and the acute respiratory distress syndrome. N Engl J Med. 2000;342:1301–8.
9. Peake SL, Bailey M, Bellomo R, et al. Australasian resuscitation of sepsis evaluation (ARISE): a multi-centre, prospective, inception cohort study. Resuscitation. 2009;80:811–8.

10. Rivers E, Nguyen B, Havstad S, et al. Early goal-directed therapy in the treatment of severe sepsis and septic shock. N Engl J Med. 2001;345:1368–77.
11. Bellomo R, Stow PJ, Hart GK. Why is there such a difference in outcome between Australian intensive care units and others? Curr Opin Anaesthesiol. 2007;20:100–5.
12. Grimshaw JM, Russell IT. Effect of clinical guidelines on medical practice: a systematic review of rigorous evaluations. Lancet. 1993;342:1317–22.

Chapter 15
Brazil: Where Are We Going?

Rubens Costa-Filho

Introduction

IBGE (Brazilian Institute for Geographical Statistics) has just completed the 2010 census which concluded that Brazil is transforming its profile of a younger country into an elderly one. The dimension of the Brazilian census is huge [1] (see Table 15.1). According to this census, Brazilian life expectancy in 1980 was 65.75 years, and 77.01 in 2009. In an analysis performed by this Institute, in 2050 life expectancy is expected to increase faster to 81.29. Thus, in a few decades Brazil will be comparable to countries such as Iceland (81.8), Hong Kong, China (82.2), and Japan (82.6), which are amongst the largest in the world today. Another peculiarity recently announced in the media regarding this census was related to the profile of female population which is in an excess of four million, and with a future projection of 14 million, signifying a triple increase of women in 2050 [2].

This picture of a rapid growing aged population will create several problems for the social health system, mainly to critical services. Based on this analysis, the health system at least would have to be more prepared in a logistic way to face more female-related problems, for example breast cancer (unless a cure has been discovered). On the other hand, the child (less than 1 year) mortality rate has decreased from 69.12 per 1,000/live births to 22.47 per 1,000/live births. This result means that Brazil had an 80% reduction in infant mortality rate in the last three decades. Furthermore, according to IBGE, in this millennium the fixed target rate would be 15 per 1,000/live births, but in fact this target has been projected to reach 18 per 1,000/live births. This was due to the efficient vaccination programs, breast feeding and follow up care of newborn babies, universal access to prenatal services, breast

R. Costa-Filho, M.D., F.C.C.P., M.B.A. (✉)
Critical Care Centre, Hospital Pró Cardíaco, Av. Gal Polidoro 192,
Botafogo, Rio de Janeiro CEP 22280-020, Brazil
e-mail: rubenscosta@me.com

D.W. Crippen (ed.), *ICU Resource Allocation in the New Millennium: Will We Say "No"?*, 113
DOI 10.1007/978-1-4614-3866-3_15, © Springer Science+Business Media New York 2013

Table 15.1 Magnitude and extension of Brazilian census 2010 [10]

Local	Quantity
Universal of census	All national territory
Municipal districts	5,565
Residences	58 million
Sector units for census	$\approx$314
Areas of coordination	222
Collection of data computerized	6,823
Population	190,732,964

feeding campaigns and, lastly, investment in basic sanitation and educational programs.

Based on the above information gathered from this deep population analysis, probably the attention will have to be moved to the elderly instead of newborns. Therefore, government should look with more emphasis at an intelligent resource allocation to a critical care service related to the older not the younger population, as well as, the investment and specialized training to health care providers in the geriatric field.

The Dilemma of Supplementary Health in Brazil

Gloomy Perspectives

Supplementary health in Brazil stemmed from the inability to put into practice the fundamental theory of universal health care such as mentioned in Chap. 1 by Carvalho, FB et cols. Although this system was born to give more resources and benefits to the public health care, the players (see Table 15.1) involved are still coping with serious problems of relationship with each other to fulfill a working system in the private sector.

Nowadays, this system is dysfunctional because they are working in an isolated way with different perspectives and so obtain different outcomes losing values and energy during trading and communication.

As previously mentioned regarding the growth of the elderly, health problems will increase and also the medical resources that will be required, as well as the technology used which is becoming even more advanced and expensive.

Users/Beneficiaries

The beneficiary, due to lack of information, has a short-term vision in which their goal is the maximum utilization of the health services even reaching the point of lending their ID to others that have not contributed to the health plan. Also the media has been stimulating the population, disseminating information about new technologies, new exams, and correlating them with the best practice. Furthermore, the beneficiary tends to forget that this account has to be paid by someone.

Suppliers

The group of suppliers can be divided into hospitals, physicians, and providers of medical materials such as orthoses and protheses. Due to the fact that they blame low prices applied by official table of prices stipulated by government agencies they are constantly developing ways to make more money. An example of this was increasing the number of consultations performed by physicians, more tests, hospital admissions, utilization of more expensive health materials, and excess of unnecessary procedures. This situation that is being adopted pressures the suppliers working on behalf of a variable remuneration to forget that the product of the industry in which they are embedded is the health care system.

HMO Groups

HMOs work in the compensation scheme, paying suppliers and charging the beneficiaries, i.e., do not add the majority of business value. The argument for this stance is the high turnover of the portfolio of lives, which would not be worth the long-term investment, as far as getting the return, this life insurance would already belong to another carrier portfolio.

The increased demand for care causes an increase in charges for the users, which is one of the main indicators of this system. The relationship between the charges such as medical consultations, procedures, exams, and revenues collected is now about 90%, which is a threat to the health of this system. According to international standards, in order for companies to recover capital invested, the charge index has to represent at the most 75%.

The backdrop of the government, where the Law 9656/98 regarding supplementary health (cited in the previous chapter) as a saving measure of health, served as a complicating factor for those involved (see players in Table 15.2) in the processes of health, establishing a reference plan where almost all procedures should be achieved, and on the other hand controlling the budget. The differentiation of this sector was to rely on the budget of only accredited network and accommodation; the competition is the price and deadline.

This obligation to offer services that meet a standard of coverage set by the regulatory agency has blocked the option to supply management as a strategic business, leaving them only the demand management, which leads many HMOs to purchase their own hospitals and laboratories (verticalization management model).

Table 15.2 To accomplish good results the players must be totally integrated [11]

Players	Definition
Users/beneficiaries	patients pay to use this health care
Suppliers	Establishments that provide health care, physicians and suppliers of medical materials
Health management organizations[a]	Managers of health care according to contracts purchased by user/beneficiaries
Regulators[b]	Government agencies that regulate HMOs and suppliers

[a]The National Agency of Supplementary Health (ANS) classified these organizations into eight modalities: Administrators, medical cooperatives, odontological cooperatives, philanthropic institutions, self-management (sponsor and non-sponsor), security specialized in health, medicine groups and odontological groups
[b]ANVISA (National Agency of Sanitary Surveillance) and ANS

The current age structure (based on bands) of health plans in Brazil serves the actuarial balance, where the increase among the age band is preestablished. Therefore, it is necessary to have the first one (younger) to offset the last (older) increase, creating a barrier to entry for new users (younger) in which the profit margins would be better, leading to concentration of risk and increase costs. The fact is that the government is not willing to take a political decision in order to change the design of this system.

Suppliers Not Yet Well Aligned

The hospitals that make up the private sector have excessive numbers of beds, as well as health care providers in some regions of Brazil, although with great heterogeneity (culture and continental dimensions). The private sector has been stressed by hospital administrators as well as physicians in order to develop greater technological modernization regardless of their costs or cash flow compromised. Their policy of obtaining best results with higher profit margins has created unethical procedures such as extra charges, unauthorized charges, hospitalizations and questionable procedures, unnecessary tests, where admissions account for about 16–18% of total costs, and in some institutions critical care services have produced more than ¼ of the hospital budget.

In this barely and badly functioning system, physicians contribute poorly ordering unnecessary exams in excess, which could contribute to 38–40% of the total costs. Also, events such as ethical–professional incompetence and recklessness has increased by the number of processes analyzed by regional councils of medicine (CRM), as well as identification of unethical behavior by doctors to ensure themselves better payment.

The third group consists of those suppliers that provide materials used by doctors in medical procedures such as implants (ortheses and protheses). In this case, doctors order extra material for a particular procedure that is not normally used, and its billing is charged to the HMOs; afterwards the medical supplier forwards his profit to the doctor in the form of a commission.

This highlighted the existence of an evil competition among the stakeholders in the health industry, fighting for conflicting goals in a relationship where all destroy value.

The results are dissatisfaction among employers, clients, patients, doctors, laboratories, and hospitals.

The vision of these agents ought to be changed in order to accomplish better outcomes and quality. Due to this, they are acting as opponents inside the system with important conflicts of interests, allowing loss of values and discontinuation of the market share of the few. The key to this scenario might be a strategic alliance in all directions in a way to preserve the private sector. Additionally, the government knows what ought to be done in order to redesign a better system, and the future of supplementary health depends on this. A new relationship should be created where in fact health care could survive and pave the way to growth of quality and efficiency.

A good scenario would be everybody working in a synergic way such as hospitals working together with the pharmaceutical industry, medical material suppliers together with health plans, and users/beneficiaries together with all health care providers, e.g., physicians. In this way, new horizons could be more fruitful through synergy and knowledge exchange along with commitment of all involved in order to develop and sustain a better construction and improvement of the system.

Who Is Responsible for Changing a Culture?

Hospital centered model is entrenched in our culture. Many people believe that the best practice medicine is that linked to high technology, sophisticated and expensive exams. Thus, this scenario pressures mainly private physicians to work on the side of money instead of efficiency and rational and evidence-based protocols. The measurements of indicators of quality and performance inside hospitals are still facing many barriers [3, 4].

Thinking about the future is a challenging task because it involves multiple aspects of the geographical and political scenario [5] (see Fig. 15.1). The private sector involves only 30% of the Brazilian population. Around 70% of the population in Brazil depends exclusively on SUS (Single Health System) [6] being followed up by 27,000 teams related to health family which attend 92% of the municipalities, which is the base of a new model of assistance. In 2006, SUS performed 2.3 billions of ambulatory procedures, more than 300 million medical consultations and two million birth assistance.

In a scenario of high complexity, SUS performed 20,156 transplants last year, 215,000 cardiac surgeries and nine million procedures related to chemotherapy/radiotherapy, and 11.3 million hospital admissions [6]. The number of transplants has grown by 71.5%. Brazil has the third largest bone marrow donor's bank of the world, only behind the United States and Germany. More than 1.2 million people are registered [7].

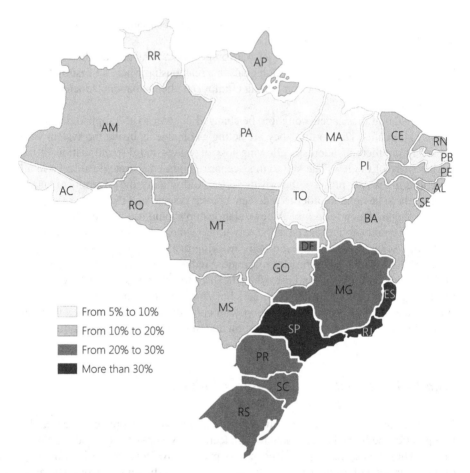

Fig. 15.1 Distribution of health care plans—private sector—among states of the federation. São Paulo and Rio de Janeiro (more developed in the country) and Espirito Santo (ES) receive more attention from this system (≥30%) [5]

Considering the economic field, health care corresponds to 20% of global spending (public and private including research and technological development) which represents around US$ 135 billion. These prospects highlight very clearly that health care is one of the most dynamic areas of the world.

Brazil Health System Is Reorganized and Increased

The SUS went through a continuous process of integration from 2003 to 2009. The service was reorganized and the assistance in the public hospitals became more agile. Functioning in the network are: the Family Health Strategy (ESF), that brings health care closer to the families; the Emergency Mobile Medical Service (SAMU), that helps in urgent cases; and the First Aid Unit (UPAs), built to receive patients in emergency situation, such as hypertension crisis and burns. This articulation frees up hospitals and reduces queues. The Ministry of Health implemented 329 UPAs from May 2009 to March 2010 [7] which has already reduced unnecessary admissions to hospitals and critical care facilities.

Population Has More Access to Medicines [7]

* Popular Pharmacy: In 2009 the Program Popular Pharmacy of Brazil, with its 529 own facilities, sold 107 medicines that cure the most common diseases in the country, such as hypertension and diabetes. The medicines, sold at cost price, helped 11.4 million people. The number of users, in 2004, was 470,000.
* Here's your Popular Pharmacy: Partnership with registered private drugstores grew by 72.6% between 2006 and 2009, from 2,955 to 10,790. The citizen buys some medicines (12 different active ingredients) up to 90% cheaper. The Federal Government pays the difference.
* Generic medicines: The offer of generic drugs in the market grew 14 times. In 2003, there were 213 medicines with lower price in the private drugstores. In 2009, there were 2,972 medicines recorded.

Reduction of Vulnerability in Strategic Sectors [7]

* Stimulus to the national production of medicines, vaccines, and health equipments: there has been an increase of investments with research and technology (R&D) transfer agreements for the pharmaceutical industry and Brazilian labs since 2008.
* Drugs for the treatment of aids—Efavirenz: Free access was guaranteed to the medicine to all that need the treatment. The Brazilian Government decreed the compulsory license of the medicine in 2007 and began the national production of the generic antiretroviral. Besides of the benefit to Brazilians, the Union has already saved R$ 154 million with this process.
* Control of sexually transmitted diseases and unwanted pregnancies—Condom Factory: In April 2008, the first factory in the world to produce condoms with natural rubber from a native rubber tree was inaugurated in Xapuri (AC). The condom is called Natex. In 2009, 34 millions condoms were distributed for free by the company, built with federal resources.

Due to the above actions well executed by government, in the field of prevention and low cost delivery of medicine for specific groups of diseases, these could decrease the excess use of hospitals and even better use of critical care in the future.

Conclusions and Final Remarks

A modern vision to sustain quality and best critical care services goes beyond the severity of the patient admission, resources or diseases related to age and comorbidities. Future prospects in critical care setting also came to the forefront globally triggered by the Institute of Medicine's (IOM) to Err is Human and Crossing the Quality Chasm reports [8, 9]. These reports had the capacity to shake perceptions from the professional health care point of view to the common citizen around the world. Also this report led to a multitude of suggestions for patient care quality and safety initiatives whose approaches were inspired by the pioneer work from Donabedian's framework for assessing quality and also focused on specific clinical care structural changes (see Fig. 15.2). In Brazil numerous national organizations are mirroring international organizations to put forward initiatives regarding quality or patient safety such as those spread by the Leapfrog group, the Joint commission on Accreditation of Healthcare Organizations (JCAHO) and National Quality Forum (NQF).

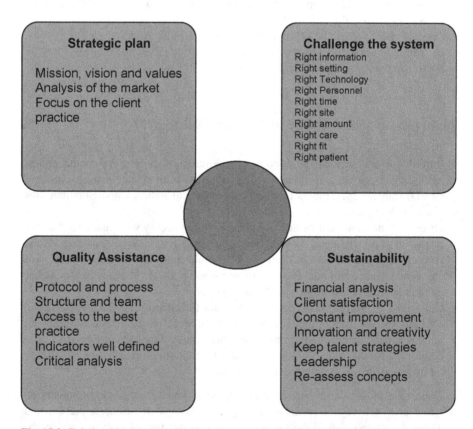

Fig. 15.2 Relationship among quality assistance, strategy, sustainability, and ideal goals (adapted from refs. [4, 12])

The ONA (National Organization on Accreditation) was created in 1999 in Brazil initiating the process to implement technical norms inside the Brazilian system of accreditation, based on international experience. Furthermore, various institutes were certified by this organization and government agencies in order to improve and audit the process adopted by the establishments that want to be accredited.

Therefore, the idea of security and quality is starting to gain force also in the Brazilian health care, mainly in the critical care field. Changing culture in a heterogeneous mixed health care system (private and universal) is a huge challenging task for all players.

It is vital for the future that governmental actions work towards the dissolution of the wrong idea held by the population of believing that it has the constitutional right to receive health care, not as an object of consumption in a society where a hedonistic model of life is dominant, but as an efficient allocation of best practices and tighten the chasm between the money gathered from taxes and contributions and the return of them in health care benefits to all the society.

Acknowledgments   I would like to thank my critical care team from Hospital Pró Cardíaco which have been supporting and stimulating my work in the ICU.

References

1. IBGE. IBGE inicia contagem regressiva para o Censo 2010. 2010. http://www.ibge.gov.br/home/estatistica/populacao/contagem2007/default.shtm. Accessed 2010.
2. Neto EV. O BRASIL ESTÁ ENVELHECENDO. 2010. http://www2.brasil-rotario.com.br/revista/materias/rev976/e976_p6.htm. Accessed 2010.
3. Cabana MD, et al. Why don't physicians follow clinical practice guidelines? A framework for improvement. JAMA. 1999;282(15):1458–65.
4. Fernandes HPS, Costafilho R. Intensive care unity quality. Rev Bras Clin Med. 2010;8: 37–45.
5. ANS. Taxa de cobertura de assistência médica por unidade da federação. Dados do site 2011. http://www.ans.gov.br/index.php/materiais-para-pesquisas/perfil-do-setor/dados-gerais.
6. Saúde Md. Mais saúde direito de todos – 2008–2011. Ministério da Saúde S-E, editor. Brasilia: MS; 2008. p. 100.
7. Government P.o.t.F. Highlights: actions and programmes of the Federal Government/Secretariat for Social Communication, S.f.S.C.o.t.O.o.t.P.o.t.R.o. Brazil, Editor. Brasilia; 2010. p. 73.
8. Kohn LT, Janet CJ, Donaldson MS. To err is human: building a safer health system. Washington, DC: National Academy Press; 2000.
9. America C.o.Q.o.H.C.i. Crossing the quality chasm: a new health system for the 21st century. Washington, DC: National Academy Press; 2001.
10. IBGE. Censo 2010: população do Brasil é de 190.732.694 pessoas. 2010. http://www.ibge.gov.br/home/estatistica/populacao/censo2010/default.shtm.
11. ANS. Entenda o setor. 2010. http://www.ans.gov.br/portal/site/dados_setor/dados_setor.asp.
12. Wakefield DS, Ward MM, Wakefield BJ. A 10-rights framework for patient care quality and safety. Am J Med Qual. 2007;22(2):103–11.

Chapter 16
Canada: Where Are We Going?

Randy S. Wax

Introduction

As medical care in Canada enters the new millennium, our basic tenet of universal access to health care is in question. The Canada Health Act, enacted in 1984, established the rules by which the federal government transfers funds to provincial governments to provide health care to their constituents. Provincial governments are required to provide all medically necessary hospital and physician services to all insured persons, without co-payments. Key aspects of The Canada Health Act include: Public Administration, Comprehensiveness, Universality, Portability, and Accessibility. This key piece of legislation has driven the health care agenda in Canada since its implementation. In the absence of clear definitions of these principles, however, many issues have been left vague and subject to considerable variability between provinces over time. In the coming years, the definitions of "comprehensiveness," "universality," and "accessibility" related to provision of critical care in Canada will likely require the added qualifier of "acceptable," given that the potential demand for critical care will progressively increase over the next decades. Recent events, such as the SARS outbreaks and the H1N1 influenza pandemic, have brought to light numerous vulnerabilities within the Canadian healthcare system. This chapter will explore factors in the Canadian healthcare environment

R.S. Wax, M.D., M.Ed., F.C.C.M. (✉)
Department of Medicine, Queen's University,
Kingston, ON, Canada

Department of Medicine, University of Toronto,
Toronto, ON, Canada

Department of Emergency Medicine and Critical Care,
Lakeridge Health, Oshawa, ON, Canada
e-mail: rwax@lakeridgehealth.on.ca

D.W. Crippen (ed.), *ICU Resource Allocation in the New Millennium: Will We Say "No"?*,   123
DOI 10.1007/978-1-4614-3866-3_16, © Springer Science+Business Media New York 2013

that will influence developing regional and national standards for what is considered "acceptable" critical care, and will provide some predictions about the future of critical care in Canada.

Anticipated Demand for Critical Care in Canada

Like many other countries, Canada has noted a progressive demand for critical care beds over the past decades [1], and anticipates a dramatic further increase in demand in the next few decades. One provincial analysis has predicted a 57% increase in demand for mechanically ventilated beds from 2006 to 2026 [2]. Many jurisdictions have attempted to mitigate the need for more ICU beds in the coming decades through strategies to reduce need for ICU beds by preventing critical illness, avoiding ICU admission when not consistent with patient philosophy on end-of-life care, and preventing factors that contribute to increased length of ICU stay and subsequent inefficient use of ICU resources. Ontario has linked efficiencies of critical care services with publically reportable wait times for surgery and delays in Emergency Department care, with high political stakes for success with the provincial government. Initiatives have included funded rapid response teams in larger hospitals [3], coaching teams to help improve patient flow, and education initiatives [4]. Other multicenter projects have used intensive knowledge translation strategies to promote critical care best practice to reduce factors contributing to increased length of ICU stay and morbidity [5]. The ICU resource burden and cost due to preventable ICU complications such as ventilator associated pneumonia are great, so the regionalization of critical care has created the opportunity for regional leadership in quality improvement across hospitals, aiming for greater efficiency of ICU resource use [6].

Regionalization

As the second largest country in the world, the geography of Canadian population density is a key factor determining the organization of healthcare services. Roughly 80% of the Canadian population lives within 150 km of the northern border of the United States, yet the country has an extremely low average population density. Providing equitable care to patients living in urban, rural, and extreme rural locations within Canada is a huge challenge. By necessity, a regionalized network approach to medical care has been established in most jurisdictions. Although one approach to resolving this inequity has involved the creation of acute care beds in rural areas, the occupancy of these beds is significantly less than those acute care beds in urban areas. Similar to observations in other jurisdictions, treatment of some critically ill patients in low volume smaller critical care units is associated with poorer outcomes compared with transfer to a larger facility [7].

Despite the creation of hospital capacity in rural areas, difficulties in recruitment of physicians, especially specialists, limits the nature of care available. Increasing infrastructure to produce more trained intensivists annually has caused a slow saturation of positions in some urban centers; however, as the population slowly spreads to the suburbs, new positions are being created that appear more desirable than jobs in more rural locations. Intensivists are less willing to take positions in rural hospitals given the lack of a critical mass of ICU beds to generate expected income, lack of specialist support, and inavailability of technology (such as renal replacement therapy) necessary to care for sicker ICU patients. Care of the critically ill patient is provided by surgeons, internists, anesthesiologists, family physicians, and others in many rural centers. Some attempts have been made to train nonintensivists to improve their ability to stabilize critically ill patients and potentially reduce transfers; however, these efforts have not been sustained or implemented on a large scale.

Regionalization of specialty critical care has already taken place, mainly in the areas of trauma, cardiovascular surgery, and pediatrics. The outcome of medical critical care patients appears to be worse when cared for in lower volume centers that are less likely to transfer patients to regional centers. There will be an increasing push towards the creation of a "hub and spoke" model of critical care, with regional centers affiliated with other smaller centers. Regional health authorities, such as the Local Health Integration Network (LHIN) system in the Province of Ontario, are responsible for developing a network of Critical Care for centers within their jurisdiction. Ontario has appointed an experienced intensivist as the Critical Care Lead for each LHIN, who can advocate for clinically appropriate distribution of services within the LHIN. The network will go beyond clinical care to include efforts to ensure the widespread distribution of critical care best practices within the LHIN through education. A recent pilot project using existing information technology infrastructure has had early success in ensuring that smaller centers can provide high quality care through knowledge translation strategies [5]. Data collection beyond the institutional level will allow better organization and distribution of resources. For example, the Critical Care Information System project in Ontario, which collects extensive data from most critical care units in the province, provides valuable administrative and quality- related data that allows data-driven decision making at regional and provincial levels [4].

The distinction between academic critical care hub and nonacademic referral facility will also become increasingly blurred over time throughout Canada. With the increased number of intensivists being trained, hospitals without academic affiliation will increasingly have the benefit of intensivist-led ICU care. In addition, the increased number of undergraduate and postgraduate medical trainees has forced academic centers to develop more satellite and rural training institutions. Regional critical hubs will not necessary be defined as academic centers, and referral centers will not necessary be defined as nonacademic. The impact of this shift will likely lead to significant changes in the way Canadian intensivists are trained, learning how to practice critical care in centers that do not necessarily have all services or technology available. This trend will potentially spread latest evidence-based care throughout regions using trainees as vectors, and encourage participation in research at more centers. The redistribution of services forced by regional health authorities has created considerable turmoil in some jurisdictions. In some cases, the regional authorities become scapegoats for making difficult decisions, and act as a buffer between an angry local population and provincial governments. These regional authorities will require increased support by their provincial governments to empower them to make difficult decisions to relocate services. Regional authorities may be restructured or dismantled based on transient political pressures. Nonetheless, the regionalization of critical care in Canada is a likely forgone conclusion required by the nature of geography, scarcity of critical care human resources, and evidence of improved quality of care and efficiency through providing higher level critical care with a sufficient critical mass of patients and resources.

Role of e-ICU

Although care will likely involve regionalized critical care throughout Canada, the nature of the country's geography will necessitate increased use of telemedicine technology to support critical care. In many areas of Canada, the distances between referral and tertiary care centers is very large. Winter conditions can create enormous transport challenges, leading to delays of hours to days for the movement of critically ill patients to definitive care. The most extreme examples can be found in the north, with care provided in nursing stations or emergency medical responder outposts with transport times of 2–3 h to definitive care in the best of weather. Limited transport resources can benefit from minimizing unnecessary transfer of patients who could be cared for adequately in outside of tertiary critical care centers. High bed occupancy rates in tertiary ICUs can also lead to delays in being able to accept patients. The role of e-ICU must be considered in the care of patients during the stabilization process, evaluation for the appropriateness of transfer, and the prolonged provision of critical care during delays in transport.

Despite the extensive use of telemedicine in Canada, e-ICU infrastructure has been limited. There are tremendous opportunities to expand e-ICU in Canada over the coming years. The formation of "hub-and-spoke" regionalized approaches to critical care lend naturally to consultation relationships than can be improved through use of telemedicine technology. Ontario has participated in a pilot Virtual Critical Care project through the existing Ontario Telemedicine Network infrastructure, which also included specific physician reimbursement for e-ICU services. However, the reliance on hard-wired solutions and expensive audiovisual installations has limited the expansion of this project. Newer and less expensive wireless technology is on the brink of revolutionizing the way in which e-ICU is provided. e-ICU extension outside of the traditional walls of the ICU onto the wards, into the Emergency Departments, and potentially into the prehospital/transport medicine environment, will extend the clinical reach of intensivists into locations historically deprived of such expertise.

More important than any clinical recommendations may be the ability to evaluate the appropriateness of critical care and the clarification of end-of-life goals for patients prior to ICU admission. Reduction in the inappropriate transfer of patients who could be cared for in a palliative manner without transport could significantly mitigate the projected shortfalls in critical care resources and reduce transport cost and risk. e-ICU will likely include family conferences bringing together distant relatives virtually, expert second opinions, and ethics consultations, modeled after existing mental health telemedicine initiatives present throughout Canada.

Legal and Ethical Issues Related to Access to Critical Care

To date, Canadian governments and courts have not provided clear support to intensivists having to make difficult decisions on how to avoid use of scarce critical care resources in circumstances they feel are not medically or ethically warranted. Guidelines for processes in handling such difficult cases are present in some jurisdictions; however, potential for controversies and conflict remain. A recent case in Manitoba involving intensivists following recommended steps to limit critical care in a case they deemed futile led to court interference with their decisions [8]. Multiple intensivists at the same institution resigned their hospital privledges rather than be forced to comply with what they felt was an unethical court order. Accusations against intensivists by family members and press about alleged unilateral restrictions on critical care have occurred recently in Ontario [9, 10]. The potential political risk for sitting provincial governments responsible for health care has likely prevented any clear declaration or legislative/regulatory direction protecting critical care clinical teams when making difficult decisions regarding approach to end-of-life care when in conflict with family members.

One strategy increasingly used by intensivists in some jurisdictions is to use governmental Consent and Capacity boards (example, Ontario's http://www.ccboard.on.ca) to consider whether the designated substitute decision maker (SDM) is making decisions in the best interest of the patient. The board can order the SDM to make certain decisions, or can relieve the substitute decision maker and replace him/her with another person that is better able to represent the wishes of the patient. This approach has had some success in cases when the patient had prior expressed wishes to limit the aggressiveness of care but the family member with decision making authority cannot abide by those wishes. When there are differing views amongst family members, the board may appoint a different SDM that has a view more consistent with the recommended plan from the critical care team. There is an increasing trend for intensivist involvement sitting as members on these boards in cases related to end-of-life care in the critical care unit, lending a better understanding to the board panel of these complicated issues. Further appeals can be made to the courts, so this approach may not definitively resolve conflicts.

The regionalization of critical care in Canada creates a poorly acknowledged mechanism for controlling critical care resource allocation. When a patient presents with a potentially futile clinical situation, the family usually has access to consulting physicians to plead their case for aggressive care. As regionalization becomes more extensive, fewer centers will have all the services required for comprehensive care. Physician-to-physician remote consultation becomes the means for accessing specialty services, including critical care.

Receiving physicians may choose not to accept the transfer of a patient given a futile situation, having minimal interaction with substitute decision makers. This leaves the sending physician in a potentially awkward situation, with inadequate resources to carry on with life support measures but no place to send the patient.

Increasingly, jurisdictions are developing mandatory "risk to life or limb" policies that mandate timely transfer acceptance by tertiary care centers responsible for designated regions. One can expect tension between the need to transfer promptly versus the need to avoid transfer when not appropriate. Further evolution of these policies must incorporate clear guidelines about the mechanisms to decide not to transfer a patient despite patient/family request when deemed medically inappropriate.

Seeking Alternative Access to Critical Care Beyond the Funded System

The ability to seek medically necessary services outside of the funded system in Canada remains a hot topic for debate. There is considerable opposition to the development of a two-tiered system for health care similar to the public/private system found in many other countries. Given the potential political risk for disrupting the concept of universal Canadian health care, governments will be very reluctant to offer a privatized second system within the borders of Canada. However, the incidence of Canadians seeking medical services outside of Canada has increased. Long wait times for certain surgeries, particularly those meant to improve quality of life but not necessary life-threatening conditions, have prompted Canadians to go elsewhere for care [11]. In some cases, Canadians will seek surgery in other countries when refused by Canadian surgeons due to excessive comorbidity, or when a procedure is not offered in Canada due to slow uptake by surgeons or due to inadequate supporting evidence. For surgeries that will require post-operative ICU care, global medical tourism creates a functional two-tiered health care system for those who can afford it, with financial resorces now diverted outside of the country. Although the decision to not offer a surgical option is not made by the intensivist, the impact on critical care resources is still important. This phenomenon creates a mechanism to bypass evolving processes for prioritizing access to critical care services. Long-term trends of global medical tourism by Canadians remain unclear.

Conclusions

Canada is not immune from the global trend of increasing demand for critical care resources, combined with tremendous financial constraints being experienced at all levels of government, health care regional authorities and hospitals. The unique geographical reality of Canada creates a huge challenge to serve patients in the rural areas and north at the same level of care as those in the densely populated south. The evolution of regionalization will almost certainly progress and will change the nature of ICU resource allocation. Critical care staff on the front lines in Canada are hungry for concrete support through governmental authorities of ethically appropriate but medically reasonable processes for "saying no" to critical care admission.

References

1. Needham DM, Bronskill SE, Sibbald WJ, Pronovost PJ, Laupacis A. Mechanical ventilation in Ontario, 1992–2000: incidence, survival, and hospital bed utilization of noncardiac surgery adult patients. Crit Care Med. 2004;32:1504–9.
2. Hill AD, Fan E, Stewart TE, Sibbald WJ, Nauenberg E, Lawless B, Bennett J, Martin CM. Critical care services in Ontario: a survey-based assessment of current and future resource needs. Can J Anaesth. 2009;56:291–7.
3. Baxter AD. Critical care outreach comes to Canada. Can Med Assoc J. 2006;174:613–5.
4. Trypuc J, Hudson A, MacLeod H. The pivotal role of critical care and surgical efficiencies in supporting Ontario's wait time strategy: part 3. Healthc Q. 2006;9:37–45.
5. Scales DC, Dainty K, Hales B, Pinto R, Fowler RA, Adhikari NK, Zwarenstein M. An innovative telemedicine knowledge translation program to improve quality of care in intensive care units: protocol for a cluster randomized pragmatic trial. Implement Sci. 2009;4:5.
6. Muscedere JG, Martin CM, Heyland DK. The impact of ventilator-associated pneumonia on the Canadian health care system. J Crit Care. 2008;23:5–10.
7. Needham DM, Bronskill SE, Rothwell DM, Sibbald WJ, Pronovost PJ, Laupacis A, Stukel TA. Hospital volume and mortality for mechanical ventilation of medical and surgical patients: a population-based analysis using administrative data. Crit Care Med. 2006;34:2349–54.
8. The Canadian Press. CBC News online. http://www.cbc.ca/canada/manitoba/story/2008/06/04/golubchuk.html. Posted 4 June 2008.
9. Cribb R. The Toronto Star. http://www.thestar.com/news/gta/article/856741--lawsuit-could-set-precedent-about-end-of-life-decisions. Posted 4 Sept 2010.
10. Editorial page. The Globe and Mail. http://www.theglobeandmail.com/news/opinions/editorials/are-we-ready-to-die-with-dignity/article1786413/. Posted 5 Nov 2010.
11. Turner L. Medical tourism: family medicine and international health-related travel. Can Fam Physician. 2007;53(10):1639–41.

Chapter 17
Germany: Where Are We Going?

Thomas Kerz

Germany's health policy, in the past, has tried to limit health expenditures by defining an overall health budget, by installing a DRG-system and by forcing patients to participate directly at health costs by demanding copayments. These measurements were somehow effective in keeping premiums stable but have led to resource allocation at the bedside and therefore on an implicit level. Yet, implicit-level decisions create ethical dilemmas for physicians as they not only have to deliver best medical care but shall also have cost in mind, are unjust because rationing criteria differ from one case to another, and create a general fear of legal uncertainty, in turn leading to defensive medicine with the overuse of inappropriate medical acts. Furthermore, they have no democratic legitimation.

As in most Western Societies, Germany's society is aging rapidly. Health costs are generally expected to rise more steeply, and this idea is often advanced to justify rationing needs in the future. Yet, many older people live in rather good health condition and diseases start only at an older age than one or two generations before. There are studies showing that the explosion in health care costs at higher ages may not be a function of age per se but of individual proximity to death, and individual cost distribution over time is primarily determined by remaining lifetime, not by chronological age [1]. Patients up to their 60s (men) or 70s (women) consume three to four times the amount of money than oldest old patients do [2]. Reasons for this could be an already existing mechanism of palliative care in the care of the oldest old or a lower demand of health care benefits from those patients as they are aware that their life expectancy is already low. In addition, ICU therapy does not need to be expensive compared to other lifesaving therapies as therapy in a German medical intensive care unit was found to cost 22,000 Euro per QALY, a sum that is within the accepted range [3]. So, the aging society by itself does not contribute to rising health costs.

T. Kerz, M.D. (✉)
Klinik und Poliklinik für Neurochirurgie, Universitätsmedizin der
Johannes Gutenberg Universität, Mainz, Germany
e-mail: kerz@uni-mainz.de

D.W. Crippen (ed.), *ICU Resource Allocation in the New Millennium: Will We Say "No"?*, 131
DOI 10.1007/978-1-4614-3866-3_17, © Springer Science+Business Media New York 2013

However, the need for ICU beds will possibly rise as interventions in older patients who generally have more concomitant diseases demand higher therapeutic intensity in the perioperative period. One German Hospital (in Nuremberg) has already created a specialized geriatric intensive care unit. On the other hand, older age is accompanied with more chronic sicknesses—whether the focus of health care will lie on acute care (more beds in hospitals and ICUs) or chronic care (focused on prophylaxis and preventive measurements) remains difficult to predict.

The development of rationing in Germany will therefore be influenced by two main factors. One is the development of the social security system: Who eventually bears the cost and how much is society ready to pay for health issues. The other one pertains to the discussion on distribution of goods, social justice, and futility. Currently, there are two major political trends that aim for different financing models of the public health service, and it is unclear which one will prevail. One trend, more social democratic and solidarity-based, tends to maintain the universal flat rate model where everybody pays a fixed percentage of his income into the statutory system and family members have free coverage. To enlarge the money input into the system, not only labor income shares would have to increase but income from interests and rents would need to be taxed, too. By doing so, medical services could be extended and rationing become less a priority. The opposite trend, more liberal, tends to bring per capita flat rates into effect. Here, every citizen (and family members, too) pays a fixed sum into the health fund, regardless of income. If more money is consumed than is within the system, copayments can be charged. To allow lower income groups access to health care, federal subsidies would be accorded. As it is, however, not clear what part of the costs would be reimbursed, there is fear that the second system would result in restricted access of low-income classes to medical benefits. Proponents of both systems sometimes call for a basic benefits model were those wanting to get better health care could pay extra premiums. This would potentially result in excluding expensive treatment options such as ICU care after having passed a certain age or when a patient has specific comorbidities. That of course would take away pressure from the ICU as well as from the single doctor because everyone would be able to make his own choice in advance.

Direct billing from doctors to patients with refunding limited to 90% has also been proposed which would increase out-of-pocket spending. Patients would be prompted to consider the necessity of each visit to a doctor—the aim being to reduce the frequency of physician-patient contacts. This frequency is extremely high in Germany with 243 patients per doctor and week whereas doctors in other industrialized countries see only between 102 and 154 patients per week [4]. Physicians, on the other hand, fear that up to 25% of private practices could close because of the reduction of patient's visits, then. Most Germans, however, feel unhappy with these suggestions, as they fear that these first steps mark the start of farewell of solidarity in the German Health Care System and the beginning of privatization of health risks. On the other hand, it is unknown what price Germans would be ready to pay for solidarity purposes.

While politicians do not advance the idea of rationing and rather talk about rationalizing resources, the central ethics committee of the German Medical Council in 2007 called for a public debate on prioritization which should primarily consider ethical and legal questions and should not exclusively take into account economic questions [5]. The paper's main point is that all citizens should continue to have access to a basic health care system instead to allow access to high-end medicine for only a part of the population. Largely unnoticed by the public, the paper calls amongst other points for transparency, evidence-based decisions, consistency, democratic legitimation, and the possibility to enter objections (for patients or health care providers who are refused treatments or market access). Main criteria, in order of relevance, should be based on medical need, individual benefit of treatment and cost-effectiveness. The paper refers also to the British NICE agency and its implicit threshold of 20,000–30,000 GBP per quality-adjusted life-year [6] and proposes a similar limit, with exceptions possible when there is no alternative treatment. A prioritization model, based on new federal laws and fundamental ethical considerations, should take into account firstly protection from loss of life and protection from severe pain, secondly protection from severe damage of body parts or body function, and, on third and fourth rank, to protect from less severe or temporary disorders. At ranks one and two, exclusion or differentiation of treatment because of financial reasons or in relation to the type of one's health insurance would not be allowed. In order to implement such categories, the constitution of a federal health advisory board is proposed to develop such criteria conjointly with all relevant social groups (physicians, patient representatives, clergy, health maintenance organizations). Within that line, many other specialist associations have demanded similar efforts such as the German Association of Anesthetists. Its president called in 2009 for a public debate of when and how intensive care therapy should be limited: "No health care system works without rationing. Effective intensive care medicine will not be possible in the future without limiting therapeutic interventions" [7]. Prioritization to avoid rationing was also proposed by the president of the German Association of Internal Medicine that same year [8].

While prioritization and rationing are often used interchangeably, they in fact have two distinct meanings: Prioritization should lead to a definition of procedures that have either a high or a low priority. If resources are scarce, medically and democratically legitimated prioritization lists should be available from which doctors could deduct which procedures to perform. Cost-effectiveness could represent a central marker that allows the development and ranking of such lists. In Germany, a first example of calculating cost-effectiveness was given by Manns et al. in 2002 for activated protein C [9]. They found one additional life-year to cost $28,000 (20,000 Euro) for patients with an APACHE II score greater than 25. In those with a lower APACHE score, costs for one live-year were $575,000 (410,000 Euro).

Rationing, as the second step, would come into effect when prioritization is no longer an option because of even more scarce financial resources. First efforts to establish such criteria were undertaken from 2006 to 2009 with assistance from the federal ministry of education. Preliminary studies analyzed the current situation of rationing in two highly expensive areas, e.g., interventional cardiology and intensive care medicine. Based on these findings as well as with data on cost and effectiveness from the literature, the project tries to create standards of care for expensive new treatments in these two fields. These standards will finally be submitted to physicians and their applicability and acceptance will be tested. Ultimately, a catalogue could be established in the statutory system, defining standard therapies (which would be paid for by health maintenance organizations) as well as supplementary therapies with a lower benefit–cost ratio where patients would have either to wait or bear costs by themselves.

As a first example calculating evidence-based benefit on life expectancy, Marckmann et al. have proposed cost-sensitive guidelines for two interventions, namely drug-eluting stents and the implantation of cardioverter-defibrillators (ICD). Additional charges per QALY for ICD-implantation are 62,000 Euro (compared to amiodarone therapy) in patients with an ejection fraction (EF) of less than 30%, 168,000 Euro when the EF is 31–40% and 479,000 Euro for an EF > 40% [10]. While European guidelines call for an ICD implantation for an EF < 35% [11], Marckmann's cost-sensitive guidelines would propose ICD-implantation only for patients with an EF of <30%. However, when published and discussed in 2009, these cost-sensitive guidelines provoked great skepticism as the German Constitutional Court already in 2005 had ruled that HMOs have to pay every treatment in case of life-threatening diseases, inclusive of those with marginal benefits and regardless of costs. Therefore, rationing because of economic reasons would require a change in the German constitution that is unlikely to happen in the near future. Furthermore, the research group considered it impossible to determine standards for all medical fields and was aware of the need for doctors to continue rationing on a case-to-case basis. Therefore, they also proposed to develop methods which doctors could apply in these cases [12]. Of note, while 92% of physicians interviewed in the preparing study agreed that treatments should be limited when benefits are limited compared to costs (without further definition what this would mean exactly), 78% of physicians said that doctors should follow such a guideline in all or almost all cases. Only 30% voted for an absolute adhesion to such protocols in order to ensure a consistent and fair utilization of resources. It therefore seems that German physicians believe on the one hand that cost-sensitive guidelines should be implemented, but on the other hand, they are all too ready to accept exemptions from this rule. Rationing by cost-sensible guidelines must also be discussed in the light of individually different judgements of cost–benefit ratios. A discussion is needed whether an individual would have the right to opt for a treatment that only confers a marginal cost–benefit ratio. One could discuss if, even when paying for such a treatment by his own, the patient would have a right to consume such a treatment or if panels would decide over the options available.

Guidelines for prioritization are currently also developed by the German Research Foundation (Deutsche Forschungsgemeinschaft, DFG). Within 11 sub-projects, the research groups explore prioritization from different perspectives (ethical, philosophical, legal, economical, psychological, empirical, and medical) in order to develop a possible guideline for priority settings. Amongst other projects and for the first time in Germany, a representative population-based survey was conducted to identify preferences and ranking orders in the general population. Three thousand citizens of Lübeck received a questionnaire, two thousand four hundred of them answered, most of them with a generally positive attitude towards prioritization. A representative group of 18 persons was then invited to debate: "What is important for us in health care—how to decide about prioritization." The panel developed criteria for prioritization and identified problematic areas such as personal fault and individual responsibility. The results were sent to national and state physician's associations as well as to leading politicians in the field. The initiators hope that civic participations like this will elicit the discussion on the subject to a broader extent.

Rationing could also come closer to Germany's medical system as it is facing severe work force shortages, which could result in closing of medical facilities. Although in 2007, 10,000 physicians more than in 2000 were registered in Germany, the actual work hours remained almost identical [13]. This is due on the one hand to more female physicians working part time, on the other hand on new working time regulations. Moreover, 70,000 physicians will retire until the year 2020 and might not be replaced timely which will possibly lead to severe staffing problems, not only in ICUs which in Germany exclusively are run as closed units. PricewaterhouseCoopers calculates 56,000 physicians missing in 2020 [14]. The same situation holds true for nursing staff where 20% is older than 50 years—those retiring actually cannot be replaced as there are not enough nurses available on the job market. Several strategies will have to be employed, beyond others recruiting of workers from outside Germany or more attractive wages and working conditions.

When discussing prioritization and rationing in the ICU, the question of treatment futility is of major importance. Futility here means avoiding unnecessary therapies, overtreatment, or therapies that will not achieve a desired goal—by avoiding those, resources could be saved, which could benefit other patients. In Germany and according to German legislation, either the physician or the patient (or his surrogate) defines futility. In its simplest form, a therapy for a given disease or a specific patient condition is judged futile by a physician and therefore not administered—when there is no indication, physicians cannot be forced legally to give therapies. However, overtreatment is still a fact and happens because of either fear of legal consequences, over-ambition or prognostic uncertainty. From the patient side, futility means that a therapy is not effectual and cannot achieve a health status the patient could accept as meaningful. More and more patients reach the ICU with either a living will or a durable power of attorney—about 1 million people out of the 80 million inhabitants are estimated to have such a directive, and 29% of patients dying in a surgical ICU died with an end-of-life decision [15]. While directives generally include only very global instructions like "I do not want to be resuscitated if there is no chance of a meaningful recovery," sometimes they comprise very specific directives such as "I do not agree to artificial ventilation" or "I do not want a gastrostomy." With the new German Law on advance directives from 2009, the patient's will, either as fixed in the advance directive or as stated by the surrogate, has to be observed, even though when patients could die because of that decision. Only in case of discordance between physician and patient/surrogate, a court has to be appealed. Nonadherence to such a directive theoretically could lead to indictment over bodily harm, but there are no known cases up to now. It can be assumed that although not at all uncommon today, more and more end-of-life decisions, treatment withholding, and withdrawals will occur as patients and their relatives realize their new power.

The decisions to forego or stop treatment are today mostly devoid of financial considerations because hospital and rehabilitation costs are covered by HMOs. Most families do not initially consider that they will have to take their share in nursing facilities where costs can easily reach 4–5,000 Euro per month and statutory long-term care insurance covers only up to 1,800 Euro. Yet, relatives have a right to keep their homes and to hold assets of 100,000 Euro when welfare comes in. As nursing care in specialized homes or even at home is so expensive, many nurses from eastern European countries such as Poland, Hungary or from the Czech Republic have been employed. This will certainly be extended in the future.

Beside these issues, futility as judged by the physician is closely related to prognosis—though scoring systems such as APACHE or SAPS can classify cohort mortality risks, they are inapplicable for prognostication of a given individual patient. We need more data similar to that after anoxic cardiac arrest where somatosensory evoked responses and neuron-specific enolase allow reliable outcome prognostication [16]. It is, however, clear that for many diseases, objective prognostic parameters will either not be available or fall short of the desired accuracy. This emphasizes the need for intensive patient–physician talks in which a patient's will is determined, prognostic factors are considered, and both are aggregated into a meaningful clinical directive.

Health care spending has to be seen in the light of its opportunity costs, that is, budget restraints imposed on other fields of social welfare. Many, e.g., Asian countries spend remarkably less money than Western Economies for health care purposes. However, they achieve similar or even better results. E.g., Japan spends 2,300 USD for health care, while the US spends 6,900 USD and Germany 3,200 USD. Strikingly, many health indices such as life expectancy or infant mortality are better in the Japanese system. Some authors therefore advocate transferring federal expenses away from the medical sector and demand increased financing of the education sector as the association between better health and higher education has been clearly established [17].

Some German centers have undergone restructurations in the way they organize their ICU care—while, traditionally, every department had its own ICU, the university hospital Hamburg in 2005 merged all ICU beds into one department of intensive care medicine, now disposing of 64 ICU and 12 IMC beds. Thereby, the occupation rate is hoped to rise as there is more flexibility in allocation of beds. This is expected to lead to cost savings and quality improvement by better standardization of care and the interplay between the departments concerned such as surgery, medicine, and anesthesiology.

Self-determined timing of death will certainly become another issue in the future. The suicide of Eberhard von Brauchitsch at the age of 83, together with his wife, received much attention in German Media in October 2010. Brauchitsch was a member of the same family as Nazi field marshal Walther von Brauchitsch and was one of the most powerful economic leaders of the country in the 1980s. He suffered from severe emphysema, his wife being severely affected by Parkinson's disease. Both could hardly walk with a walking frame. His daughter stated that both had discussed the issue in the past and, with declining health status, decided to commit suicide as long as it was still possible. As assisted suicide is not illegal in Germany but drugs cannot be prescribed for this indication, both traveled to Switzerland where they were assisted by "Exit," one of several associations to assist suicide. An estimated 80–100 non-Swiss citizens choose this option every year. Assisted suicide has been debated within the national German Ethic Council. While most members said it was inconsistent with the duty of doctors, a minority favors legalization of physician-assisted suicide [18]. The Association of German Jurists adopted a resolution in 2006, stating that assisted suicide is ethically acceptable [19]. It seems conceivable that, in the medium term, assisted suicide will be legalized in Germany and end-of-life discussions will be influenced by this option.

In conclusion, Germany has a considerable delay in discussing rationing issues when compared to other European and non-European countries. A public debate on rationing has not been held in Germany up to now and most Germans seem unwilling to start it. Politicians focus on keeping premiums stable by limiting the amount of money put in the health system. Implicit rationing leads to increased out-of-pocket spending for those who want more and better benefits. The German principle of solidarity in health care therefore will be challenged. Privatization of the medical sector will progress. A debate on prioritization and/or rationing (or of radical changes in health care delivery) will happen only in the future when and if consumers will become aware of shortcomings of the current system and the extent of implicit rationing. Several already existing legal bodies could see their influence rising and find themselves entitled to propose (cost-sensitive) guidelines. Quality indicators and cost-effective ratios will certainly play a major role in these guidelines. This would, however, need a change of the German constitution that will be difficult to obtain. Use of standard operation procedures will increase as hospital administrations will demand evidence-based therapies to cut costs. Clear and nationwide uniform admission and discharge guidelines will have to be formulated. Workload for intensive care personnel will increase, perhaps nurses and critical care practitioners with shortened training could replace more and more physicians at the bedside. Informatics and telemedicine will probably assist in these changes, more so, as larger areas in eastern Germany will see their population density decrease. As German citizens place a high priority on solidarity, a conceivable evolution of the security system could consist in a 3-tier system: The first tier or basic level will probably be a compulsory insurance, covering life-threatening and acute diseases. The second-tier will demand more copayments and will leave extent of coverage and premiums to every insured. The third-tier will round off coverage for even marginal health problems. Saying no to infinite health care demands will then be achieved by a mix of limitations set by the organization of health care system itself and by personal preferences. Federal agencies in Germany in 2050 will probably only set minimal requirements of health care, benefits and hospital care to which all HMOs will have to adhere to. The predominantly state-run system, as known today, will have been disappeared.

References

1. Lubitz J, Beebe J, Baker C. Longevity and medicare expenditures. N Engl J Med. 1995;332(15):999–1003.
2. Brockmann H. Why is less money spent on health care for the elderly than for the rest of the population? Health care rationing in German hospitals. Soc Sci Med. 2002;55(4):593–608.
3. Graf J, Wagner J, Graf C, Koch KC, Janssens U. Five-year survival, quality of life, and individual costs of 303 consecutive medical intensive care patients—a cost-utility analysis. Crit Care Med. 2005;33(3):547–55.
4. Koch K, Gehrmann U, Sawicki P. Primärärztliche Versorgung in Deutschland im internationalen Vergleich. Dtsch Ärzteblatt. 2007;104(38):A2584–91.
5. Zentrale Ethikommission bei der Bundesärztekammer. Priorisierung medizinischer Leistungen im System der Gesetzlichen Krankenversicherung (GKV) Stellungnahme der Zentralen Kommission zur Wahrung ethischer Grundsätze in der Medizin und ihren Grenzgebieten (Zentrale Ethikkommission) bei der Bundesärztekammer 2007; Available from: http://www.zentrale-ethikkommission.de/page.asp?his=0.1.53.
6. NHS. Measuring effectiveness and cost effectiveness: the QALY. 2010 2010-20-04 2012-11-06; Available from: http://www.nice.org.uk/newsroom/features/measuringeffectivenessandcosteffectivenesstheqaly.jsp.
7. Boldt J, Schollhorn T. Ethics and monetary values. Influence of economical aspects on decision-making in intensive care. Anaesthesist. 2008;57(11):1075–82. quiz 1083.
8. Rabbata S, Stuewe H. Die heimliche Rationierung muss endlich aufhören. Deutsches Aerzteblatt. 2009;106(20):957–9.
9. Manns BJ, Lee H, Doig CJ, Johnson D, Donaldson C. An economic evaluation of activated protein C treatment for severe sepsis. N Engl J Med. 2002;347(13):993–1000.
10. Marckmann G, Reimann S. Kostensensible Leitlinie zum Einsatz eines implantierbaren Defibrillators (ICDs) 2009 2012-11-06; Available from: http://www.iegm.uni-tuebingen.de/images/pdf/ksll%20icd.pdf.
11. Dickstein K, Cohen-Solal A, Filippatos G, et al. ESC Guidelines for the diagnosis and treatment of acute and chronic heart failure 2008: the Task Force for the Diagnosis and Treatment of Acute and Chronic Heart Failure 2008 of the European Society of Cardiology. Developed in collaboration with the Heart Failure Association of the ESC (HFA) and endorsed by the European Society of Intensive Care Medicine (ESICM). Eur Heart J. 2008;29(19):2388–442.
12. BMBF Forschungsverbund "Allokation." Ethische, ökonomische und rechtliche Aspekte der Allokation kostspieliger biomedizinischer Innovationen: Exemplarische Untersuchungen zur expliziten und impliziten Rationierung in der interventionellen Kardiologie und der Intensivmedizin. 2012-11-06; Available from: http://www.mm.wiwi.uni-due.de/fileadmin/fileupload/BWL-MEDMAN/Forschung/BMBF-ForschungsverbandAllokation_Information.pdf.
13. Kopetsch T. Dem deutschen Gesundheitswesen gehen die Ärzte aus! 2010 2010-10-11; Available from: http://www.bundesaerztekammer.de/downloads/Arztzahlstudie_03092010.pdf.
14. PwC. Fachkräftemangel Stationärer und ambulanter Bereich bis zum Jahr 2030. 2010 2012-11-06; Available from: http://www.pwc.de/de/gesundheitswesen-und-pharma/fachkraeftemangel-stationaerer-und-ambulanter-bereich-bis-zum-jahr-2030.jhtml.
15. Meissner A, Genga KR, Studart FS, et al. Epidemiology of and factors associated with end-of-life decisions in a surgical intensive care unit. Crit Care Med. 2010;38(4):1060–8.
16. Zandbergen EG, Hijdra A, Koelman JH, et al. Prediction of poor outcome within the first 3 days of postanoxic coma. Neurology. 2006;66(1):62–8.
17. Schmidt V. Priority setting at the macro level. Health care in relation to other fields of social policy. Ethik Med. 2010;22:275–88.
18. German National Ethics Council. Self-determination and care at the end of life. 2006 2012-11-06; Available from: http://www.ethikrat.org/files/Opinion_end-of-life_care.pdf.
19. Deutscher Juristentag. Beschlüsse des 66. Deutschen Juristentags. 2006 2012-11-06; Available from: http://www.djt.de/fileadmin/downloads/66/66_DJT_Beschluesse.pdf.

Chapter 18
India: Where Are We Going?

Farhad Kapadia and J.V. Divatia

Introduction

It is always difficult to predict future trends. A few generalizations and some educated guesses given below may serve as an indicator as to what the future of Critical Care will be in India. One can confidently state that more and more ICUs will be primarily staffed by Intensivists. One can also state, with some certainty, that many newer tertiary institutes, in addition to the existing ones, will focus on medical tourism. Market forces and regulatory and accreditation bodies will probably focus on ICU services ensuring adequate resource allocation to critical care. Insurance and other third party payees will slowly increase as out of pocket payment becomes beyond the reach of an increasing proportion of the population. Litigation, still rare, may influence medical decision making and cost. End of life considerations will have to be addressed as empty ICUs beds become increasingly difficult to find in emergencies.

Trends from the Last Decade

Govt spending has been miniscule in terms of percentage of gross domestic product or total budget expenditure, but encouragingly this is increasing. The government allocated Rs 25,154 crore (one crore = ten million) for the financial year 2010–2011 (0.36% of GDP or 2.3% of the total budget expenditure). This is significantly higher

F. Kapadia, M.D., F.R.C.P. (✉)
Department of Medicine and Critical Care, P.D. Hinduja Hospital
and MRC, Mumbai, India
e-mail: fnkapadia@gmail.com

J.V. Divatia, M.D.
Department of Anaesthesia, Critical Care and Pain, Tata Memorial Hospital,
Mumbai, India

D.W. Crippen (ed.), *ICU Resource Allocation in the New Millennium: Will We Say "No"?*,   139
DOI 10.1007/978-1-4614-3866-3_18, © Springer Science+Business Media New York 2013

that the budget for 2004–2005 (Rs 8,086, 0.26% of GDP and 1.6% of total budget expenditure). Critical Care services in the country are rapidly changing and evolving. There is minimal analysis of cost considerations in this relatively new field. It is estimated that approximately 10% of Critical Care services are from government hospitals and institutes. The rest is from the private sector. Despite the economic growth and increasing government health expenditure, it is likely that private sector will continue to offer the majority of critical care services to the population.

As discussed in the earlier chapter, the actual costs of ICU and that of resource allocation for critical care services by the government are poorly documented. For the limited published data one noted that the ICU expense per day in large government hospitals in the 1990s was estimated to be Rs 2,000–5,000. In the Mid-2000, the data from a private–public specialized cancer hospital suggested that ICU expenses approximated Rs 12,000 daily. A survey done from a large private hospital in the end of 2010 noted that daily direct costs varied from Rs 4,600 to 10,400 but this represented a small percentage (~20%) of the total hospital bill (till death of ICU discharge). It is apparent that there is marked variation in resource availability and utilization in different health systems, but one could reasonably infer that costs will similarly escalate in next decade as India becomes more developed and prosperous and more of critical care services are delivered from the private sector.

Another noticeable trend over the last decade has been a slow increase in "third party" cover, with a correspondingly decreasing need for "out of pocket" direct payments by the patient. This trend has been necessitated by increasingly large hospital bills, and it is conceivable that direct payments of these hospital bills will soon be beyond the means of most of the population.

Payment Structure

Given the rapidly increasing cost of hospitalizations in general and of critical illness in particular, it is pragmatic to assume that direct out of pocket payments will become less frequent over the next decade. Indirect payment has lead to conflict with hospitals and there is no reason to believe that this conflict will be amicably resolved in the near future. Currently the hospital has three modes of payment: First, directly from the patient. Second, the patient pays the initial deposit and the final bill and later collects the money from the insurers. Third, the employer or the insurer "preauthorizes" the hospital to treat the patient to a given "sanctioned" amount. In elective surgery, this is obtained prior to admission and the patient does not pay the initial deposit or the final bill. (Often a relatively small amount is collected as a deposit in case the insurer subsequently denies part of the claim.) For emergencies, the patient pays the initial deposit, and then gets the sanction or authorization from the insurance company. After this is received by the hospital the patient does not have to pay the final bill, and is refunded the initial deposit. As this administrative burden on insurers increased, a set of "TPAs or third Party Assurance" evolved which were designed to smooth the interaction between patients, insurance companies and hospitals. In reality the system has lead to major conflict between the insurers (direct or indirect TPAs) and the hospital. TPAs have started dictating to patients which institutes they recognize and which they do not, limiting patients' choice. Hospitals have had problems with delayed or incomplete payments from insurers and TPAs, resulting in many being blacklisted. The patient and family is caught in-between this tussle. They first have to borrow or break saving deposits to pay the bills, and then have to run form pillar to post to get back the money. The system is clearly not working to the patients, advantage and it is difficult to predict as to how this will resolve over the next decade.

Intensivists Lead ICUs

The last couple of decades have already witnessed these changes and it is distinctly unusual for newer large tertiary institutes no to have intensivists in charge of ICUs. One can confidently state that more and more ICUs will be primarily staffed by Intensivists. As has already happened in the more developed world, this trend will be strengthened by increasing numbers of officially qualified intensivists emerging from the newly established training and certification programs. Also, the younger generation of non-ICU specialist consultants will themselves be less competent general physician or surgeons and less confident of their ability to manage critically ill patients in modern ICUs.

The financial aspect of this will be twofold. Hopefully, appropriate level of monitoring and care will result in a decrease investigations and unnecessary medications. Improved care may also result in shorter length of stay in ICU and hospital, and thereby bring down costs overall cost. Lower mortality may translate to lower cost per life saved. On the other hand, costs may increase to pay the salaries or fees of the intensivists. A "fee for service" billing may unfortunately lead to excess interventions and costs.

Market Forces and Medical Tourism

The last decade has witness a large number of newer modern hospitals in the major cities and smaller towns. This trend is clearly going to increase over the next decade or two, and these hospitals will probably offer the majority of standard medical services to the population. Smaller nursing homes will probably occupy a niche role only, either offering specialized services or serving a local community. These newer hospitals labeled as "tertiary referral" and described as offering "state of the art" services, will inevitably compete with each other. In this scenario, ICU services will play a major role. Hospitals which offer suboptimal ICU services will be seriously disadvantaged regardless if the patient is paying out of pocket or if the insurer or employer is covering the costs. Currently, despite the rapidly increasing number of newer hospital with large ICUs, there is still a significant scarcity of ICU beds for medical and surgical emergencies. In this context there will be a symbiotic financial growth between ICU and other hospital services. Operation theater, laboratory and imaging services will benefit in terms of turnover and revenue due to the coexisting ICU services. Hospital accreditation is gaining recognition and optimal ICU services will be needed to ensure ongoing accreditation.

The country is already witnessing a large amount of "medical tourism." Earlier, this essentially consisted of People of Indian Origin choosing to come to India for medical care. This was seen mostly in those Indian residents from the Far East, the Middle East and Africa. Modern medical tourism is attracting residents, not of Indian origin, from all parts of the world including the developed countries in Western Europe and Northern America. The main attraction of medical care in India is based on lower costs and shorter waiting periods. Additional advantages included a well trained medical community fluent in the English language. A perceived disadvantage is the weaker infrastructure, but this is rapidly improving as medical care becomes more professionalized in larger modern hospitals. Medical tourism for people of non-Indian origin initially focussed on relatively simple elective procedures like ophthalmology and infertility. However, there are an increasing numbers of patients seeking more complex medical procedures like coronary artery bypass graft surgery, joint replacement and plastic surgery. These procedures need reliable ICU services, either due to the complexity of the surgery or due the increasing age and comorbidity of the patients undergoing these procedures. Patients and insurers will expect reliable ICU services for these procedures. Hospitals and institutes which are not accredited by national or international bodies will not be able to attract these patients. This service from ICUs will probably increase in the future, and the reliable income generated from this will further strengthen the ICU services for the general population.

End of Life Care

End of life care (EOL) is become a major issue in ICUs across the world. There are many difficult issues to deal with ethically and legally. Governments and civil society are still a long way from having a clear and transparent approach to these difficult issues. In the foreseeable future, EOL issues will significantly impact ICU spending and resources. Once again finances become a major issue. Either because it is too costly, or because the family is spending money on futile prolonged terminal care without due discussion and effective palliation, or because individual family members do not want to be perceived as denying their family member medical care on financial grounds. Encouragingly, more doctors, especially intensivists and other consultants are beginning to work on explaining the limitations of intensive care, and the options of effective palliative care outside ICU. There is a significant lack of clarity on the legal status of this kind of limitation of care. The Indian Society of Critical Care Medicine in 2005 attempted to assist decision making by releasing a

Position Statement "Guidelines for Limiting Life-Prolonging Interventions and Providing Palliative Care Towards the End of Life in Indian Intensive Care Units." This was followed by the Law Commission of India presenting a draft bill to the Parliament and Government of India "196th Report on Medical Treatment to Terminally Ill Patients (Protection of Patients and Medical Practitioners)." Should this become law, decision making at the end of life will become easier for medical professionals and for patients and families.

Conclusion

It is difficult to predict the direction of Critical Care and the financial implication of this over the next decade. Private services will probably continue to offer the majority of the service, and intensivists will play an increasing role in this. Costs will continue to escalate and indirect payments will be the only feasible option. Conflict between insurers and hospitals will hopefully be resolved and the patient and family will hopefully no longer be caught in-between the conflict between hospitals and payees. Market Forces and Medical tourism will ensure that ICUs which maintain higher standards of service are accredited and attract revenue, while end of life care will hopefully become more pragmatic and less defensive, sparing unnecessary expense yet ensuring protection of patients' families and of profession care-givers and institutes.

Chapter 19
Italy: Where Are We Going?

Marco Luchetti and Giuseppe A. Marraro

Introduction

The Italian Hospital, as the institution at the source of health care delivery, has been operating under crisis conditions for a long time because of an increasing demand for health care from a rapidly and constantly changing population. In Italy, child birth survival has increased threefold [1], infantile mortality has fallen to 4.3 per 1,000 live births while the general population has basically aged [2]. It is estimated that the ageing index will double before 2050 and thus future public health expenditure will be more and more subordinate to long-time care [3]. In today's Italy, health care is universally provided under the law to all Italian citizens and the state, in various guises, sustains the bulk of the expenditure [4]. The last decades have witnessed the development of a private health care sector, although state as well as insurance contributions still account for most of the expense. The Italian health system is characterised by marked regional differences. Italian economic growth has never been uniform and the gap between the wealthiest northern and central regions and the less affluent southern ones has never been bridged. Regional socio-economic disparities have caused wide differences in the quality and efficiency of health services in this country, especially in the south where a high percentage of low-income families mostly live [5]. In the course of time, the evolution of the health care system has been marked by various attempts at reform [6] in the face of ever more stringent budgetary limits on the one hand, and an ever increasing process of decentralisation on the

M. Luchetti, M.D. (✉)
Department of Anaesthesia, Intensive Care, & Pain Management, A. Manzoni General Hospital, Via dell'Eremo 9/11, Lecco 23900, Italy
e-mail: m.luchetti@ospedale.lecco.it; m.luchetti@fastwebnet.it

G.A. Marraro, M.D.
Department of Anesthesia and Intensive Care, Fatebenefratelli and Ophthalmiatric General Hospital, University of Milan, Milan, Italy

D.W. Crippen (ed.), *ICU Resource Allocation in the New Millennium: Will We Say "No"?*,   145
DOI 10.1007/978-1-4614-3866-3_19, © Springer Science+Business Media New York 2013

other. This devolution is ongoing and could lead to the creation of as many diverse local health systems as there are regions in the country [7].

Forty years ago, the concept of health care rationing was not widely perceived or mentioned. During the last 15 years, health care costs have become a major focus of public policy. While looking for ways of controlling costs, an awareness of economic trade-offs in the health care decision process has increased. Allocating resources to one branch of the service has meant leaving less available for other services. Allocating resources to one patient means that fewer resources will be available for others. Rationing has thus gained more visibility among the public and has now become explicit at all levels of the health care system. Moreover, in many intensive care units rationing still remains a taboo issue—thoroughly carried out but insufficiently discussed.

Physicians' responsibility in carrying out the rationalisation of health care expenditure still meets with considerable resistance of a cultural nature. Doctors are wont to have a direct personal relationship with their patients and thus feel culturally alien with intervention costing which, compared to the life and health of the patient, is felt to be not commensurable. Thus, physicians may take a moral stand against considerations of cost in medical acts for the benefit of the health of all who trust themselves to their care [8–10].

Italy, too, has adopted a management cost reducing strategy for hospital-based facilities and has begun to introduce cutbacks in hospital bed capacity. Such a reduction should have slashed inappropriate admissions, made inter-hospital care more efficient and effective, as well as facilitated early discharge procedures. This strategy, however, has not been well accepted politically because of the strong dissatisfaction it causes among health care users and because it endangers the survival of local politicians. Moreover, the planning of alternative facilities and medical services, even with a reduced inter-hospital bed capacity, does not necessarily translate into an overall reduction in health care spending [11].

Despite the fact that emergency and intensive care facilities have always been considered beyond the reach of such a reductionist rationale, there has been a covert attempt at reducing the overall number of critical beds indiscriminately, since a controlled and targeted cutback is up against the vested interests of small-size hospitals.

Critical wards, whether providing specialist care or not, have always represented a show-piece for any hospital without, however, actually taking into account their utility within the context in which they operate, the efficiency of their role in specialist care delivery to critical patients and the results they achieve.

Italian intensivists as a profession belong to different sub-specialties of medicine and have now come to realise that intensive care unit beds are a limited and expensive type of resource [12, 13]. The shortage of beds is an everyday issue in many an intensive care unit [14, 15] and the allocation of beds is viewed as one of the most crucial aspects of the Italian health care system [16].

The National Health Care Plan and Intensive Care

The 1998 National Health Care Plan, and subsequent Regional Plans, recommended a ratio of three critical beds to one hundred hospital beds. These estimates, however, not only fail to reflect the existing actual Italian situation but they do not even taken into account the impossibility of new bed increments at zero cost. Delays in the application of the law and a lack of transparency about its purpose have resulted in the creation of intensive care units with a much reduced number of beds with the result that, on the one hand, shortages in critical patient care have not been made up and, on the other, that health care spending has actually increased.

There are marked differences in Italian health care regarding the number of beds each Region deems necessary to ensure intensive care to all its citizens. In 2005, a poll carried out by the Italian Association of Hospital-Based Anaesthesiologists and Intensive Care Specialists (AAROI) revealed that Italy has 3,814 beds available in intensive care units, which represented 50% of needs as assessed by international standards [17]. This investigation confirmed, moreover, the existence of notable differences between the various Regions as well as the existence of intensive care units with a scarce number of beds—in general two—which, in small hospitals situated in marginal areas, did not justify their own existence and actual utility.

To compound matters even further, there are in Italy a polymorphic variety of intensive care units which, in some instances, hardly qualify as emergency care facilities (although they are allotted specific resources for this purpose) and which should be classified under the heading of sub-intensive therapy, given the type of admissions and treatments they provide. Any attempt at making clear which type of patients they admit, the specificity of the care they provide or the appropriateness of in-patient care has invariably met with failure, as does any bid which had come up against the counter-claims of these varied health care operators who, with local politicians, have a vested interest in upholding the status quo at all costs in order to ensure their survival. To plot thickens even further when one considers that, up to the present day, there has only been one system of Anaesthesiology and Intensive Care Specialist Schools to cover the whole theoretical and practical basis of the intensive care curriculum. This could have induced to think that all areas of intensive care could have been managed and carried out by the professional figures who had been graduates of these schools. In actual fact, the picture is somewhat chequered as it falls to the various heart surgeons, neurosurgeons, paediatricians, neonatologists, lung specialists, etc.—who sometimes lack an actual cultural background and specific training—to manage on their own, or with the help of an "on-call" consultant intensive care anaesthesiologist/resuscitation specialist, an intensive therapy which requires professionalism of the most expert kind.

Last but not least, the government will have to choose where it wants to go, what to do and what services to offer citizens in matters of health care, within the public or private sector, or a mixture of the two. For the time being, the only firm point is the continuing requirement on members of the public in need of care to incur the expense of additional charges on medical services in order to shore up the national health care spending, thereby levying a larger amount of "ad hoc" tax than what would derive from fixed fiscal revenue apportionment.

Critical and Intensive Care Departments in Italy

General intensive care is usually under the management of anaesthesiologists and intensivists and caters to both general medicine and surgery patients. It often deals with multiple trauma cases and sometimes accepts paediatric patients above 3–5 years of age.

Intensive care in cardiosurgery is mainly managed by cardiosurgeons who may collaborate with anaesthesiologists trained in resuscitation techniques as cases demand. It admits acute and sub-acute cardiosurgery cases.

Intensive care in neurosurgery is managed either by neurosurgeons working collaboratively with anaesthesiologists trained in resuscitation techniques as cases demand or by the anaesthesiologists themselves. It admits acute and sub-acute neurosurgery cases as well as general cranial trauma cases uncomplicated by multiple injuries.

Intensive care in cardiology is managed by cardiologists who rarely avail themselves of consults from anaesthesiologists trained in resuscitation techniques. It admits acute cardiology patients.

Trauma centres are managed by specialists with various cultural backgrounds and, in particular, by emergency care physicians, neurosurgeons and anaesthesiologists with resuscitation training. They manage patients with multiple trauma. The number of trauma centres is currently limited in Italy and where they exist they are distributed among the major cities. Their area of operation is not clearly defined.

Intensive care in neonatology is managed by paediatricians with training in neonatology. Only three centres (Turin, Genoa and Novara) are under the direct management of anaesthesia-resuscitation specialists with neonatal expertise. They operate within their respective neonatal pathology divisions and mainly deal with the management of the premature infant, either because of intensive care requirements or because only supportive care is required.

Intensive care in paediatrics is managed mainly by anaesthesiologists with resuscitation training and experience in paediatrics as well as, in some instances [2, 3] paediatricians with resuscitation expertise. It caters to all paediatric patients and may care for newborn children as well as adolescents over 14 years of age.

Sub-intensive care is ill-defined and the presence of intermediate care on the national territory is thus difficult to tally. The most frequently found units include:

1. Post-surgery recovery rooms.
2. Cardiosurgical intermediate care.
3. Neurosurgery intermediate care.
4. Sub-intensive therapy pneumology unit.
5. Sub-intensive unit for long-stay patients.

Optimistic Horizon Forecast for the Near Future

What we may expect to happen in the next 10 years in order to improve the care of the truly critical patient, utilise existing resources appropriately and reduce overall management costs is difficult to forecast since it is conditional on several internal as well as external factors, e.g. political, which lay outside the sphere of the health care institution itself. On the other hand, the rationalising of health care expenditure, meaning the best exploitation and the correct limitation of available resources for intensive care, has become indispensable and legitimate from both a legal and ethical point of view, although limitations of themselves are still insufficient to reach these goals.

The most useful course of action is to act on the unjustified expectations, bordering on miracle-working, which the public entertains regarding the efficacy of medicine—and of intensive care in particular—by providing accurate information on the inevitability of death and the persistence of serious disease which remain without a cure.

Moreover, acting on the physician's culture is necessary, with particular emphasis on the actual possibilities of treatment in intensive care and on the risks incurred by patients because of inappropriate admission. The intensive care unit admission allows the survival of patients who may not otherwise do so but this can also expose patients to risks of complications linked to the very treatments applied, such as ventilation-associated pneumonia or ventilation-induced lung injury. Furthermore, it may let patients with severe disabilities, later requiring prolonged rehabilitative therapy, to survive and usher the prolongation of a terminal state attended by ethically unacceptable suffering on the part of the patients and their family [18–22].

Currently, developing countries have had to choose *not* to invest out of all proportion in intensive medicine, being aware that they would have allowed a smaller number of patients to survive than if an equivalent economic investment had been diverted towards providing basic medical care (e.g. vaccinations, prevention, etc.) to a wider segment of their population.

The main problems against which intensive care units have to struggle on a daily basis can be summarised thus:

- The definition of the requirements for beds for the hospital needs and within its catchment area (in the absence of effective regional planning).
- The number of available beds and possibly available beds of immediate activation according to ISO resource management, unavailable because of the operational rigidity of health care personnel.
- The appropriateness of admission and the clear definition of treatable patient categories.
- Difficulties with the discharge of chronic patients for the lack of adequate facilities to care for them.
- The need to deal with pandemic emergencies, especially those of a respiratory nature.

It is thus foreseeable that in the next 10 years, with the progression of a continuing long-term trend, there will be a necessity to act on different levels to attempt a cost reduction in order to rationalise existing resources and improve the performance of intensive care.

Action at Institution Level

1. It is necessary to actually distinguish between first-, second- and third-level hospitals in relation to the type of care they can deliver without their personnel feeling frustrated for being assigned to the lower tiers. What should be aimed at is the final outcome of the treatments applied and the patients' satisfaction with the care provided.

   Adequate resources will have to be allotted to each level of the intensity of care (I, II or III) by carrying out expense reviews and by reducing squandering. Waste can be reduced by monitoring departmental performance, the type of patients admitted and expected outcomes, which must include fatality following long-term hospital stay or prolonged rehabilitative care because of iatrogenic issues deriving from treatment received.

2. Only intensive care units with at least six beds should remain operative and those units with few beds (2–4) should be closed, especially if allocated to small-size hospitals on the outskirts, or outside cities.

3. Among the benefits accruing from such closures are: improved quality of care, as small-size unit personnel acquire experience on a number of pathologies insufficient to increase their expertise and manual skills, and an economy of scales, deriving, e.g. from inferior acquisition costs of large quantities of consumables in comparison to small-size purchases.

4. Patient-centred care with the more severe cases admitted to specialist care units, which necessitate devolving to Regions the responsibility of intensive care medicine, taking into account the number of inhabitants, the ease of access to the facilities, etc. An example to emulate is the centralisation of at-risk pregnancies in a limited number of highly qualified centres. This has allowed not only the reduction of both maternal and neonatal mortality. It has also markedly reduced the management costs of this type of medical condition. The centralisation process has been made possible by the implementation of emergency services and of both road and airborne transportation of at-risk patients.

5. The creation of intermediate facilities (intermediate care units, step-down units, transitional care units, etc.), which provide less intensive care, but of a level consistent with the patients to be treated. Intermediate care reduces hospital costs by decreasing staffing to coincide with the needs of the patients. Since personnel costs may comprise up to 80% of total ICU expenses, the savings afforded by staff reduction (dealing with patients with an illness of intermediate severity) can be substantial. Intermediate care units reduce costs, ICU length of stay and do not impact negatively on patient outcomes [23].

6. Expansion of long-stay patient facilities in order to find the correct placing for patients once their hospital career (especially following intensive care) is terminated. It is important to be aware of the fact that many intensive-care patients require long rehabilitation periods and that some may be remain with seriously invalidating disabilities [24–28].

Action at Management Level

1. Awareness of what should be understood by intensive care, how many beds should be attributed to the unit, what type of care the unit must provide and what type of patients should be admitted to it. There should be the utmost clarity in stating the mission of the intensive care unit and well-defined protocols should be applied to patient admission, taking into account the possibility of the certainty of the diagnosis. As long as a reasonable doubt—or an uncertainty—remains about the irreversibility of the clinical condition, it is appropriate to initiate or continue intensive care. Conversely, if there is a reasonable certainty regarding the irreversibility of the diagnosis, it is also appropriate not to initiate or to suspend intensive measures so as not to unduly prolong the process of dying [29, 30].

   This is not an easy duty to perform, and recent studies in the United States have shown that most academic MICUs do not strictly apply ICU admission and restriction guidelines, as recommended by the Society of Critical Care Medicine and by the American Thoracic Society [31].

   Over-treatment is ethically reprehensible and unanimously condemned, since it determines an inappropriate use of health care means, is uselessly painful to the patient by causing physical and mental injuries and fails to respect the patient's dignity in death. Excessive care is also morally wrong in that it further increases the suffering of family members, care-givers frustration and generates an unfair distribution of resources by detracting them from other patients [32].

2. Reduction in inappropriate admissions. Unfortunately, inappropriate ICU admissions are perceived as a frequent occurrence, mainly attributable to the difficulties inherent to the assessment of the appropriateness of the admission itself. Physicians are naturally aware that their decisions are based purely on medical necessity, but their decisions are also often influenced by extraneous factors such as pressure from their superiors or the referring physician, by the patient's family or by the threat of litigation. The main perceived problems are of a clinical nature and the economical influences are poorly recognised, even if they are present [9].

3. More attention should be paid to the "do not resuscitate" order on patients' forms, with special reference to terminal cases or patients who may reach the hospital in a deep coma or who suffer from injuries incompatible with survival. It is wise to always keep clearly in mind the intensive care mission: to maintain and support vital functions in patients with an elevated potential for recovery from acute illness.

4. To disconnect the patient from the means of support in cases of assessed brain death, as allowed under Italian law. In general, the procedure can be carried out

without particular problems if the patient is eligible for organ donation, while greater reluctance should be exercised in other settings. Beds in intensive care are a precious commodity, and thus incorrectly admitting to, or keeping in the unit one patient may preclude another also in dire need from benefiting from the same right to treatment [33, 34].

5. To implement the indispensable requisites for high-grade and up-to-date intensive care, the following should be considered:

- A specialist in intensive care medicine should be on duty at all times.
- A nursing team should on duty at all times: one nurse for two patients on average and in relation to the level of the intensity of care.
- A minimum of six beds should be available at all times with the possibility to add two or more.
- The capability of dealing with an occasional increase in bed requirements in accordance with ISO resource management [35].
- The possibility to assess cardiac output (e.g. in case of cardiac decompensation, sepsis, etc.).
- The use of ultrasound for routine assessment, also in respiratory pathologies.
- The availability of ultra-filtration systems, dialysis equipment, and in general of filtering devices.
- A facilitated access to radiology (conventional, CAT, NMR, ultrasound) at all times.
- The availability of laboratory findings and facilities and of the blood bank at all times.
- The availability of beds equipped with ventilator, heart monitor and an adequate number of pumps.
- A centralised monitoring and alarm control system.
- Medical personnel adequate for the level of intensive care delivered to the individual patient.

Additional Action

1. Choosing as much as possible those treatments appropriate to the specific patient in order to avoid the onset of complications such as ventilation-induced lung injury, ventilation-associated lung injury and ventilation-associated pneumonia, even if these problems are often considered almost as if they belonged to the natural course of intensive care patient's illness [36–38].
2. Minimising the use of invasive techniques of direct (intubation or mechanical respiration) or indirect (central catheter placement and in-dwelling for longer periods than necessary for therapy) treatments to the amount of time strictly necessary for diagnosis and acute phase treatment [39].
3. Bringing forward the initiation of the best possible treatment for the specific pathology, even if it could appear at first to be the more invasive course and for the

specific patient, thus avoiding intolerable waiting times or inefficacious alternative interventions. Once the pathology is established and the patient's clinical status reaches its acme (respiratory insufficiency becoming cardio-respiratory insufficiency, onset of multiple organ failure), it becomes necessary to resort to more invasive forms of treatment with subsequently increased iatrogenic risks [40].

4. Early disconnection of invasive means of support (e.g. in-dwelling tubes) and increased use of non-invasive means (e.g. non-invasive ventilatory support delivered through a mask following extubation).

5. Early but protected discharge from intensive care as the unit is a constant source of infection. The patient not in need of active monitoring of her vital functions should be transferred to an area with a lesser risk of infection within the shortest possible time.

Final Considerations

It is hardly imaginable to think that there may be no more resources available to dedicate to the care of the intensive care patient, regardless of considerations of age and acute pathology, in the near future. However, for want of extensive resources to dedicate to intensive care medicine, it is necessary to consider a model of such care in different terms with respect to the past. Physicians must continue to be the defenders of the patient's health, but in this role they must take into account the resources they use and oversee their utilisation with circumspection. Furthermore, it is necessary to consider new operational models of intensive care in order to guarantee optimal resource utilisation by regrouping critical patients with different acute pathologies in large-size intensive care facilities where there will always be a need to ensure both isolation and the differentiation of their treatment. Some attempts are already in progress, for instance by incorporating acute neurosurgery patients into general intensive therapy units. The benefits that may accrue include: (1) improved resource use; (2) economy of scale; (3) greater expertise and improved treatment for patients with complex pathologies.

References

1. Istat. Italia in cifre 2005. Rome: Istat; 2006. http://www.istat.it. Accessed 5 Oct 2010.
2. Organisation for Economic Cooperation and Development. OECD health data 2005. Paris: OECD; 2005.
3. Italian Ministry of Economics and Finance, Department of General Accounts. Mid-long term trends for the pension and health care systems summary and conclusions. The forecasts of the Department of General Accounts Updated to 2005. Rome: Ministry of Economics and Finance; 2005. Available in English at http://www.rgs.tesoro.it/ENGLISH-VE/Institutio/Social-exp/Forcast-ac/_____mid-long-term-trends-for-the-pension--health-and-long-term-care-systems. pdf. Accessed 10 Oct 2010.

4. Constitution of the Italian Republic, Art. 32. http://www.governo.it/Governo/Costituzione/1_titolo2.html. Available in English at http://www.senato.it/documenti/repository/istituzione/costituzione_inglese.pdf. Accessed 10 Oct 2010.
5. World Health Organization. World health report 2000. Geneva: WHO; 2000. http://www.who.int/entity/whr/2000/en/index.html. Accessed 12 Oct 2010.
6. Decreto Legislativo 19 Giugno 1999, n. 229. "Norme per la razionalizzazione del Servizio sanitario nazionale, a norma dell'articolo 1 della legge 30 novembre 1998, n. 419". Pubblicato nella Gazzetta Ufficiale n. 165 del 16 luglio 1999 – Suppl. Ord. n. 132. http://www.parlamento.it/parlam/leggi/deleghe/99229dl.htm. Accessed 12 Oct 2010.
7. Giannoni M, Hitiris T. The regional impact of health care expenditure: the case of Italy. Appl Econ. 2002;14:1829–36.
8. Marraro G. Choices that have to be made by a doctor. Int J Clin Monit Comput. 1993;10:163–6.
9. Giannini A, Consonni D. Physicians' perceptions and attitudes regarding inappropriate admissions and resource allocation in the intensive care setting. Br J Anaesth. 2006;96:57–62.
10. Ward NS, Teno JM, Curtis JR, Rubenfeld GD, Levy MM. Perceptions of cost constraints, resource limitations, and rationing in United States intensive care units: results of a national survey. Crit Care Med. 2008;36:471–6.
11. McKee M. Reducing hospital beds: what are the lessons to be learned? European Observatory on Health Systems and Policies. Policy Brief No. 6, 2004.
12. Szalados JE. Access to critical care: medical rationing of a public right or privilege? Crit Care Med. 2004;32:1623–4.
13. Cook D, Giacomini M. The sound of silence: rationing resources for critically ill patients. Crit Care. 1999;3:R1–3.
14. Vincent JL. European attitudes towards ethical problems in intensive care medicine: results of an ethical questionnaire. Intensive Care Med. 1990;16:256–64.
15. Metcalfe MA, Sloggett A, McPherson K. Mortality among appropriately referred patients refused admission to intensive-care units. Lancet. 1997;350:7–11.
16. Coomber S, Todd C, Park G, Baxter P, Firth-Cozens J, Shore S. Stress in UK intensive care unit doctors. Br J Anaesth. 2002;89:873–81.
17. Associazione Anestesisti Rianimatori Ospedalieri Italiani (AAROI). Censimento nazionale dei posti letto di rianimazione attivi al 30 giugno 2005. http://www.aaroi.it/Pagine/iniziative/iniziative_2005/censimento_nazionale_01.pdf. Accessed 15 Oct 2010.
18. Gallesio AO. Improving quality and safety in the ICU: a challenge for the next years. Curr Opin Crit Care. 2008;14:700–7.
19. Manser T. Teamwork and patient safety in dynamic domains of healthcare: a review of the literature. Acta Anaesthesiol Scand. 2009;53:143–51.
20. Reader TW, Flin R, Mearns K, Cuthbertson BH. Developing a team performance framework for the intensive care unit. Crit Care Med. 2009;37:1787–93.
21. Garrouste Orgeas M, Timsit JF, Soufir L, Tafflet M, Adrie C, Philippart F, Zahar JR, Clec'hC, Goldran-Toledano D, Jamali S, Dumenil AS, Azoulay E, Carlet J, Outcomerea Study Group. Impact of adverse events on outcomes in intensive care unit patients. Crit Care Med. 2008;36:2041–7
22. Levy MM, Rapoport J, Lemeshow S, Chalfin DB, Phillips G, Danis M. Association between critical care physician management and patient mortality in the intensive care unit. Ann Intern Med. 2008;148:801–9.
23. Nasraway SA, Cohen IL, Dennis RC, Howenstein MA, Nikas DK, Warren J, Wedel SK. Guidelines on admission and discharge for adult intermediate care units. American College of Critical Care Medicine of the Society of Critical Care Medicine. Crit Care Med. 1998;26:607–10.
24. Jackson JC, Mitchell N, Hopkins RO. Cognitive functioning, mental health, and quality of life in ICU survivors: an overview. Crit Care Clin. 2009;25:615–28.
25. Hopkins RO, Jackson JC. Short- and long-term cognitive outcomes in intensive care unit survivors. Clin Chest Med. 2009;30:143–53.

26. Adhikari NK, McAndrews MP, Tansey CM, Matté A, Pinto R, Cheung AM, Diaz-Granados N, Barr A, Herridge MS. Self-reported symptoms of depression and memory dysfunction in survivors of ARDS. Chest. 2009;135:678–87.
27. Cheung AM, Tansey CM, Tomlinson G, Diaz-Granados N, Matté A, Barr A, Mehta S, Mazer CD, Guest CB, Stewart TE, Al-Saidi F, Cooper AB, Cook D, Slutsky AS, Herridge MS. Two-year outcomes, health care use, and costs of survivors of acute respiratory distress syndrome. Am J Respir Crit Care Med. 2006;174:538–44.
28. Rubenfeld GD, Caldwell E, Peabody E, Weaver J, Martin DP, Neff M, Stern EJ, Hudson LD. Incidence and outcomes of acute lung injury. N Engl J Med. 2005;353:1685–93.
29. Task Force of the American College of Critical Care Medicine. Guidelines for ICU admission, discharge, and triage. Crit Care Med. 1999;27:633–8.
30. Bioetic Commission SIAARTI. SIAARTI guidelines for admission to and discharge from intensive care units and for limitation of treatment in intensive care. Minerva Anestesiol. 2003;69:101–18.
31. Walter KL, Siegler M, Hall JB. How decisions are made to admit patients to medical intensive care units (MICUs): a survey of MICU directors at academic medical centers across the United States. Crit Care Med. 2008;36:414–20.
32. Irone M, Parise N, Bolgan I, Campostrini S, Dan M, Piccinni P. Assessment of adequacy of ICU admission. Minerva Anestesiol. 2002;68:201–7.
33. SIAARTI – Italian Society of Anaesthesia Analgesia Resuscitation and Intensive Care Bioethical Board. End-of-life care and the intensivist: SIAARTI recommendations on the management of the dying patient. Minerva Anestesiol 2006;72:927–63.
34. Bertolini G, Boffelli S, Malacarne P, Peta M, Marchesi M, Barbisan C, Tomelleri S, Spada S, Satolli R, Gridelli B, Lizzola I, Mazzon D. End-of-life decision-making and quality of ICU performance: an observational study in 84 Italian units. Intensive Care Med. 2010; 36:1495–504.
35. Iapichino G, Radrizzani D, Bertolini G, Ferla L, Pasetti G, Pezzi A, Porta F, Miranda DR. Daily classification of the level of care. A method to describe clinical course of illness, use of resources and quality of intensive care assistance. Intensive Care Med. 2001;27:131–6.
36. Dries DJ, Adams AB, Marini JJ. Time course of physiologic variables in response to ventilator-induced lung injury. Respir Care. 2007;52:31–7.
37. Vincent JL, de Souza Barros D, Cianferoni S. Diagnosis, management and prevention of ventilator-associated pneumonia: an update. Drugs. 2010;70:1927–44.
38. Diaz E, Lorente L, Valles J, Rello J. Mechanical ventilation associated pneumonia. Med Intensiva. 2010;34:318–24.
39. Kress JP, Marini JJ. Acute respiratory distress syndrome: adjuncts to lung-protective ventilation. Semin Respir Crit Care Med. 2001;22:281–92.
40. Marraro GA. Protective lung strategies during artificial ventilation in children. Paediatr Anaesth. 2005;15:630–7.

Chapter 20
The Netherlands: Where Are We Going?

Frank H. Bosch

Introduction

There are many problems in modern healthcare in The Netherlands.

The current combination of an increasingly older population, the increasing use of medical technology and many others are putting severe stress on the system. The costs are increasing and the willingness to continue to pay for these increasing costs is diminishing. The Netherlands will face difficult choices in the future. For 2012, only the health sector is allowed some growth by the government, but the predictions are that in the future, more productivity will have to be accomplished by an increase in efficacy only.

Future Trend: Concentration of Power

Hospitals and Quality of Care

Most hospitals in The Netherlands offer a broad array of services. The all have a First Aid Department and an intensive care department. This is changing quickly. Most medical associations are now imposing strict guidelines for hospitals. For instance: a hospital is only allowed to do perform lung cancer surgery if they perform more than 50 thoracotomies every year. Similar numbers have been agreed on Whipple surgery, breast cancer surgery, etc. This is the beginning of a new trend. Other conditions are following. Cardiology: every cardiologist should perform at least 150 Percutaneous Coronary Interventions; to be recognized as an intervention

F.H. Bosch, M.D., Ph.D. (✉)
Department of Internal Medicine, Rijnstate Hospital Internal Post 1241,
PB 9555, Arnhem 6800TA, The Netherlands
e-mail: fhbosch@rijnstate.nl

D.W. Crippen (ed.), *ICU Resource Allocation in the New Millennium: Will We Say "No"?*,   157
DOI 10.1007/978-1-4614-3866-3_20, © Springer Science+Business Media New York 2013

center, at least 600 interventions need to be done every year. These agreements will lead to patient shifts from hospitals that only perform a limited number of procedures to hospitals that have special programs for these diseases (focused factory).

The Netherlands is a densely populated country and people are wondering whether it is necessary that every hospital has a First Aid Department. Some are suggesting that the number can be decreased by as much as 40% and that this will lead to cost reduction and quality improvement. This will lead to a new landscape in which a number of smaller hospitals will not have a First Aid Department anymore, at least during nighttime. Agreements have to be made upon how much delay is acceptable for an ambulance to travel to a hospital with a fully functional First Aid Department. Doctors and insurance companies are beginning to draw guidelines with regard to these issues.

Professional societies who are involved in the care of critically ill patients are drawing up new guidelines for intensive care units. It is foreseeable that they will advise that only large (mostly teaching and university) hospitals have an ICU and that these ICU's should have regional functions, i.e., they admit sick patients from smaller regional hospitals as well. This can only be done in a situation where reimbursements follow costs.

Patients will have to travel further for the treatment of acute and chronic diseases. Fortunately, the Netherlands is mostly a densely populated country, so travel distances are never very far. The treatment of chronic diseases will mostly be done in a regional setting.

Hospitals and Their Representation

The university hospitals have formed a very powerful union that looks after their interests. They have a strong political lobby. The large teaching hospitals are also working together. In the near future new alliances will be formed, for instance regional to organize the health care in new ways with clear agreements about which hospital provides which care.

Insurance Companies

In the last years there has been a tremendous consolidation in the number of insurance companies for health care. In 2007 there were 15 companies who were active in this market; in 2011 there are only six large health insurance companies. This consolidation had led to an increase in buying power and an increase in the influence that they are having in the health care field. One of the companies was instrumental in the start of the discussion how many breast cancer surgeries should be done at least in a certain hospital to get reimbursed for these procedures. It is foreseeable that insurance companies will demand more information about the quality that they are getting for their patients and that they will buy care more selectively. The insurance companies have together formed an organization "zorgverzekeraars Nederland". This organization has a lot of lobbying power in the political arena.

Because the costs of healthcare keep increasing, new choices about what is covered by insurance and what should be paid by the patients will have to be made in the future.

Doctors

Doctors in The Netherlands have been rather badly organized. They are several organizations that have not been working together very well. Due to increasing external pressures, this will probably change in the near future. Doctors are beginning to realize that it is extremely important to have a powerful lobby at the government and a powerful organization that is capable to be a trustworthy partner in the discussions about where health care should go, but also to defend their monetary and other interests.

E-health

E-health will play an important new role in the future. The internet/computer structure and density are very high and people are willing to do more and more online. Care providers will give advice increasingly with the aid of computer technology, enabling patients to participate more actively in their own treatment and limiting the number of doctors office visits for chronic diseases.

Conclusion

Health Care will change tremendously in the near future in The Netherlands. Hospitals and doctors will see many changes in their daily practice. Concentration of services will be a movement that will not go away in the near future. Many believe that the quality of care will improve and that more patients will be treated at lower costs per patient. The insurance system will remain largely unaltered, except for the loss of certain reimbursements from the basic insurance. People will have to pay more themselves or seek additional insurance.

Chapter 21
New Zealand: Where Are We Going?

Stephen Streat

The Current State of Intensive Care Medicine in New Zealand

New Zealand is a small remote country with 4.4 million increasingly ethnically diverse inhabitants. There is a well-entrenched tradition of publicly funded social services including healthcare, education and welfare. Public expectations of the capabilities of the welfare state have become tempered in recent years by the legacy of the free-market reforms which began in the late 1980s and by the almost-stagnant economy which has yet to recover from the 2008 global recession. Per capita GDP at purchasing power parity is just over half that of the US. Health expenditure [1] is around 75% of that in Australia, Canada and most of western Europe but only a third of US health expenditure. Publicly funded healthcare occupies a dominant position in health care with 81% of expenditure being Government-funded [2]. Intensive care medicine (as it is known in New Zealand) is almost completely confined to the 25 ICUs in public hospitals [3]. Most of these are multi-disciplinary ICUs—only one is a speciality ICU (cardiovascular) and there is only one paediatric ICU. There are small intensive care units confined to post-surgical (mostly cardiac) patients in three private hospitals.

Intensive care resources in New Zealand have increased a little over the last few years, largely in availability of high-dependency beds. National data for 2008 onwards are not yet available but there was no increase in the number of "ventilated beds" between 2000 and 2007 (defined as "A single patient care location fully configured to ICU standards, plus a ventilator") [3].

S. Streat, B.Sc., M.B., Ch.B., F.R.A.C.P. (✉)
Department of Critical Care Medicine, Auckland City Hospital,
2 Park Road, Grafton, Auckland 1023, New Zealand
e-mail: stephens@adhb.govt.nz

D.W. Crippen (ed.), *ICU Resource Allocation in the New Millennium: Will We Say "No"?*,   161
DOI 10.1007/978-1-4614-3866-3_21, © Springer Science+Business Media New York 2013

Predicting the Future

It remains difficult to predict the future, especially prospectively! History is littered with examples of spectacular failure to predict even the immediate future, particularly in areas where quantum technological innovation (or even slow incremental development) has fundamentally destroyed the validity of underlying assumptions—especially the seductive assumption that "the future is a parametric extension of the past". Will the urban myth, allegedly uttered about 30 years ago by Bill Gates (which he denies), that "No one will need more than 637 kb of memory for a personal computer. Six hundred and forty K ought to be enough for anybody" ever be forgotten? As robotic aircraft are poised to take over from remote-controlled drones in the prosecution of war, what should we make of the imperious dismissal of a delegation of aircraft builders in 1910 by the British Secretary of State for War "Gentlemen, much as we would like to help you by placing orders, we regret we cannot do this as we are trustees of the public purse and we do not consider that aeroplanes will be of any possible use for war purposes" [4].

Learning from the Past

The history of intensive care medicine is very brief indeed. I was born only a few months before Bjørn Ibsen's famous young patient with polio was treated with prolonged positive pressure ventilation [5]. We have been privileged to have been part of what now seems like the golden age of modern intensive care medicine, yet half a century on it is clear that the complex interplay of expectations and preconceived prejudices, technological capability, societal ills, case mix and financing that determine intensive care resource utilisation has evolved in very disparate ways around the world and is far from stable in any country, including my own. The long history and records of our own department allow reflection on some of these various drivers of demand and supply and illuminate the limitations of predicting the future by extrapolating from the past.

Initially, intensive care was seen as a treatment of last resort—an almost experimental fusion of anaesthesia, surgery and the animal physiology laboratory. Intensive treatments were often begun late in the course of an illness—when deterioration was clearly immediately life-threatening—indeed sometimes only at the point of respiratory arrest [5] or during trans-tentorial herniation. Expectations were not high, as previously most of these patients had died. What were then seen as heroic and desperate measures were applied first to small numbers of children and young adults in previously good health whose physiology was both resilient and reminiscent of the animal laboratory. Individually simple treatments (e.g. high inspired oxygen concentrations, endotracheal intubation and tracheostomy, mechanical ventilation [6], blood volume expansion, continuous catecholamine infusions, opioids, sedatives and neuromuscular blockers, antibiotics, enteral feeding of homogenised hospital meals, drainage of sepsis, debridement of dead tissue, closure of open wounds, intrapleural drainage, stabilisation of fractures, and even haemodialysis [7]) were applied in novel combinations by dedicated nursing and medical staff and produced good survivors. For example [6]—over the first 4 years of intensive care medicine at Auckland Hospital (1958–1961), there were only 141 admissions, only ~35 admissions per year—most commonly for postoperative respiratory failure (24), croup (23), tetanus (21) poliomy-elitis (20), barbiturate intoxication (19), traumatic brain injury (9) and Guillain-Barre syndrome (7). Other conditions accounted for only 15 cases. The median age of all admissions was 26 and a quarter of them were children (under the age of 15). Because of spectacularly long ICU stays (median 14, mean 73.9 days) mean ICU bed occu-pancy was 7.0. With the notable exception of patients with traumatic brain injury (all nine of whom died), survival was 106/132 (80%) in the others.

This good result rapidly altered (everyone's) expectations and drove substantial growth in resource utilisation. A decade later the number of admissions had increased eightfold—to 283 patients in 1969 [8]. Intensive care was offered to young patients—the median age of all admissions was still 26 and a third were under 15 years old. Poliomyelitis had gone, but tetanus and Guillain-Barre syndrome remained and there was an enormous increase in admissions for trauma, drug intoxication, men-ingitis, epiglottitis, pneumonia and other causes of sepsis. Length of stay had decreased dramatically (median 3, mean 10.9 days) so that mean ICU bed occu-pancy remained only 8.3 and survival (77%) was similar to what it had been a decade earlier. By this time intensive care medicine had come to be seen as effective and no longer "quasi-experimental" and in New Zealand it diffused quickly into most public hospitals.

My personal involvement in the specialty coincided with another substantial increase in intensive care resource allocation which took place during the 1970s. This era saw the beginning of effective therapy for multiple organ failure [9, 10] as well as rapid development of modern medical devices and equipment (including ventilators, infusion pumps and electronic monitors). I think that it also marked the beginning of what has recently been called "gizmo idolatry" [11] in medicine—a fascination with technological devices, accompanied by "magical-thinking" of their talismanic power to effect cure. By the end of the 1970s, intensive care admissions had again doubled over a decade previously, due in large part to substantial increases in admissions for trauma, sepsis and tricyclic poisoning as well as a new cohort of patients—admitted postoperatively "for monitoring" after major elective surgery—comprising 11% of admissions in 1977. That year there were 655 admissions, and mean ICU occupancy was 16.5, but fewer patients were ventilated (44%) than had been the case a decade earlier (58%). ICU length of stay fell further (median 2, mean 9.4 days) and ICU survival was 81%.

During the 1980s the number of ICU admissions rose dramatically—reaching 1,169 by 1987. The "first peak" passed in a wave of elective admission of what we would now call "postoperative high dependency" patients. By 1987 these comprised only 1% of the 1,169 admissions that year (918 of which were in adults). The increase in ICU admissions during the 1980s was of young people with blunt trauma following road crash, others with self-poisoning (largely with tricyclic antidepres-sants) and a mysterious epidemic of severe life-threatening asthma [12] which saw

New Zealand have at that time the world's highest asthma mortality [13]. We also saw the treacherous syndrome of slow-release oral theophylline poisoning [14] as this drug was being widely prescribed for asthma at that time. More patients (68% in 1987) were being ventilated but length of stay fell sharply (median 1, mean 3.9 days) and as a result mean ICU occupancy also fell (to 11.3).

Over the next decade, New Zealand society changed dramatically and some of these changes were reflected in the use of ICU resources. Tariffs on the importation of cheap late-model used cars were abolished and as a direct result the number of motorcycle registrations fell sharply. Improvements in vehicle safety and implementation of injury prevention measures (e.g. improved road design and construction, compulsory use of seat-belts, motorcycle and bicycle helmets, and child-restraints, roadside breath alcohol testing and increased police enforcement of traffic offences) saw road fatalities fall sharply during the 1990s. This was accompanied by a fall in the number of ICU admissions for trauma (by a third). Tricyclic antidepressants were being replaced by SSRIs and inhaled steroids were widely prescribed for asthma (and slow-release oral theophylline was no longer). Admissions for poisoning dropped by 30% and for asthma by 72% while there were large increases in admissions for sepsis and subarachnoid haemorrhage. A paediatric intensive care unit opened in the new children's hospital on the same site in late 1991 and all critically ill children were subsequently admitted there. The number of adult ICU admissions continued to rise, to 1,006 by 1997 but length of stay continued to fall (median 1, mean 3.1 days) and mean occupancy (now of adults only) was 8.6 while ICU survival was 87%.

Over the next decade admissions remained near-constant (1,036 in 2007) but the case mix continued to evolve in the direction of greater complexity. The median age rose from 47 to 53, while admissions for trauma, asthma and particularly poisoning declined substantially. Over this decade, admissions for subarachnoid haemorrhage remained constant while those for sepsis and cardiovascular failure increased and we undertook liver transplantation and simultaneous kidney–pancreas transplantation. Despite increasing case complexity, ICU survival increased to 92% while length of stay continued to fall (median 1, mean 2.6 days) and as a result mean occupancy also fell to only 7.7.

Looking to the Future

Advances in surgery, anaesthesia, coagulation support, monitoring and investigative technology have emboldened surgeons and anaesthetists to offer increasingly heroic surgery to increasingly elderly co-morbid patients with malignancy, degenerative conditions and cardiovascular disease. Intensivists' gloomy predictions of the effectiveness and efficiency of intensive treatments being increasingly applied to such patients has been countered by the contrary evidence of our own improved performance.

An additional high-dependency unit (HDU) was created within our ICU in 2009, specifically for postoperative admission of "high-risk" (non-cardiac) surgical patients—but where invasive ventilatory support is not provided. This additional resource allocation was driven by "external customer demand" despite a lack of evidence of effectiveness or cost-effectiveness [15] and it is probably too early to speculate on the durability of this popular model of care delivery. Similar HDU expansion has occurred in many other New Zealand ICUs over the same period, as it has in other countries and it may provide a means of helping to constrain inordinate "demand" for intensive care services.

Intensive care medicine continues to show increasing therapeutic efficacy [16] as well as efficiency, although this is now by increments rather than by steps. Even slow progress can be enough to sustain an optimistic expectation that in time every problem will yield to meticulous research, application and development. This may not be true as such an optimistic expectation is also common in other areas of human activity (e.g. resource depletion, climate change, ideological conflicts) which appear as intractable now as they have at any time.

Intensive care admission has come to be seen as a "moral good" by virtue of its evident capability to effect dramatic rescue of individuals in crisis. This "virtue status" does I believe underlie a sense of "entitlement to the opportunity for rescue" that intensive care medicine might offer. Conversely, rationing of access to intensive care medicine can be seen (or portrayed) as a dereliction of a general moral duty viz. a "…duty to attempt an easy rescue of an endangered person" [17]. While such a view may have some support in other countries it is likely to have less resonance in New Zealand where intensive care medicine is almost exclusively a public funded endeavour and shows no sign of changing in this regard.

It is clear that throughout the developed world there will be a substantial rise in the number of elderly co-morbid persons who *could* potentially be treated in intensive care units. Using linear models of predictable relationships between inputs and outputs (e.g. ref. [18]) many authors have recently made projections of ICU demand based largely on anticipated demographic changes. In many countries with similar population demographics to New Zealand, these have resulted in large projected increases in demand over the next 5–20 years e.g. Australia [19, 20] 50 and 72%, Canada [21] 57%, and the US [22] 35%. Similar results would occur also in New Zealand but testing the sensitivity of such projections to underlying assumptions is difficult when at least one unpredictable "quantum event" (such as an emerging disease, a fundamentally efficacious therapy or an effective prevention strategy) seems likely to occur over the period of those projections. Noting the chaotic (in the sense of non-linear) and unpredictable forces which have demonstrably defined and driven intensive care medicine over the last 50 years, modelling the future as a projection from the status quo and the existing relationships between demography and resource utilisation is surely simplistic.

Summary and Conclusion

It seems likely that despite projected increases in the number of elderly patients who could be treated in intensive care units, that other *force majeure* factors, particularly societal willingness or ability to pay, changes in medical technology and changes in the values, beliefs and expectations of clinicians, patients and their families, will unpredictably confound projection models based on demography. I suspect that over the next 20 years the balance of these factors in New Zealand, in particular fiscal limitations, will not result in anything more than minor increases in ICU resource availability and utilisation.

References

1. International Monetary Fund. World economic outlook database. Washington DC: International Monetary Fund. 2010. http://www.imf.org/external/pubs/ft/weo/2010/02/weodata/index.aspx. Accessed 16 Nov 2010.
2. The New Zealand Health and Disability System: Organisations and Responsibilities. Briefing to the Minister of Health. Wellington: Ministry of Health. 2008. http://www.moh.govt.nz/moh.nsf/pagesmh/8704/$File/nz-health-disability-system-briefing2008.doc. Accessed 16 Nov 2010.
3. Drennan K, Hart GK, Hicks P. Intensive care resources and activity: Australia and New Zealand 2006/2007. Melbourne: ANZICS. 2008. http://www.anzics.com.au/. Accessed 23 Nov 2010.
4. Verdon-Roe A, Gilbert J. The world of wings and things. New York: Arno Press; 1979.
5. Trubuhovich RV. August 26th 1952 at Copenhagen: 'Bjørn Ibsen's Day'; a significant event for anaesthesia. Acta Anaesthesiol Scand. 2004;48(3):272–7.
6. Spence M. The emergency treatment of acute respiratory failure. Anesthesiology. 1962;23: 524–37.
7. Irvine RO, Montgomerie MB, Spence M. Assisted respiration, noradrenaline infusion and the artificial kidney for glutethimide ("Doriden") poisoning. Med J Aust. 1963;2:277–9.
8. Trubuhovich RV. From respiratory support to critical care: my early days in intensive care medicine. Crit Care Resusc. 2007;9(2):123–6.
9. Eiseman B, Beart R, Norton L. Multiple organ failure. Surg Gynecol Obstet. 1977;144(3): 323–6.
10. Routh GS, Briggs JD, Mone JG, Ledingham IM. Survival from acute renal failure with and without multiple organ dysfunction. Postgrad Med J. 1980;56(654):244–7.
11. Leff B, Finucane TE. Gizmo idolatry. JAMA. 2008;299(15):1830–2.
12. Zimmerman JE, Galler LG, Judson JA, Streat SJ, Trubuhovich RV. Severity stratification in life threatening asthma. J Intensive Care Med. 1990;5(3):120–7.
13. Pearce N, Beasley R, Crane J, Burgess C, Jackson R. End of the New Zealand asthma mortality epidemic. Lancet. 1995;345(8941):41–4.
14. Parr MJ, Anaes FC, Day AC, Kletchko SL, Crone PD, Rankin AP. Theophylline poisoning – a review of 64 cases. Intensive Care Med. 1990;16(6):394–8.
15. Bellomo R, Goldsmith D, Uchino S, Buckmaster J, Hart G, Opdam H, Silvester W, Doolan L, Gutteridge G. A before and after trial of the effect of a high-dependency unit on post-operative morbidity and mortality. Crit Care Resusc. 2005;7(1):16–21.
16. Streat S. Improving the effectiveness of our practices. In: Flaatten H, Moreno R, Putensen C, Rhodes A, editors. Organisation and management of intensive care. Berlin: Medizinisch Wissenschaftliche Verlagsgesellschaft; 2010.
17. Peters DA. An individualistic approach to routine cadaver organ removal. Health Prog. 1988;69:25–8.
18. Finarelli Jr HJ, Johnson T. Effective demand forecasting in 9 steps. Healthc Financ Manage. 2004;58(11):52–6.
19. Corke C, Leeuw E, Lo SK, George C. Predicting future intensive care demand in Australia. Crit Care Resusc. 2009;11(4):257–60.

20. Bagshaw SM, Webb SA, Delaney A, George C, Pilcher D, Hart GK, Bellomo R. Very old patients admitted to intensive care in Australia and New Zealand: a multi-centre cohort analysis. Crit Care. 2009;13(2):R45.
21. Hill AD, Fan E, Stewart TE, Sibbald WJ, Nauenberg E, Lawless B, Bennett J, Martin CM. Critical care services in Ontario: a survey-based assessment of current and future resource needs. Can J Anaesth. 2009;56(4):291–7.
22. Angus DC, Kelley MA, Schmitz RJ, White A, Popovich J Jr; Committee on Manpower for Pulmonary and Critical Care Societies (COMPACCS). Caring for the critically ill patient. Current and projected workforce requirements for care of the critically ill and patients with pulmonary disease: can we meet the requirements of an aging population? JAMA. 2000;284(21):2762–70.

Chapter 22
South Africa: Where Are We Going?

Ross Hofmeyr

Introduction

South Africa is a country proud of its diversity, epitomised by the nickname "Rainbow Nation", outgrowing a political past which may euphemistically be called colourful. The spectrum of diversity is apparent in the topography, cultures, languages, skin tones, and in the great disparity in wealth. Areas of First World affluence and service availability are found in the major cities while much of the country can be described as the Third World; properties with the greatest value per unit area in Africa are within a short drive of rambling and squalid shanty-towns.

More than 15 years since emerging as a democracy, the legacy of Apartheid is fading away as a target for blame for the challenges faced by the nation, and public demand for realisation of revolutionary promises becomes more insistent. Along with economic growth, employment, education and housing, access to quality health care is one of the most frequent concerns. As the leading emerging country and economy in Africa [1] with the most developed and sophisticated medical services on the continent [2], many eyes further north follow our progress closely as a model for the rest of Africa.

To theorise upon the future of resource allocation in intensive care in South Africa, one must understand the setting and challenges that exist in the country as a whole. Healthcare in South Africa is subject to the First/Third World dichotomy in several ways. First, there is the problem of wealth distribution and population density. Most of the country's 50 million inhabitants [3] live either in large cities or deep rural areas, with a very significant disparity in medical services. Often rural patients live remote from the simplest primary clinics and must travel or be referred great distances to secondary or tertiary care hospitals. Second, there is no equality

R. Hofmeyr, M.B.Ch.B. (Stell), Dip.P.E.C. (SA), D.A. (SA) (✉)
Department of Anaesthesia and Critical Care, GF Jooste Trauma
and Emergency Hospital, Cape Town, South Africa
e-mail: wildmedic@gmail.com

D.W. Crippen (ed.), *ICU Resource Allocation in the New Millennium: Will We Say "No"?*,   169
DOI 10.1007/978-1-4614-3866-3_22, © Springer Science+Business Media New York 2013

between the services offered in the nine provinces of the country, with some having almost no tertiary services (at which level ICU care is provided) whatsoever. The health departments of several provinces are woefully in debt, and as such they are unable to maintain staffing levels required to provide adequate care. In other provinces (most notably the Western Cape and Gauteng) within the large urban areas, health services are well established: despite continuous budgetary and staffing pressure, a very high level of care is achieved with standards and training to rival the best centres in the developed world. It is not by chance that South African-trained doctors are well respected internationally, and significant contributions to medical development (with the most oft-quoted example of the first heart transplant [4]) have been made.

Each and every South African is entitled by constitutional right [5] to State-provided health services, which are provided at no charge to the indigent and via a scaled system to those with income. For the affluent minority who do not wish to rely on State healthcare, there exists a parallel system of private practitioners, clinics and hospitals across the country. Herein lies the third dichotomy: those that can afford private care (for most, this implies some form of health insurance or medical aid plan) can rapidly access First World resource-rich facilities with a large number of specialists across the medical disciplines, while the poorer majority rely on State facilities staffed primarily by medical officers at various levels of experience, often with scanty specialist cover. However, there is a curious reversal of this situation at the "quaternary" level: almost all academic, research, medical and nursing training, and sub-speciality services are to be found in State hospitals. Indeed, there is no registrar (resident) training outside of the State hospitals linked to the major universities, and almost no training in critical care in the private sector.

If we are to consider the future of intensive care in South Africa, it must be with reference to the great diversity of locations and levels of care, the influence of cultural differences, the greatly different public and private systems which exist, and the socio-political climate.

Locality and Levels of Critical Care in South Africa

A globally young discipline, critical care has only been formally established in South Africa for a little more than a decade. Data delineating the quantity and distribution of high and/or intensive care units (HCUs and ICUs respectively) in South Africa was not available until this millennium, with the first thorough audit performed by Bhagwantee and Scribante in 2004–2005 and published in 2007 [6]. This form of data is essential in understanding and planning the future of critical care.

The audit revealed while the number of beds in the public and private sector is numerically fairly balanced (57% of 4,168 total being found in private hospitals), the public-sector beds are concentrated in relatively few facilities. Indeed, while 84% of private hospitals have critical care beds, only 23% of state hospitals can claim these capabilities. This disparity is even starker when it is recalled that only

18% of the population has access to private care through medical insurance. Furthermore, there is a very significant regional variation in the number and concentration of beds; 86% are to be found in only three of the nine provinces (Gauteng, Western Cape and KwaZulu-Natal). Bed to population ratios vary from 1:14,000 in the Western Cape to 1:82,000 in the Limpopo Province. The inequality is more pronounced for the private sector but distribution is similar, which challenges the concept of public–private partnerships to alleviate load (see below).

First World standards dictate an ICU:total bed ratio within a hospital to lie between 5 and 12% dependant on the level and nature of work. Private hospitals in SA have a ratio of 8.9% (well within this range) while for public facilities the number falls to 1.7%. If only hospitals with critical care facilities are considered, this increases to 9.6 and 3.9% respectively. The audit found a paucity of beds dedicated to paediatric and neonatal patients (19.6% of critical care beds). A similar dearth of burns beds exists. Interestingly, there is a relative lack of HCU beds (only 18% of the total). As high care units can both decrease the admission rate in for ICUs as well as providing an intermediate step-down alternative at less cost than ICU admission, this is an unfortunate state of affairs.

Another unequal facet of critical care medicine is that all training—whether undergraduate or postgraduate, for doctors and nurses alike—occurs in the State system. Exposure to intensive care for medical students is limited across all institutions. Interns may gain some exposure, either as part of their anaesthesia or internal medicine training, but this is not the norm. Frequently, it is junior medical officers (often in their first year of unsupervised practice as part of their community service rotation) who are left alone in peripheral/regional HCUs and even ICUs with minimal specialist cover.

Formal exposure and teaching in critical care is usually left to a few months' rotation as a registrar, depending on specialty. Specialists in disciplines such as surgery, orthopaedics, internal medicine, obstetrics and gynaecology, anaesthesiology, etc. usually have no more than 3 or 6 months of ICU training at best, but are expected to care for ICU patients in private hospitals after obtaining their fellowship. Only those wishing to pursue sub-speciality intensivist training will proceed to completing the 18–24 months full-time programme and fellowship examinations in Critical Care. Still, even within this programme there is significant regional variation in experience and expectations. Hopefully, the increasing attention and energy devoted to critical care in South Africa will encourage the numbers enrolled in these programmes to swell.

Intensive Care Units within public hospitals are often considered by medical administrators as a necessary evil. ICUs consume huge quantities of resources—staffing, medications, equipment, and disposables—for a small number of patients, often with "dismal" outcomes. Indeed, even in the very expense-cognisant private hospitals, critical care spending accounts for about 30% of SA medical insurance expenditure, with mortality rates at an equal figure (in line with international trends). The State ICU is thus a major financial drain for the hospital, while its benefits such as facilitating advanced surgeries, supporting severe medical admissions, etc. remain more hidden. However, the ICU is an essential resource to enable these profit-generating activities to continue. It has been suggested that units with a full-time critical care specialist medical director can both improve patient outcomes and decrease mortality and costs [7,8], which would clearly benefit ICUs in both settings. This thinking has not yet penetrated the private institutions. The national forum (under the auspices of the Critical Care Society) must drive the establishment of an integrated critical care plan within the framework of the existing health services in a proactive fashion.

Cultural Influences on Critical Care

One legacy of the human rights struggle in South Africa which resulted in the collapse of Apartheid and subsequent ascension to a democracy with one of the world's most liberal constitutions has been the emergence of a rights-based culture. Particularly amongst the previously-disadvantaged population groups, revolutionary pre-democracy promises of the emerging political parties have generated a steadfast anticipation of service delivery, free housing, employment and health-care. While an essential component of any Constitution, the enshrined right to care has been warped into an expectation that *all*, rather than *basic* care, is a human right. This is compounded when the affluent can access expensive private facilities that the poor cannot, leading to a further sense of disenfranchisement with the system. Fortunately, within most of the spectrum of South African cultures—Afrikaans, Muslim, Xhosa, Zulu, Indian, and so forth—there still exists a significant respect for medical professionals. Doctors are still viewed both as very learned professionals and frequently as "elders".

A strong argument has been made that intensivists should not be seeking to gain *consent* for changes or withdrawal of therapy (as it is impossible to encapsulate 15 years of training and experience into a bedside discussion) but should rather seek *assent* to the decision which has to be made. Patients and families often afford great value to the advice of critical care doctors, and issues regarding treatment, futility of care and end-of-life decisions are more readily accepted in South African communities (especially within State hospitals) than is apparent in the international literature. Conversely, however, while decisions are more easily approached in many of SA's cultures, there is great cohesion of the family unit. Frequently there is a request to delay assent to medical decisions until the senior family member is present.

South Africa has thus far been spared the ravages of litigation that have become prevalent in some First World medical settings, but this is beginning to change [9]. Still, critical care doctors practise with a fair degree of autonomy in their decision-making, allowing choices with regard to care to be directed by clinical criteria rather than societal expectation. Considerable variation exists, however, between population groups. Affluent patients are more likely to have higher expectations on ICUs and make demands on care-givers than those utilising State hospitals. Indigent patients typically come from so-called African ethnic groups, have lower levels of education (especially the older generations) and lower expectations. Medical malpractice suits in State ICUs are very rare and successful litigation almost unheard of.

For the meantime, it seems that South Africa has escaped many of the difficulties of patients demanding or seeking to control therapies which are not medically indicated. As global cultures become homogenised and the level of education and access to media (particularly the Internet) improves in the country, critical care providers should watch these trends carefully lest we emulate the less popular quirks of First World practice.

Public Versus Private Health Care

While the roles of intensivists in tertiary academic hospitals in South Africa are well established and do not appear to be under threat, there is a natural limit to the quantity of services that can be provided within each facility due to fiscal concerns. Typically, bed occupancy in tertiary units is consistently at capacity, with continuous demand for further admissions from lower levels of care. Surgical procedures requiring ICU admission are frequently deferred due to lack of beds, and patients deserving of ICU admission are treated for extended periods in wards, resuscitation areas and high care units (HCUs). The reflex response to the problem is to increase unit capacities, but critical care clinicians and leaders in the field should employ a more diverse response, including:

- Improving and integrating the critical care systems, triage and patient flow, so that patients can be treated at the most appropriate level in the system at all times during their disease course. The Critical Care Society of Southern Africa has published guidelines to guide admission and appropriate levels of care which are freely available [10]. Furthermore, increasing the number and improving distribution of HCU beds may dramatically reduce both the need for ICU admission as well as providing an intermediary step-down for patients no longer needing full ICU care. Effective critical care transport systems are needed to facilitate this process.
- Extending the influence of intensivists beyond the ICU. In a country such as South Africa where junior and unspecialised doctors are so often responsible for the initial and emergency care of patients, ICU specialists have much to offer through training, advice and guidance which will help arrest disease progression before ICU admission is required. Similarly, the overwhelming success of outreach teams within the hospital should encourage integration of ICU skills into the general hospital setting. Another avenue that must be pursued is telemedicine: improved communications systems and enhanced contact between peripheral/regional units and academic ICUs will allow the junior personnel to consult early, gaining more educational opportunities and reducing transfers.
- Increase, standardise and objectify critical care training across the country. Formulation of the Society and subspecialty more than a decade ago, as well as the introduction of the Critical Care Fellowship has already built a strong foundation, but unit and programme accreditation, as well as the specifics of training and exposure must be formalised.

- Promote and guide research to evaluate the local needs and requirements for critical care. Simple imitation of the solutions instituted by other developed nations will not ensure success; recognising the parameters of our systems while learning from established examples should lead to ideal solutions. This in turn can act as a leadership example to other developing nations.

Considerable concern exists regarding the future (and indeed, current prospects) of the intensivist in private healthcare in South Africa. This has been demonstrated by intensivists who have attempted to establish themselves in private hospital ICUs only to be met with resistance, from a number of sources. Many medical or surgical specialists prefer to continue managing their own patients in open ICUs even if this is not the focus or forte of their practice; this also allows them to continue to bill for consultations. For surgical intensive care patients, the expectation often falls to the anaesthetic staff to continue to provide care, whether or not they have interest in this practice. Despite the evidence in favour of intensivist-run ICUs, hospitals baulk at the concept of employing additional staff, especially if this runs against the wishes of their existing staff (as described above). Finally, medical insurers question the necessity for intensivists when many training programmes include a modicum of exposure to critical care.

The interventions proposed above with relevance to State hospitals are equally desirable in private facilities, especially if the profile and influence of critical care specialists is to be increased. However, at this time there is no indication that a change towards increased use if intensivists in the private sector is likely to occur. An active leadership role is required to further establish and promote the speciality, for the ultimate benefit of individual patients as well as the system through increased efficiency.

Socio-political Climate for Medicine in South Africa

Uncertainty exists within the health care industry in South Africa with regard to the proposed implementation of a National Health Insurance (NHI) plan in the near future. While government spending on health as a percentage of GDP has increased from 3% in 1994 to 3.4% in 2009, barely keeping pace with inflation and population growth, medical scheme expenditure per beneficiary has doubled over the last decade. This is seen as an indication that State spending has not increased to meet the increased demands placed by the HIV pandemic in the country. The greatest focus of the proposed NHI is understandably on primary and preventative medicine, and increasing access to the health system for all citizens. The exact mechanisms involved remain nebulous—strategies from outlawing private health insurance in favour of participation in the NHI through to increased taxation with public access to private facilities have all been proposed.

In this uncertain political climate, it is challenging to prognosticate on the future of critical care facilities. However, there are some potential advantages of the NHI for critical care in the future:

- Improved primary health care and access to basic and diagnostic facilities will reduce the burden of disease and allow earlier intervention, thus reducing critical care admissions.
- Better prevention and management of HIV/AIDS related illness—as well as improved rollout of antiretroviral medications—will decrease the need for ICU care.
- Public-private partnerships have been proposed, to function in several ways:
  - Patients without private insurance can be admitted to private hospitals under specific usage agreements. For example, the Walter Sisulu Paediatric Cardiac Surgery Centre [11] operates with a partnership agreement which allows children of indigent families to receive necessary surgery in a private facility, decreasing waiting times and reducing the pressure on the strained State hospital. During the recent H1N1 influenza epidemic, several South African provinces utilised partnership agreements to admit the overflow of intensive care patients in public hospitals to private ICUs.
  - Private hospital groups can be contracted by the State to provide services for an entire area or community at a set rate. This carries the potential benefit— and risk—of care being carefully rationed to contain costs, and requires careful setting of goals and subsequent monitoring. This model is being tested by the government of neighbouring Lesotho [12] and remains a contentious proposition.
- Mounting pressure exists within the NHI proposals to train medical staff in private hospitals; given the fact that the majority of ICUs and ICU beds in the country are outside the State service, this would greatly increase the number of intensivists and the profile of critical care in South Africa.

Economic uncertainty within the health service is currently a crippling factor in the development of critical care services. Much rests upon the decisions taken by the National Department of Health, and thus clinicians, unit heads and the Society must be active in influencing the choices and ensuring the provision of good intensive care and protection of clinical decision-making at a bedside level.

Looking to the Future

Clearly, critical care medicine in South Africa has emerged against a colourful background and faces many challenges moving forwards. The creation of both a society and sub-speciality in conjunction with the efforts of a dedicated few has ensured a rallying point for the discipline, but there are many uncertainties. The developing nature of the country as a whole should not be forgotten as actions are chosen; interventions to improve intensive care medicine in SA must be grounded in simple and universal principles, and the little money available must be focused where it is most valuable.

Triage principles must be universally adopted, targeting care to those patients who will benefit most. Protocols and agreement on withdrawal of therapy (never withdrawal of *care*) must be standardised and propagated. Intensivists must reach out into their hospitals and back down the referral tree to educate practitioners and improve practices, reducing the need for admissions.

Data on the local needs and requirements must be gathered, as well as data demonstrating the value of intensivist-led units. Clinicians must take interest, action and leadership in creating and managing the health care funding systems in the country as advocates for their patients, rather than as servants of the medical insurers. Increasing the number of HCU beds available may alleviate the pressure faced by ICUs at less cost.

Finally, we must find means to engage and educate communities with regards to critical illness and end-of-life issues in order to maintain their respect and prepare them for recognising the limits of the abilities of critical care medicine.

References

1. Wikipedia.org. List of African countries by GDP (nominal). http://en.wikipedia.org/wiki/List_of_African_countries_by_GDP_%28nominal%29. Accessed 9 June 2012.
2. Lipman J, Lichtman, AR. Critical Care in Africa: North to South and the future with special reference to Southern Africa. Critical Care Clinics. 1997;13(2):255–265.
3. Statistics South Africa. Mid-year Population Estimates 2012. Statistical release P0302, published July 2010. http://www.statssa.gov.za/publications/P0302/P03022010.pdf. Accessed 10 June 2012.
4. Cooper DK. Christiaan Barnard and his contributions to heart transplantation. J Heart Lung Transplant. 2001;20(6):599–610.
5. The Constitution of the Republic of South Africa, 1996, Chapter 2 (Bill of Rights), Section 27. South African Government Information. http://www.info.gov.za/documents/constitution/1996/96cons2.htm#27, last modified 19 August 2009, Accessed 10 June 2012. (Available in booklet form: ISBN 0-620-20214-9).
6. Bhagwantee S, Scribante J. National audit of critical care resources in South Africa – unit and bed distribution. SA Med J. 2007;97(12):1311–4.
7. Brown JJ, Sullivan G. Effect on ICU mortality of a full-time critical care specialist. Chest. 1989;96:127–9.
8. Hanson CW, et al. Do intensivists make a difference? Crit Care Med. 1996;24:A56.
9. Pepper MS, Slabbert MN. Is South Africa on the verge of a malpractice litigation storm? SA J Bioeth Law. 2011;4(1):3.
10. Richards, G, Mer M, Levy B et al. Best Practice Guidelines. Critical Care Society of Southern Africa/Nesipopho, 2010.
11. Kleinschmidt A. Hope on the Cape. Medical Solutions, December 2010, http://www.medical.siemens.com/siemens/en_GB/rg_marcom_FBAs/files/news/2010_11_cape/S.Africa_Med_Sol_Dec_10.pdf>. Accessed 10 June 2012.
12. Wearden G. A public-private approach to healthcare. Mail & Guardian, 21 June 2009, http://www.mg.co.za/article/2009-06-21-a-publicprivate-approach-to-healthcare. Accessed 10 June 2012.

Chapter 23
United Kingdom: Where Are We Going?

Andrew Thorniley

There is always a problem in predicting the future—some events have the propensity to alter the course of events and foul the whole concept of prediction.

First prize to Apple and the iPad—finally a system that gives portability to medicine.

Systems like Remote Presence (RP6) Robots (http://www.intouchhealth.com) that trundle around on wheels cost a small fortune to purchase, update software and demand expensive service contracts. Just pick up the iPad, use the built-in video camera and link to someone elsewhere in the world to discuss the case with. And it is cheap and off the shelf.

A quick return to the past is needed.

In 1948 the Labour Government established a publicly funded healthcare system that was to provide for the entire nation, free at source and based on clinical need and not the ability to pay.

But, by 1949 there was a recognition that the "dream" was already outstripping the funds available. The Minister of Health, Aneurin Bevan acknowledged the increasing demands on facilities and staff and the rising costs.

And to sound a warning note for the rest of this essay, in 1951 a proposal that a prescription charge of 5p be levied and the patient to bear half the cost of a pair of spectacles and dentures split the Labour party and caused it to lose the election.

The expectations had outstripped financial resources and the same continues unabated today and into the future. Woe betide the intensivist who fails to heed the bottom line in the accounting ledger.

Not only does the NHS have an insatiable appetite for money, but no political party is immune from an entity that can make or break an election.

Promises made in electioneering campaigns often fail to materialize or alter drastically when bottom line accounting shows impracticality.

A. Thorniley, M.D. (✉)
Department of Anaesthetics, The Hillingdon Hospital NHS Trust,
Uxbridge, Greater London, UK
e-mail: andrew@aaybkt.demon.co.uk

D.W. Crippen (ed.), *ICU Resource Allocation in the New Millennium: Will We Say "No"?*, 177
DOI 10.1007/978-1-4614-3866-3_23, © Springer Science+Business Media New York 2013

In the year 2010, rocked by global recession, banking crises and severe economic downturn the NHS still faces problems in resourcing care: Care that is now soaring away in costs and availability; Pressures of an aging population, advanced and more expensive drugs and operations and the need for more funding for the chronically sick and disabled.

Money and what is available will dominate the future of ITU provision in the UK more than before. In 2007 the per capita expenditure on health in the UK was $2992.00 compared with USA—$7290.00, France $3601.00, Australia $3137.00.

http://www.commonwealthfund.org/Content/Publications/Fund-Reports/2007/May/
    Mirror--Mirror-on-the-Wall--An-International-Update-on-the-Comparative-
    Performance-of-American-Healt.aspx

As money makes the world go around the next 10 years will be lean ones for intensivists. Budgets will be either held at the current fiscal year or cut. These financial binders will hinder equipment procurement both for replacement or new purchases. New builds and refurbishments, staffing and drug usage likewise will be throttled back.

This may well be a nexus point in remodelling where patients will be treated—as usual in the District General Hospital (DGH) or taking to large central ITUs with closure of small units.

Intensive Care in the Future

Intensivists—they will be much more financially and politically aware clinicians. Money as the bottom line will play an ever increasing role as cash strapped hospitals look to curb costs.

Regional health authorities will be analyzing consultancy reports on whether ITUs should be centralized or certain types of patient transferred to specialist hospitals. Currently in my area of north-west London DGHs treat the standard medical/surgical adult population and transfer to specialist centres, e.g. neurosurgery, cardiothoracic when indicated.

As an ongoing development and gazing into the future, London will have two major polytrauma units in the future—based on future costs and superior treatment of life threatening conditions. This will entail closing A&Es as well as reclassing others to deal with general emergency admissions only.

The ambulance services are already triaging at point of contact and bypassing nearby hospitals to transport patients directly to the appropriate unit. As an example, patients diagnosed by the paramedics as having myocardial ischaemic events are immediately taken to the nearest specialist cardiac unit for stenting. Our coronary care unit has seen a drastic alteration in patient type. Gradually the unit has started taking other sick medical patients and repatriation of post-stented patients from the main centre for rehabilitation.

Similarly the management of leaking aortic aneurysms has altered dramatically. Expertise has been moved to large units that can provide a 24 h vascular, radiological and stenting service. Citing lower death rates and complications they now are fed directly from the peripheral hospitals.

All this alteration in the landscape means that ITU occupation will change as similar concentrating of expertise occurs. The sting in the tail also means that there is redundancy in the workforce, i.e. less vascular surgeons are needed than before.

Training of intensivists—currently the majority are anaesthesiologists who have subspecialized. Most units used the closed model in that the admitting team surrenders patient management to the anaesthesiologists and play an observer role until discharge. There are few physicians and surgeons entering the speciality. This is an unhealthy state of affairs and the balance will need to be redressed.

Failure to recruit a variety of expertise will cause a stagnation of the discipline. One of the tough challenges will be to break the hegemony and bring in specialists other than anaesthesiologists. However, it is unlikely that in the next 10–20 years there will be any great changes to the ratio of anaesthetist to others.

Specialist Nurses

There will be more specialist nurses who will join the ranks and be an integral part of the unit. Currently our unit has two outreach nurses who patrol the wards looking for trouble. There remit is to review post-ITU discharged patients to ensure continuity of care and prevent relapses. They also examine, treat and advise on patients who trigger alerts on a scoring system of risk-based nursing assessment system that is simple to use, reduces unnecessary paperwork and reduces the risk of harm to patients.

http://www.nursingtimes.net/developing-a-risk-assessment-tool-to-improve-patient-safety/1833916.article

Use of the patient at risk scores in the emergency department: a preliminary study, J E Rees, C Mann Emerg Med J 2004;21:698–699 doi:10.1136/emj.2003.006197.

With contractions in the numbers of doctors and reductions in costs, these nurses will gradually take over more responsibilities and decision making. Is it not too revolutionary to predict that there may be specialist nurse run ITUs with doctors providing a consultative role?

ITU real estate now—the majority of ITUs are not made for the year 2010 onwards. Considerable investment will need to be found to either build newer units or refurbish current ones. Given the enormous expense, current units will soldier on in their present state for many years to come.

Amalgamation of a hospitals fragmented ITU/HDU services—changes in treatment of patients and newer technology and patient–doctor expectations have led to a differential development of services scattered throughout the hospital.

Currently a general hospital will have an A&E unit and an attached emergency medical/surgical admission unit, a set of respiratory beds for non-invasive ventilation, a coronary care unit and possibly a haematology unit.

With an aging patient population and increasing numbers of operations that require a minimum of 24 h of HDU monitoring—hip and knee replacements, laparotomies—an area needs to be set aside for their management.

I predict that by 2020 some will have seen the light and allocation of resources and space for managing all the above patients will be the new ITU/HDU model.

This model would necessitate a larger style ward for patients who are expected to stay 24 h to 3 days before discharge to a normal ward, but who need a more constant care that an HDU or post-op step-down unit supplies.

Currently the joint replacement centre at our hospital has an early fore-runner of this idea. The ward provides 24 h post-op fluid, pain and vital sign management and coupled into the recovery plan is early mobilization.

These types of ward would have specialist areas for various treatments.

Current DGHs have small CCUs and respiratory beds that have non-invasive ventilation facilities. This model gives a fragmentation of care delivery to sick patients and may require transfer from one section to another.

In the future these disparate needs will be brought into the specialized area where a respiratory physician would have his set of beds and likewise the cardiologist. Specialist nurses will come more into their own providing support and services in the renal, respiratory, diabetic areas, to name a few.

In most units there will be a seamless integration between the acutely ill patient arriving in the A&E and being treated al la Emanuel Rivers in the golden 6 h

Early goal-directed therapy in the treatment of severe sepsis and septic shock, Emanuel Rivers, M.D., M.P.H., Bryant Nguyen, M.D., Suzanne Havstad, M.A., Julie Ressler, B.S., Alexandria Muzzin, B.S., Bernhard Knoblich, M.D., Edward Peterson, Ph.D., and Michael Tomlanovich, M.D. for the Early Goal-Directed Therapy Collaborative Group. N Engl J Med 2001; 345:1368–1377 with transfer to the ITU during treatment for continued monitoring and team specialist treatment.

The Major Regional Hospitals

These hospitals are the "disease palaces" where the more complex patient can be cared for. They will be the designated polytrauma units with all the supporting specialities of neurosurgical, trauma, paediatric, vascular and cardiothoracic surgeons. The ITU/HDU set-up will out of necessity be compartmentalized. This will be a weakness in comparison to the DGH working plan of specialist team-working. However, that will be offset by the necessary subspecialization needed to achieve best results.

Data and Information Technology

Many billions of pounds have been spent trying to develop an electronic record that would allow doctors anywhere in the country access patients' medical records. Currently the whole system is treading water with a distinct possibility of failing completely. The only real entity from that has emerged from the morass is the PACS (Picture Archiving and Communication system).

What the future ITU will have is a variation for each doctor of the iPad device and a Google search engine. The physician can access the patient status anywhere in the world, anytime and in real time. (Sure the security of information and encryption, etc. will have to be sorted out.) But the intensivist will have on the rounds all current results, drug therapy checked for correct dosing and timing, alerts to events, echocardiogram, MRI,CT, ultrasound imagery and specialist reports. The integrated PADD (Personal Access Display Device) video-telephone will allow the patient's general practitioner to participate by supplying patient information as well as planning for discharge back home with support from community nursing teams. As information flows bi-directionally, community support teams can notify doctors of changes and request assistance.

The great opportunity raised by this vast database is to be able to participate in clinical trials and treatment assessments globally. Data collection becomes easier and more focussed.

The future is digital—the next generation intensivists and nurses will be carrying "mobile platforms" with them. The eMedic (c.f. iPad/PACS/anything) will allow instant access to all patient records, ongoing treatment and other consultations. The patient can in some instances be reviewed by a consultant elsewhere in the world. Information will be fed back from renal haemofiltration machines, ECMOs, ventricular support devices and a myriad of electronic support machines and diagnostic vehicles. Multi-disciplinary discussions can occur at the bedside by the attending physicians and those elsewhere on the planet. This will free up the burden of delayed consultations because of distance to be travelled and remoteness of specialist centres. Note that although digital access will be universal, the human element of clinical diagnosis and expertise will be at the bedside.

The digitally-enhanced ITU will allow the major hospitals to extend their bed base out to the district hospitals—a hub and spoke model. As a scenario, a head injured patient who is transferred for surgical removal of a subdural haematoma and intracranial monitoring can when stable be transferred back to the originating hospital. Further treatment and recovery progress can be followed up by the neurocritical care. This expertise will allow closer monitoring, management and improved outcomes.

If one has ever seen the Star Trek series where the Borg are encountered, some idea of what will happen becomes apparent. For those non-aficionados, the Borg are all linked to each other digitally—each individual can access all others at any time. Transfer of information is rapid and universal. As each individual encounters a problem, their method of solving it or searching for a solution from the collective knowledge allows selection of best approach.

Already some ITUs are either contemplating or using crude devices to achieve this concept. Acceptability and achievability and affordability have only just arrived. Wireless technology is now at a stage where it is reliable, fast and cheap and easily deployable. Any hospital can set up a wireless link anywhere in the world. Apple's iPAD and iPhone have demonstrated how a device can be used portably and to perform operations involving software and digital images via the Internet.

We are currently exploring a simple, cheap and user friendly method of ITU ward rounds. The device has to merely access the hospital Internet to avail the user of laboratory results, digital images and their reports, templates to admit, treat, manage and discharge a patient. It has to be reasonably rugged and light and individual training has to be intuitive and rapid. Today's human wanders around texting, surfing the Internet on mobile phones and has in most cases little training on the use of, or even bothers to read the manual. Software comes off the shelf. And it works! The next generation of doctors will have the patient and world store of information in their pockets wherever they go.

An Alternative Reality

Politics, money and the NHS get together and decide that best treatment is centralized, in large units and under the aegis of university centres.

Possible—the idea is in use in other specialities. Paediatric critical care has already headed down that path with district hospitals providing a way station role in handling the acutely ill child. Paediatric retrieval services work by serving a catchment area with telephonic advice, templates for a standardized management of the child, help in finding a bed if none is available and providing a trained set of teams.

http://image.thelancet.com/journals/lancet/article/PIIS0140-6736%2810%2961113-0/fulltext

Other specialities such as neurosurgical referrals provide telephonic help, report on digitally transmitted scan images and on acceptance of a patient, the referring hospital dispatches a doctor and nurse team.

Reality check—no one has brought in the accountants to cost an alternative that would need to build large units, staff them and work out methods of transfer and repatriation of patients. In all probability the expenditure involved would prove unsustainable.

So what remains is the current structure of district hospitals providing a general ITU service for adult patients: Large central hub hospitals providing the polytrauma department and all the speciality critical care needs; Plus accepting transfers from the periphery. The infrastructure is present, functions well and will not cost large sums of money to refurbish in the future. In the current economic climate there are no finances available to rebuild hospitals and intensive care units in particular. This state of affairs will last for several years with some hospitals being downsized or even closed. The surviving ITUs will have to struggle to maintain sufficient beds.

But this model of super-ITUs falls apart with the above-mentioned technological advances of mobile data access of high quality.

Disasters and Responses

SARS, Avian flu, pandemic flu and situations with mass casualties have shown that current infrastructure will not cope. Planning for the theorized H1N1 pandemic of winter 2008/2009 involved hospitals drawing up plans for triage, staff management, supply of additional beds to ventilate patients and attempts to provide emergency services simultaneously left no doubt anyone's mind that the intensive care system would fail miserably. The future looks bleak for the response to such happenings. ITU services of the future will have established plans to survive in these acute circumstances based on a national response.

Summary

The future of UK ITU services will be highly dependant on who has the money and how it is allocated. Most ITUs will in 10 years time need refurbishment or complete rebuilds to remain viable. For DGHs the current hub and spoke model will remain to feed the hungry university conglomerates referring and repatriating of patients back.

The brighter side of ITU is in the digital revolution that will allow patient data and specialist access to be instantaneously available at the bedside. What was thought to be expensive and unobtainable is now available and ready for all. Patient care will be much improved with a rationalization of care pathways.

Patient care will go global with physicians being able to talk to a world community of expertise and all from a small handheld personal eMedic.

Disease management—progressively standardized methods of caring for illnesses will permeate the ITUs. The National Institute for Health and Clinical Excellence (NICE) (http://www.nice.org.uk) publishes and provides clinical evidence for treating patients. This form of unified patient management is now a standard part of medical life and its influence will strengthen in the future. Even today the NICE method drags clinicians, nurses, hospitals and managers into conformity. The advantage will be that ITU researchers will be able to access larger numbers of patients undergoing standardized care and use the information to change for the better.

Extra

After all non-invasive respiratory support is already part of ITU/HDU management, why not bring everyone under the same wing.

I predict that there will be an amalgamation of fragmented HDU type services. For instance the advent of non-invasive ventilation allowed the respiratory physician to manage patients on the ward instead of admitting to an ITU.

Chapter 24
United States—Private Practice: Where Are We Going?

John W. Hoyt

Introduction

There are two competing models in the United States of America (USA) for delivering critical care services in private practice to intensive care unit patients with life-threatening illnesses. First is the model of consultation practiced by Pulmonary/Critical Care. Patients are admitted to the ICU in the private practice environment by a primary care physician or surgeon and various services are consulted to manage various organ systems in failure. Normally pulmonology is asked to manage respiratory failure and will take care of mechanical ventilation and other respiratory issues. They usually do not manage other organ systems so cardiology takes care of heart problems and nephrology takes care of kidney problems, etc. Under most private practice situations the pulmonologist does not take administrative responsibility for the ICU and does not see all patients in the ICU. The consultation on the ICU patient is only part of the pulmonologist's day since they have many other clinical responsibilities including consults outside the ICU, office practice, sleep lab, bronchoscopy suite, pulmonary function lab, etc. This model represents 85–90% of private practice situations where the intensivist model is uncommon. The private practice intensivist model places physicians who are fellowship trained in critical care in the ICU as their full clinical responsibility without other clinical obligations outside the ICU. The intensivist usually works 7 days in a row to preserve continuity of care. A team of intensivists cover 16 hours/day with a daytime intensivist and an evening intensivist and take night time pager call for the remaining 8 hours/day so that there is a ready response from an intensivist 24 hours/day. In very busy ICU environments with a high level of severity of illness, there is a dedicated intensivist in the ICU 24 hours/day.

J.W. Hoyt, M.D. (✉)
Department of Critical Care, Pittsburgh Critical Care Associates, Inc.,
Critical Care Medicine, University of Pittsburgh School of Medicine, Pittsburgh, PA, USA
e-mail: HoytJ@pccaintensivist.com

D.W. Crippen (ed.), *ICU Resource Allocation in the New Millennium: Will We Say "No"?*,   185
DOI 10.1007/978-1-4614-3866-3_24, © Springer Science+Business Media New York 2013

In the private practice model of healthcare delivery it is reasonable to predict that the intensivist model will prevail and the consultation model with Pulmonary/ Critical Care managing respiratory failure and mechanical ventilation will disappear over the next ten years. Emergency Medicine went through a similar transition in the 1970s. In the old emergency rooms there were nurses but no dedicated physicians. When a new patient arrived in the ER the nurse would call the patient's private practice physician and get consults to cover needed services. In the span of ten years, hospitals went from the fragmented care of emergency room consults to a dedicated staff in the Emergency Department. There was much medical staff turmoil during the transition from emergency room to emergency department as physicians resisted this change because of economic reasons. Many members of the medical staff were making substantial parts of their annual income by doing consults in the Emergency Room. The same issues exist in private practice critical care where many physicians, particularly pulmonologists, are very invested in the ICU consultant model. The changes in emergency medicine in the 1970s were driven by improved quality of care, better and faster service to patients, less fragmented care by the involvement of multiple physicians, and improved outcomes.

In a 2005 article in Healthcare Executive written by Peter Pronovost M.D., Ph.D. he notes that 30% of acute care cost come from the intensive care unit approximating $180 billion dollars a year. In 2011 that number is even larger and has probably exceeded $200 billion annually. This is 30% of acute care costs and is the largest and most expensive single element of healthcare in the USA. In 2005 over five million patients had intensive care unit stays. With the approaching surge in patients over 65 because of the aging of the baby boomers, the number of ICU patients and the cost of ICU services will be astronomical. Patients over the age of 65 have a statistically much higher use of ICU services. In the critical care consultation model the ICU length of stay is high, time on mechanical ventilation is long, and the cost of ICU services is high. There are more infections, more complications, and higher hospital mortality rates. The fragmented care of the private practice consultation model in the ICU produces poor results in many studies of the delivery of ICU services. Dr. Pronovost has a landmark paper in the Journal of the American Medical Association where he showed statistically poorer outcomes after elective abdominal aortic surgery in Maryland hospitals without intensivists. What we saw with the consultation model in private practice emergency room consultative care in the 1970s is now true in the intensive care unit consultative model but it is now more expensive.

Cost of care will be a primary driver for the conversion of the ICU consultative model to the intensivist model. A second driver will be quality of care. In the 1990s the Institute of Medicine (IOM) report "To Err Is Human" published information claiming over 100,000 unnecessary deaths annually in hospitals because of poor quality of care. A November 2010 article in the New England Journal did a 10 year followup on the IOM report and found no improvement in medical errors which lead to poor quality and outcomes. In fact this information is not surprising. The largest number of deaths in hospitals occurs in the ICU. There has been little progress in the private practice environment in replacing the ICU consultative model with the ICU intensivist model. Only by having full time intensivists who see and manage all ICU patients and take full responsibility for the day-to-day operation of the ICU will there be an improvement in quality of care and outcomes.

Other first world countries have had dedicated intensivists for years. In fact they never went down the path of the fragmented consultative model that evolved in the USA. Patients in Canada, most of Europe, Australia, and New Zealand benefit from the presence of a dedicated intensivist who sees all ICU patients and writes all orders on all patients and is responsible for ICU outcomes. These intensivists regulate ICU admissions and discharges. They use consultants for specialty situations such as a cardiac cath. The minute-to-minute decisions and treatment course corrections are made by the intensivists. There is much less futile care and patients with an opportunity for recovery get aggressively treated in an intensivist coordinated model. It is just inconceivable that the USA could continue with its current private practice model of ICU services. It is too expensive with mediocre outcomes.

There are barriers to the evolution of the private practice intensivist model. Right now there is a meager number of fully trained intensivists with good life support skills (intubation, mechanical ventilation, insertion of arterial, central venous, dialysis, and pulmonary artery catheters, insertion of chest tubes, lumbar puncture, etc.) coming out of fellowship programs. There are more intensivist positions than there are available intensivists. There will have to be substantial changes in intensivist fellowship programs to keep up with growing demand for intensivists if the transition from ICU consultative care to intensivist care is to happen. Organizations like the Society of Critical Care Medicine need to define and model intensivist practice and work with Medicare and insurance companies to fund intensivist services. Emergency Medicine had all the same challenges in the 1970s and rose to the challenge to create the private practice model of Emergency Medicine. Intensivists have the ability of controlling health care costs for 30% of the acute care market. There will be a significant return on investment in the area of improved quality of care if the parties that fund health care rise to the occasion and embrace the intensivist model in private practice medicine in the USA.

Chapter 25
United States—Academic Medicine: Where Are We Going?

Mark Mazer

Critical care providers in the United States are frustrated by their participation in the most costly medical delivery system in the world, nonetheless plagued by a dismal return on investment in terms of value. The current pace of increasing cost, without budgetary restraint and without adequate guarantee of quality, is a recipe for economic disaster. However, given proper direction and stewardship, the American medical system is capable of providing high value, cost effective healthcare.

Healthcare reform does not occur at a discrete moment in time; rather healthcare is constantly being remodeled, transformed and reorganized. The Patient Protection and Affordable Care Act, as amended by the Health Care and Education Reconciliation Act, March 23, 2010 (PPACA), is a distinct point on the timeline of healthcare reform. Whether one agrees or not with the provisions of the PPACA, at least we can now move beyond merely articulating the problems and have a contemporary tool to start the active process of transforming healthcare delivery. The tool can be refit and reforged as we move forward.

To understand why reform is necessary, one must examine the value equation, Value = Quality/Cost. Both in an absolute sense and adjusted for gross domestic product, the USA spends more money per capita on healthcare than any other member of the Organization for Economic Co-operation and Development (OECD). Unfortunately this massive per capita expenditure on healthcare does not translate into the delivery of high quality healthcare for the population at large or for individuals. The USA has higher cost per hospital discharge, more pharmaceutical expense per capita, and spends more on medical technological capacity per capita by comparison with other OECD countries [1]. Despite this prolific spending, some 17% of the population has no healthcare insurance, and 50% have inadequate insurance [2]. This is partly due to the unbridled rise of health insurance premiums,

M. Mazer, M.D. (✉)
Department of Pulmonary, Critical Care and Sleep Medicine,
Pitt County Memorial Hospital, East Carolina University,
Greenville, NC, USA
e-mail: mmazer@suddenlink.net

D.W. Crippen (ed.), *ICU Resource Allocation in the New Millennium: Will We Say "No"?*, 189
DOI 10.1007/978-1-4614-3866-3_25, © Springer Science+Business Media New York 2013

which have increased 114% in the last decade [3]. Those lacking adequate insurance are 25% more likely to die prematurely [4], which translates into about 45,000 avertable deaths per year in the USA [5]. On an equally somber note, Americans enjoy among the lowest life expectancies at birth and the poorest outcomes in terms of preventable mortality among the OECD members [1]. Thus, healthcare providers and consumers currently function in a system which has little to no fiscal restraint and is woefully lacking in benefit defined as quality healthcare in return on investment.

Central to the disequilibrium of the value equation is a virtually unfettered fee-for-service reimbursement system heavily reliant on the service of specialists. This formula is the basis of spiraling costs and less than optimal outcome benefit [6]. Academic critical care physicians are specialists par excellence, and understanding the future landscape in the context of the new evolving paradigm of healthcare delivery is crucial for the economic viability of academic critical care specialists.

To stimulate healthcare reform, the PPACA delineates three broad areas, which must be orchestrated simultaneously: access to care, quality of care, and cost of care. These generic categories will serve as the framework for the discussion of how healthcare reform may affect the practice of critical care medicine in American Academic Healthcare Centers (AAHC), particularly in American Academic Intensive Care Units (AAICU).

Whether through individual mandates, health insurance exchanges, expansion of state health insurance programs, subsidies for low wage earners, or tax incentives for employers a core goal is to increase access to healthcare. It is hoped that the current large population of uninsured and underinsured Americans will be covered by an essential healthcare benefits package, and will receive preventable services. The evolving philosophical shift is to maintain health, and deemphasize sick care. In any event, we can expect a large expansion of more insured patients actively demanding and competing for medical resources. Unless we proactively plan, the demand for critical care beds and resources may very well spiral out of control.

At present, many patients allow their condition to deteriorate to a critical point before presenting for care. Even with an emphasis on well care, there will always be an inevitable personal decision concerning the propriety of critical care. It is difficult to predict how this will affect future demand for critical care beds. However, an optimistic assumption is that there will be a significant effect on the demographics of the ICU population, with less patients presenting with preventable complications of diseases such as diabetic ketoacidosis and hypertensive intracranial hemorrhage. A key component of well care is end-of-life planning. As a nation and people we have opportunity to evolve in this regard, to overcome pervasive angst concerning end-of-life issues, and responsibly plan for it. Hopefully the renewed emphasis on preventive care will help erode the aversion to dealing with matters relating to death, and mitigate the catastrophic political and religious polemic with attempts to equate euthanasia to orthothanasia. AAHCs will play a leading role in planning for end-of-life care by expanding the training of palliative care providers, encouraging research into the delivery of palliative care, and amassing the multidisciplinary platforms needed to research the social engineering needed to help Americans mature in this regard.

The second essential pillar of healthcare reform is emphasis and focus on quality improvement. It is imperative to define the characteristics of quality and develop appropriate metrics and reporting methodologies as they pertain to critical care medicine. A considerable gap often exists between the acquisition of knowledge from research, and the implementation of this knowledge into routine medical practice. As few as 55% of patients benefit from recommended medical care, and it takes as long as 17 years for as little as 14% of new knowledge to become ensconced into routine clinical practice [7–9]. AAICUs must take the lead in comparative effectiveness research and translational research. The direction of critical care scientific research will undoubtedly shift towards analyzing the relative effectiveness of interventions such as the various modes of mechanical ventilation, preload assessment and renal replacement therapies and how to effectively implement these interventions at the bedside. The general philosophical shift towards preventative medicine will also intensify the demand for interventions and techniques to assure patient safety and lessen the occurrence of preventable complications of care in the ICU. There will be little tolerance for avertable complications of critical care procedures and treatment processes such as ventilator associated pneumonia (VAP), central line associated blood stream infections (BSI) and catheter-associated urinary tract infections (CAUTI). These reportable conditions will severely impact quality scores, and will also adversely affect reimbursement. AAHCs will redouble research efforts towards what might be phrased as, "preventative ICU medicine." Integral to this research will be an increased focus on safety, with the goal of not only reducing errors of commission, but also the much more pervasive problem of errors of omission. For example, despite evidence suggesting that mechanical ventilation strategies using lower tidal volumes have beneficial effects not only for patients with acute lung injury (ALI) and the acute respiratory distress syndrome (ARDS), but as well as for patients without underlying lung disease, evidence also exists that use of lower tidal volume lung protective ventilation is underutilized in routine critical care practice [10–15]. Failure to adopt such evidence-based concepts into routine practice in AAICUs will not be easy to excuse in the very near future.

Undoubtedly AAICUs will be held to the task of helping to define which critical care interventions are most cost effective. Interventions which are of marginal benefit or do not impact outcome will come under increasing scrutiny. More studies will be done comparing the actual and projected cost and financial implications of interventions. This will require refocus and redefinition of acceptable outcomes in terms of length of survival and quality of life. Myopic, narrow end points such as 28-day survival or simply surviving the ICU stay, without more temporal expanse and quality of life analysis will no longer serve to underwrite the use of expensive therapies such as activated protein c in sepsis [16]. The near future will also witness a paradigm shift in how critical care research is funded and coordinated. It is paramount for AAICUs to form robust interinstitutional alliances in order to define critical research direction, increase the chances for funding, increase the enrollment of subjects and resultant power of studies in the ICU, and reduce the number of redundant studies. The US Critical Illness and Injury Trials Group (USCIIT) has already made strides in this regard and serves as a national liaison between investigators from different institutions [17].

The third pillar of healthcare reform is cost containment. Purchasers and consumers of healthcare will demand higher value critical care services which means not only improving quality but also reducing costs. This will require a critical reappraisal of how intensive care services are delivered and financed. New models for the efficient delivery of high value critical care will need to be developed. The current paradigm of unregulated fee for service reimbursement bereft of quality accountability will be curtailed, and replaced with new models of reimbursement for critical care professional services [18]. At the core of evolving plans is the principle of value-based payment. Reimbursement for intensivists will be linked to adherence to guidelines and quality outcomes. Variations in practice which fall outside of guidelines and common practice, with consequent suboptimal outcomes and avoidable complications such as BSIs, will carry financial penalty.

Pay for performance incentive programs are not new, and current iterations will certainly focus on enhanced ways to incentivize changes of clinical behavior which result in better outcomes at less cost [18, 19]. Academic intensivists providers may find themselves integrated into an Accountable Care Organization (ACO) [20]. In this model, a group of providers agrees to offer overall coordinated quality care for a defined cadre of patients, in return for reimbursement based on meeting defined quality standards. Conceivably intensivists may still receive fee for service remuneration, but with bonuses or withholds based on cost savings and quality metrics. Integral to the ACO are processes to promote the use of evidence-based medicine, and AAICUs should be well poised to develop guidelines and methodologies to enhance the use of guidelines at the bedside. AAHCs have large pools of multidisciplinary providers often working in hospitals and centers allied with the primary academic institution. Given competent administrative management with input from critical care physician leaders, AAHCs and AAICUs may hold a competitive advantage over other less loosely knit groups of providers and may be able to coordinate systems of care able to flourish in the ACO model. The academic intensivists will need to develop internal systems of quality and performance accountability as the AAHC affiliated with the ACO will have little tolerance for outcomes or behaviors which jeopardize the financial stability of the institution.

Another payment model gaining traction is the episode-of-care or bundled payment plan [18, 20]. In this model the critical care physicians will be part of a group of physicians willing to equitably share a predetermined global fee for discrete episodes of care spanning a fixed period of time. For example, a patient hospitalized for a community-acquired pneumonia may spend time in the emergency room, intensive care unit and then on the general medical ward with a follow-up visit with his primary care physician after discharge. The physicians will divide the fixed reimbursement in a predetermined fashion. Elective surgical procedures such as hip replacement may carry a fixed price, and methods of compensating for critical care services will need to be considered in the event of complications. AAHCs and AAICUs will have the opportunity to focus research efforts on how to optimize and efficiently coordinate care of patients in these evolving models of healthcare delivery and reimbursement.

AAICUs will lead the way in developing methodologies for benching marking performance and quality of care based on volume, individual severity of illness and the population at risk. This will be important not only in terms of improving outcomes, but also for financial viability. Critical illness carries risk not only because of the underlying disease but also because of issues related to the process of care. A major challenge facing AAICUs centers on the plethora of providers and learners at every level. It will be crucial to standardize evidence-based practices in all domains and to monitor compliance in a more arduous manner than is current practice. Enhancing compliance with standardized protocols for invasive procedures such as central line placement and site care, as well as treatment bundles for disease states such as sepsis is a major initiative for AAICUs. Use of robust functional safety checklists addressing common ICU procedures, and for use during patient care rounds and handovers of care will be developed and implemented. Developing the elements of these protocols and bundles, ensuring implementation at the bedside and demonstrating beneficial effect on outcomes and cost are fertile areas for academic creativity and research.

The Accreditation Council for Graduate Medical Education has taken steps to significantly decrease residents' duty hours [21]. This has had the effect of creating a multiplicity of hand offs of patient care, with an increased risk of error as a result of gaps in continuity of care. AAICUs will have to adapt to this perilous circumstance with measured development of a team approach to patient care. This will necessitate a cogent and effective means of guaranteeing safety and continuity by the use of effective and comprehensive tools during handovers of care.

In effect healthcare reform will dramatically increase the pool of insured patients. Unless coordinated properly, this threatens to further strain and perhaps even cause the collapse of the current healthcare delivery infrastructure as demand for service rises faster than resources can be marshaled. Implicit in controlling the impending *tsunami* of demand is the concept of triage. Conceptually, increased access to primary care would allow for enhanced end-of-life planning and use of palliative care services. This could serve to mitigate unreasonable use of intensive unit beds. However, given the present political climate, where responsible end-of-life planning evokes accusations of rationing and death panels, there is little hope that such measures will prove fruitful in the foreseeable future. However, healthcare reform with emphasis on value based purchasing and payment for deliverables may eventually have the salutary effect of curtailing funding of unrealistic demands for resources on the part of patients and or their families.

Were healthcare reform to result in the regrettable situation where demand for intensive care beds and resources routinely outweighs supply, then strong consideration must be given to applying the utilitarian principles of general public health ethics. In effect this would mimic an ongoing, chronic state of public health emergency. As such, consideration of individual autonomy and requests for unreasonable care and resource consumption would be considered secondary and even antithetical to the common good. Some have suggested integrating use of scoring systems such as the Sequential Organ Failure Assessment (SOFA) into decision-making concerning allocation of intensive care resources use during influenza epidemics. Unfortunately decisions integrating the SOFA score have not proven successful in accurately prioritizing patients in this circumstance [22]. To meet the challenge of these situations, AAICUs must be on the forefront of efforts to develop equitable, accurate, and transparent means of triaging intensive care resources. Hopefully the state of affairs will equilibrate long before a relative penury of intensive care resources devolves into a perpetual state of public health emergency.

At present triage for ICU admission and discharge is heavily dependent on pathophysiology, the need for specialized procedures and aggressive monitoring. While outcome prediction is important, current prognostic models such as Acute Physiology and Chronic Health Evaluation (APACHE) and SOFA are population-based and are not applicable to the individual patient. A major practical focus of clinical and research activity in the AAICU will be on developing acceptably precise outcome prediction tools applicable for triage on an individual basis. Generic risk assessment based on genetic predisposition, and proteonomic codification of illness severity based on cytokine and biomarkers levels, while of immense interest, are not sufficient to forecast outcome with reasonable certainty. Studies of the RNA transcriptome during critical illness and the development of immune cartography as a system of blood immunomonitoring is of great potential interest in this regard. There is hope that soon we will be able to monitor individual patient's molecular vital signs, and track the immunoinflammatory response to disease and therapy [23]. Hopefully the ability to adjudicate demands for intensive care resources in questionable and difficult circumstances will be bolstered by rapid, specific, individual gene expression signatures highly associated with death or survival. AAICUs will focus on melding physiologic, genomic, proteonomic and transcriptomic data so that triage decisions based can be rapidly individualized with a high degree of accuracy in terms of outcomes.

Ultimately AAICUs will be accountable to develop, implement, monitor, and report the results of these initiatives and results as reimbursement will depend heavily on predefined metrics of efficiency, quality, and outcomes. Variations of clinical behavior which negatively affect these metrics will demand accountability to the academic institution's administration and physician practice plan. In the not too distant era of bundled payments and reimbursement based on adherence to quality standards there will be increased accountability to one's colleagues. There will be little willingness to tolerate deviation from evidence based medicine and quality guidelines as remuneration will be adversely affected. For example, failure to wash one's hands may carry a significant financial disincentive which will not be countenanced. In regard to unrealistic demands for care, acquiescence will dissipate as fee for service reimbursement is limited and payments are linked to performance and outcomes. Sadly, monetary and financial concerns will eventually succeed in positively effecting behaviors such as hand washing and reasoned ethical decision-making at the bedside where scientific evidence, common sense, and logic have thus far failed.

In summary, AAICUs will be on the forefront of evolving innovative paradigms to deliver state-of-the-art critical care to patients in the new millennium. But what will the state of the art be? In the context of this discussion, the critical assumption is that demand for medical resources and service will increase as the population expands yet resources lag in comparison. Framed within the myopic viewpoint of the current state of the art of critical care medicine and the health care delivery system, this seems to presage extremely unsettling prospects. However, future advances in critical care medicine will eventually allow more rapid and precise diagnostic modalities and prognostication, in parallel with more effective therapeutic options. The combined force of social, legislative, and political pressures coupled with future advances in the science of critical care medicine will eventually cause the bedside delivery of quality critical care to be more cost effective and efficient, and in due course supply and demand will reequilibrate. Healthcare reform notwithstanding, the sanctity of the patient-critical care physician relationship will ultimately be preserved, because fundamentally no one really wishes to see this bond dissolve.

References

1. http://www.oecd.org/home/. Accessed 30 Aug 2011.
2. US Census Bureau. Income, poverty, and health insurance coverage in the United States: 2009. US Census Bureau. 2010. http://www.census.gov/prod/2010pubs/p60-238.pdf. Accessed 30 Aug 2011.
3. Kaiser Family Foundation. Employer health benefits 2010 annual survey. Kaiser Family Foundation, Health research and Educational Trust, National Opinion Research Center. 2010. http://ehbs.kff.org/pdf/2010/8085.pdf. Accessed 30 Aug 2011.
4. Franks P, Clancey CM, Gold MR. Health insurance and mortality. Evidence from a national cohort. JAMA. 1993;270:737–41.
5. Wilper AP, Woolhandler S, Lasser KE, McCormick D, Bor DH, Himmelstein DU. Health insurance and mortality in US adults. Am J Public Health. 2009;99:1–7.
6. Meunnig PA, Glied SA. What changes in survival rates tell us about US health care. Health Aff. 2010;29(11):2105–13.
7. McGlynn EA, Asch SM, Adams J, Keesey J, DeCristofaro A, Kerr EA. The quality of health care delivered to adults in the United States. N Engl J Med. 2003;348:2635–45.
8. Woolf SG. The meaning of translational research and why it matters. JAMA. 2008;299(2): 211–3.
9. Westfall JM, Mold J, Fagan L. Practice-based research: "blue highway" on the NIH roadmap. JAMA. 2007;297(4):403–6.
10. Brower RG, Matthay MA, Morris A, Schoenfeld D, Thompson BT, Wheeler A. The acute respiratory distress syndrome network. Ventilation with lower tidal volumes as compared with traditional tidal volumes for acute lung injury and the acute respiratory distress syndrome. N Engl J Med. 2000;342(18):1301–8.
11. Gajic O, Dara SI, Mendez JL, Adesanya AO, Festic E, Caples SM, Rana R, St. Sauver JL, Lymp JF, Afessa B, Hubmayr RD. Ventilator-associated lung injury in patients without acute lung injury at the onset of mechanical ventilation. Crit Care Med. 2004;32(9):1817–24.
12. Sakr Y, Vincent JL, Reinhart K, Groeneveld J, Michalopoulos A, Sprung CL, Artigas A, Ranieri M. The Sepsis Occurrence in Acutely Ill Patients Investigators. High tidal volume and positive fluid balance are associated with worse outcome in acute lung injury. Chest. 2005;128:3098–108.
13. Kallet RH, Jasmer RM, Pittet J-F, Tang JF, Campbell AR, Dicker R, Hemphill C, Luce JM. Clinical implementation of the ARDS network protocol is associated with reduced hospital mortality compared with historical controls. Crit Care Med. 2005;33(5):925–9.

14. Weinert CR, Gross CR, Marinelli WA. Impact of randomized trial results on acute lung injury ventilator therapy in teaching hospitals. Am J Respir Crit Care Med. 2003;167:1304–9.
15. Young MP, Manning HL, Wilson DL, Mette SA, Riker RR, Leiter JC, Liu KS, Bates JT, Parsons PE. Ventilation of patients with acute lung injury and acute respiratory distress syndrome: has new evidence changed clinical practice? Crit Care Med. 2004;32(6):1260–5.
16. Bernard GR, Vincent J-L, Laterre P-F, LaRosa SP, et al. Efficacy and safety of recombinant human activated protein c for severe sepsis. N Engl J Med. 2001;344:699–709.
17. http://www.massgeneral.org/research/researchlab.aspx?id=1262\. Accessed 2 Sept 2011.
18. Rosenthal MB. Beyond pay for performance – emerging models of provider-payment reform. N Engl J Med. 2008;359:1197–200.
19. Baker G, Haughton J, Mongroo P. Pay for performance incentive programs in healthcare: market dynamics and business process. 2003. http://www.leapfroggroup.org/media/file/Leapfrog-Pay_for_Performance_Briefing.pdf. Accessed 2 Sept 2011.
20. Shomaker S. Commentary: health care payment reform and academic medicine: threat or opportunity? Acad Med. 2010;85:756–8.
21. http://www.acgme.org/acwebsite/dutyhours/dh_index.asp. Accessed 2 Sept 2011.
22. Khan Z, Hulme J, Sherwood N. An assessment of the validity of SOFA score based triage in H1N1 critically ill patients during an influenza pandemic. Anaesthesia. 2009;64:1283–8.
23. Polpitiya AD, McDunn J, Burykin A, Ghosh BK, Cobb JP. Using systems biology to simplify complex disease: immune cartography. Crit Care Med. 2009;37(1) Suppl:S16–21.

Chapter 26
Reflections on the Demand for Critical Care in the Global Medical Village

Timothy G. Buchman and Donald W. Chalfin

Beliefs and expectations surrounding critical care form a global mosaic. The contemporary reports collected in Part 1 of this book point to cultural foundations, individual and group expectations, and political and economic constructs as the primary drivers of the critical care allocation and delivery systems. Understanding this interplay globally illuminates the options available to transform critical care delivery locally.

Individual Right vs. Collective Good

The most striking contrasts among the critical care delivery systems described in Part 1 lie in cultural perceptions of health care and the willingness of populations to accede to public standards and consider, if not accept, potential limits. It is apparent that "universal health care" and "socialized medicine" are inadequate descriptors of comprehensive delivery systems and the forces that underlie allocation of critical care services also represent terms that are reductionist at best and politically charged at worst. Compare and contrast Israel and New Zealand. Both countries support universal access to health care funded by taxation and government distribution of services. Both countries have a heterogeneous population. Yet Israel, or at least her inhabitants, support critical care as an individual right that cannot be refused regardless of efficacy or resource availability.

T.G. Buchman, Ph.D., M.D. (✉)
Department of Surgery, Emory University School of Medicine,
Emory Center for Critical Care, Emory Healthcare, Atlanta, GA, USA
e-mail: tbuchma@emory.edu

D.W. Chalfin, M.D., M.S., F.C.C.M.
Ibis Biosciences, Carlsbad, CA, USA
e-mail: dchalfin@applied-decision.com

D.W. Crippen (ed.), *ICU Resource Allocation in the New Millennium: Will We Say "No"?*,   197
DOI 10.1007/978-1-4614-3866-3_26, © Springer Science+Business Media New York 2013

As a consequence, demand for critical care services is often overwhelming and frequently outpaces supply. Care is provided as best as possible in any available location, and there often appears to be little relationship between the care delivered, its likelihood for meaningful benefit, and the inevitable clinical effectiveness likely to be attained. As an extreme example, consider Israel's celebrated general Ariel Sharon who sustained a devastating stroke more than 6 years ago that left him in a persistent vegetative state. He continues to receive hospital-based care without expectation of meaningful recovery.[1] The ethical principle of autonomy trumps beneficence, nonmaleficence, and the notion of distributive justice.

In New Zealand, critical care is viewed as a public good. Control over the allocation of critical care services is ceded to critical care units and their physicians, along with the responsibility of managing that allocation within the national budget. These privileges and responsibilities are enshrined in law, and the courts consistently uphold the authorities of the critical care units in making decisions about critical care delivery, to whom and for how long. As a consequence, previously healthy patients with acute illnesses and a high likelihood of reversibility and return to normal function have priority, while those with significantly reduced chances of benefit with either be denied access to critical care or else offered a brief "therapeutic trial" and hence an opportunity to demonstrate physiologic resilience under the constraints of an "indicative contract." (Indicative contracts specify time and resource limits beyond which critical care with curative intent will no longer be pursued.)

The contrasts are evident. New Zealand manages critical care with fewer beds and per capita resource consumption than Israel, yet there appears to be no apparent dissatisfaction with the critical care services provided. The cultural foundation in each country simply supports different expectations and a very different social contract that does or does not contain the individual expectations of highly technical care for any and all.

The Ambiguous Role of the Physician

A second profound contrast exists in physicians' perceptions of their roles.

Before exploring this contrast, it is worth remembering Jerome Frank's famous observation that almost the entire history of medicine is the history of the placebo effect.[2] As Postman and Weingartner pointed out, if the entire history of medicine is

---

[1] Sharon was briefly sent home and promptly returned to the hospital. Although his son "is sure that he hears me," he admits Sharon remains in a coma-like state, which is unchanged over the past few years. Bronner E. 6 years after stroke, Sharon still responsive, son says. The New York Times. 20 Oct 2011.

[2] Frank was a psychiatrist discussing the power of belief in healing. Many experienced critical care clinicians have observed a strong mind–body relationship, in which a patient and family seem to have an outcome. A scientific basis for this is emerging with the identification of neurological inputs to the immune system.

mapped to 60 min on a clock, the antibiotics that ushered in the modern era appeared about a minute ago.[3] Critical care appeared about 30 s ago. Ventilators and dialysis machines appeared about 20 s ago. Put simply, survival after failure of even a single organ is, from the perspective of medical history and its traditions, brand new. Such survival has led physicians and their patients into hitherto uncharted territories of expectations and responsibilities.

Time-honored exhortations to unflinchingly and unfailingly advocate for the patient regardless of cost, resource constraints, budgetary restrictions, and even risk to the physician now run headlong into the reality that the function of nearly every organ can be manipulated and improved and, if necessary, replaced (mechanically, pharmacologically, or via transplantation) indefinitely. Yet replacement of function is different from functional reconstitution. The birth and growth of long-term acute care (LTAC) hospitals in the USA and elsewhere speak to that distinction. LTACs are often "waystations" for those enervated patients who depend on life-supports for physiologic survival yet have demonstrated little capacity for functional survival and often significantly reduced objective quality of life.

Against this background, it is useful to compare and contrast the perspectives and travails of Canadian and Italian critical care physicians. Both subscribe to similar values and principles. Both have access to similar critical care technologies. Our Italian colleagues believe that larger social systems ought to determine any limitations to critical care on a population basis, and until such limitations are applied it is the responsibility of the Italian physician to forcefully advocate for the patient's access to the most advanced technologies irrespective of cost and marginal benefit. To do otherwise is seen as a violation of the covenant that binds patient with physician.

Canadian intensivists are positioned somewhat differently. In selected Canadian provinces, critical care physicians have the authority and responsibility to limit and even withdraw life-supporting services over the objections of next-of-kin when the local medical community opines that persistence or escalation of care will be ineffective. In these situations, professional judgment is weighted more heavily than patient/family preference. Different from New Zealand's social contract, many Canadians perceive that they are entitled to eventually get all the services they deem necessary provided they queue up and wait. Those Canadians who "jump the line" either by paying for private care in Canada or becoming medical tourists to the USA and elsewhere are seen by some as skirting a shared sacrifice.

With respect to critical care, the problem arises in Canada that professional judgment is never quite uniform, and families appeal to any physician who will come rescue their loved one from inevitable death as life-supports are curtailed. If the courts decline to be engaged, the conflicts at least become fodder for the press as

---

[3] The observation was offered in 1968. Although a half century has elapsed, the approximation still holds.

several recent cases have demonstrated.[4] The issue is rarely (if ever) that the patient has an illness that will quickly lead to death. Rather, the issue is whether the medical community or the patient/family unit has authority over the resources to be used to prolong the dying process.

Political and Economic Constructs

Once a nation acknowledges that the allocation of health care resources generally, and critical care resources specifically, must be subject to certain potential standards and perhaps constrained, that nation must identify a mechanism that balances political and economic realities. In impoverished countries, the mechanism is simple, albeit draconian: critical care is unaffordable and health care monies may be best expended primarily on public health projects. In developed countries, the mechanisms that balance politics and health care economics are specific to nations and peoples and hence, wide variation is noted.

When Great Britain created the National Health Service at mid-century, it was quickly apparent that the costs of the program were unsustainable. Two responses ensued. First, systems of private insurance arose that enabled faster access to care for those who chose to subscribe to the schemes. (Interestingly, public servants were and remain heavy subscribers.) Second, there gradually emerged an understanding that health care had to be rationed, and rationed transparently. A National Institute of Clinical Effectiveness (NICE) was commissioned to establish and apply a method for evaluating the cost-effectiveness or the objective and quantitative value of health care interventions.[5] The metric chosen was the QALY (Quality Adjusted Life Year) and the National Health Service determined the value of a treatment based on return on investment. At the time of this writing, treatments costing less than approximately $35,000 per QALY are covered under National Health Insurance, while those that cost more are denied. Two inevitable consequences have emerged. First, treatment providers have re-priced their treatments "inside" the cost per QALY threshold. Second, special interest groups have formed demanding exceptions in order to access more costly treatments without out-of-pocket expense.

Germany's economic success sustained rising health care costs for several decades. Utilization of hospitals generally and critical care specifically exceeds that of the USA, and cost controls were often subtle and local. For example, many German ICUs were given an annual budget and expected to manage all patients within that global figure. So long as the post-war population remained young, the expanding economy could absorb the additional costs. The economic collapses of the past decade (beginning with the burst of the millennial technology bubble and

---

[4] A Web search for "Isaiah James May" and "Sam Golubchuk" yields recent illustrations.

[5] Commissioned in 1999, the name was changed to the National Institute for Health and Clinical Excellence in 2005. The acronym ("NICE") remains.

more recently with the global recession) is forcing Germany to consider a future of tiered services. In such a scheme, every citizen would be guaranteed basic services, while those who subscribe to additional tiers can access progressively advanced services. In such a scheme, clinical effectiveness data will assume a supporting role while individual economics and preferences dictate service availability.[6]

Normative Beliefs and Expectations

At the core of the three contrasts described in the preceding sections lie sets of normative beliefs and expectations about health care generally and critical care specifically. We emphasize that those norms are not biological but rather social, cultural, political, and economic. The expectations of access to critical care emerge from a Web of beliefs about individual rights and collective responsibilities. In order to understand the peculiar case of the USA, it is necessary to explore its history in more detail.

The USA: Background

Health care in the USA is expensive, inefficient, variable, and inequitably distributed. The aggregate return on the national annual investment of $8,362 *per capita*[7] is inferior to that of other developed nations, reflected both in objective standard measures (e.g., infant mortality and longevity) and subjective indices (e.g., perceived quality of life). The unbridled growth in health care expenditures is unsustainable: projections by its trustees suggest that Medicare will exhaust its trust in 10 years' time. The number of Americans who have lost their employment-based health insurance is substantial.

In his 2009 inaugural address, President Barack Obama stated that "our health care is too costly." Two decades ago, the USA experimented with cost controls through expansion of managed health care. The experiment faltered as patients complained they were being denied needed care, doctors asserted that they were drowning in bureaucracy, and politicians attacked the managed care organizations as evil profit centers more interested in returning shareholder value than in the public good. With contraction of managed health care, the fraction of the gross domestic product consumed by health care resumed its ascent.[8]

---

[6] Germany's public health insurance companies remain fiscally strong, seemingly because of expense controls. In aggregate, they made €4 billion in profits in 2011, bringing their total reserves to more than €10 billion according to end of year results published in the *Frankfurter Allgemeine Zeitung*.

[7] These are the most current data available from the World Bank and reflect 2010 calculations. See http://data.worldbank.org/indicator/SH.XPD.PCAP.

[8] Cost controls are being resuscitated and appear under the guise of "accountable care organizations," a phrase invented 6 years ago.

President Obama continued: "We will restore science to its rightful place and wield technology's wonders to raise health care's quality and lower its costs." Although modern medical technologies have great power to relieve suffering and to foster healing, the manufacturers of medical drugs and devices are vilified for their aggressive marketing and pricing. While technology has great potential to raise quality, there is little basis for a prediction of lowered costs. Of greater concern is the perpetuation of the myth that health is obtainable, maintainable, and sustainable primarily through science and technology.

These two ideas—first that health results from the proper application of science and technology and second that health care is too costly—have polarized Americans along economic lines (those that do not have health insurance and those who do). The "health is a public good" advocates claim that health care is a fundamental human right and support varying extents and levels of taxpayer-financed universal health care. The "health is a personal benefit" adherents are skeptical in either the notion that universal health care will lower overall costs (their skepticism appears well-founded) or that they have a responsibility to underwrite the costs of caring for those who do not have employment-linked health insurance. While the two camps skirmish, existing government support remains stretched and distorted. For example, Medicaid, which is the safety-net program administered by the individual states originally intended to protect children and poor single parent families, has increasingly been co-opted by elders spending down their assets to qualify for long-term nursing home care.

While payment strategies are important, the current emphasis on health care financing is at best inadequate and at worst potentially misguided. Comparisons with other developed nations' investments in health care suggest that Americans adequately and disproportionately fund—in fact, overfund—health care. Where we Americans appear to fall short is in recognizing and managing the applications of medical science and technology towards those patients who will benefit the most.

The Rise of Science and Loss of Faith in American Medicine

As recently as the nineteenth century, the outcome of a medical treatment was a matter of faith and chance. Despite the rapid expansion of new biological thought (including Charles Darwin's theory of evolution, Claude Bernard's founding of physiology, Louis Pasteur's germ theory of disease, Robert Koch's discovery of the tubercle bacillus and his eponymous postulates) clinical practice remained largely disconnected from biological science. For example, the medical community so ridiculed Ignaz Semmelweis for suggesting that the cause of puerperal fever was being carried on doctors' hands from the autopsy suite to the birthing suite that he sustained a nervous breakdown.

Alarmed that the teaching of clinical medicine remained so separate from science, the Carnegie Foundation sponsored Abraham Flexner's tour through each of the 155 schools of medicine operating in the USA and Canada in the first decade of the twentieth century. His scathing critique, first published in 1910, included comprehensive recommendations for reform in medical education and, by extension, in medical practice. Science would describe the anatomy, physiology, and pathology of disease; clinicians would make specific diagnoses and patients would respond to treatments aimed at the underlying cause. The modern 4-year medical curriculum that blends 2 years' grounding in medical science with 2 years of supervised practice has remained standard for a century.[9]

The then-novel paradigm of accurate diagnosis and specific therapy proved successful beyond all expectations. Frederick Banting (a surgeon and Nobel laureate) isolated insulin, offering hope and longevity to diabetics. Peyton Rous showed that a cell-free filtrate could transmit cancer in birds, identifying viruses as a new class of etiologic agents. These ideas, coupled with emerging philosophies of public health, led to the eradication of polio and smallpox and prevention of more prosaic illnesses such as influenza and measles. Concurrent with the rise of public health measures and specific treatments for acute illness, American life spans increased from age 49 (around 1900) to current estimates of age 75 or more, further cementing the faith of patients in their physicians to provide technical solutions (pharmaceutical and surgical therapies) for acute illnesses. With treatments from penicillin to open heart surgery, Americans' confidence in American medicine to address American illness grew unbounded.[10]

"Screen and Prevent" vs. "Diagnose and Cure" vs. "Recognize and Manage"

As a consequence of prevention and of effective treatment of previously lethal acute illnesses, Americans started to accumulate chronic health problems. Whether the accumulation of chronic illnesses is an unavoidable result of our longer life spans or whether the definition of "well" has been so narrowed to that of an unattainable prototype continues to be the subject of scientific and social debate. What is not debatable, however, is that chronic illnesses have taken on the characteristics of modern plagues.[11]

---

[9] The national agency that accredits medical schools, the Liaison Committee on Medical Education, specifies that *"the curriculum of a medical education program must incorporate the fundamental principles of medicine and its underlying scientific concepts; allow medical students to acquire skills of critical judgment based on evidence and experience; and develop medical students' ability to use principles and skills wisely in solving problems of health and disease."* See http://www. lcme.org/functions.pdf.

[10] The apparent success of antibiotic treatments for acute infection, and the subsequent rise of promiscuous prescribing of those drugs for mild self-limited disease, is discussed elsewhere. See: McDonnell Norms Group Antibiotic overuse: the influence of social norms. J Am Coll Surg. 2008;207(2):265–7.

[11] 27 of Americans have multiple chronic illnesses. Together, they account for 66 of health care spending. Anderson G. Chronic care: making the case for ongoing care. Princeton, NJ: Robert Wood Johnson Foundation; 2010. Available at http://www.rwjf.org/files/research/50968chronic.care.chartbook.pdf.

Mass screening was implemented as an initial approach to preventing the most threatening of chronic illnesses, namely, cancers. Such screening programs are now viewed with increasing skepticism by statisticians, actuaries, and health care economists. Although occasional patients benefit from such programs, American clinicians have only recently come to appreciate that there are significant harms following the inevitable false-positive results that arise in large populations. The USA Preventive Services Task Force currently (as of 2012) recommends delaying routine mammography to age 50, against routine screening for colon cancer after age 75, against routine prostate-specific antigen (PSA)-based screening for prostate cancer at any age, and so on.

"Screen and Prevent" and "Diagnose and Cure" are being thus replaced by "Recognize and Manage." Our longer lives are increasing tainted by incurable conditions. It is the management of chronic illness that is prolonged, costly, and increasingly unsatisfying to American patients and their providers. Nowhere is this "recognize and manage" paradigm so oddly instantiated as the USA intensive care unit, where revolving door admissions of patients with incurable conditions near the end of life have become commonplace.

Expectations of and Incentives in Intensive Care in the USA

The jargon surrounding ICU admissions with chronic and incurable illness reflects a remanufacturing mindset: the patient is there for a "tune-up," much as one would adjust the carburetor on older automobiles. Implicit in this notion is the idea of "presentation prototype"—how the system performed when it was new. With this remanufacturing mindset, "evidence-based medicine" is used to justify a search for out-of-tolerance deviations from the prototype image and their correction.[12]

In the ICU, physiology is monitored continuously, laboratory values are checked on a schedule, and small deviations "corrected" with a precision that defies common sense and a zeal that may cause harm.[13] While such aggressive measures are perhaps understandable when applied to the previously healthy patient with sudden and life-threatening derangements, the virtue of complete correction of chronically deranged physiology is questionable.

---

[12] Herbert Simon remarked that "Engineering, medicine, business, architecture and painting are concerned not with the necessary but with the contingent—not with how things are but with how they might be—in short, with design." Simon HA. The sciences of the artificial. 2nd ed. Cambridge: MIT Press; 1981. That perspective has led to extraordinary advances in bioprostheses enabling rehabilitation of wounded soldiers. It has also enabled extraordinary preoccupation with aesthetic surgery. See, for example, www.cindyjackson.com.

[13] For example, replacement of potassium to commonly recommended levels is associated with increased mortality following acute myocardial infarction. Goyal A, Spertus JA, Gosch K, Venkitachalam L, Jones PG, Van den Berghe G, Kosiborod M. Serum potassium levels and mortality in acute myocardial infarction. JAMA. 2012;307(2):157–64.

What is not questionable are the financial incentives for the USA health care system, which have been satirically encapsulated as "Alfred E. Neuman's Cosmic Health Care Equation," namely, that Every Dollar of Health Spending = Someone's Health Care Income.[14] Patients who cycle into the ICU for a tune-up, leave, deteriorate, and are then readmitted to the ICU to begin the next cycle generate significant revenues under traditional fee-for-service models of payment. Those revenues sustained hospitals and their medical staffs.

Changing compensation models can change hospital utilization patterns in chronic illness. When CMS (the payor for care of USA's elderly population) announced financial penalties for avoidable readmission for congestive heart failure, hospitals nationwide quickly organized programs to ensure more effective outpatient treatment of those at highest risk.

Health and Well-Being

We have illustrated how the current American care paradigm relies on recognition and management of deviations from some arbitrary state of perfect "health." Thus health is something to be "restored" by medical science. Critical care, of course, restores nothing. It merely gives patients time and protected space to heal—and that healing depends on internal resilience.

The international goals of health care are equally unrealistic. The preamble to the Constitution of the World Health Organization famously defined health as "a state of complete physical, mental and social well-being and not merely the absence of disease or infirmity."[15] The breadth of this definition is apparent, as is its implicit unattainability.

The Pursuit of Health

Earlier, we asserted that the economic focus on health care was incomplete, if not misguided. We now confront the questions directly: why is health prized, and why should we Americans believe that we are entitled to good health? Neither church nor state makes such a promise. The Constitution of the USA does not assert a right to

---

[14]Uwe Reinhardt, Ph.D. U.S. Senate Finance Committee hearing on "Health care reform: an economic perspective". November 19, 2008.

[15]Preamble to the Constitution of the World Health Organization as adopted by the International Health Conference, New York, June 19–22, 1946; signed on July 22, 1946 by the representatives of 61 States (Official Records of the World Health Organization, no. 2, p. 100) and entered into force on April 7, 1948.The Definition has not been amended since 1948.

health care, and there is no evidence that the founding fathers had any intention of guaranteeing health care, much less unbridled access to highly technical critical care.

What the 14th Amendment to the USA Constitution does guarantee, and what Americans continue to hold as a fundamental right, is the right to equal protection under the law.[16] Generally accepted interpretations have transformed this right into expressions of equal opportunity, and those expressions underlie the guarantee of an education, the guarantee of nondiscrimination on the basis of age, gender, and so on.

Without reasonable health, Americans cannot learn in school, compete for employment, or exercise their right to pursue happiness. While there may not be a right to unlimited health care, it seems reasonable to assert the existence of a threshold of health below which equal opportunity is compromised. This assertion was echoed by President Obama in remarks accompanying his signature of the SCHIP legislation: "No child should be falling behind at school because he can't hear the teacher or see the blackboard. I refuse to accept that millions of our children fail to reach their full potential because we fail to meet their basic needs."[17] A level of health that enables education does not guarantee success. It merely ensures that young Americans have the opportunity to become successful.

Thus health might be desired by the individual for its own sake, but that is not the justification for public investment. The latter are made not for health's sake, but to promote the opportunities that health provides. This creates an explicit rationale for considering both costs and benefits of the public investment to individuals and to society. Young people with acute and reversible illnesses have greater opportunity and therefore greater claim on health care resources than the elderly during their inexorable decline. Preventive strategies in early and mid-life that sustain opportunity have priority over highly technical interventions applied near the end of life.[18]

Currently only about 5 of Medicare beneficiaries die each year, but those who die consume more than 27 of all Medicare costs for the elderly.[19] Much of that "last year of life" cost pays for technically advanced care (imaging, laboratory testing, sophisticated medicines and surgeries and hospitalization including ICU admissions) that yields comparatively little "quality time" for each dollar invested.

---

[16]This expands on a position initially articulated by Turka LA, Caplan AL. The right to health care. J Clin Invest. 2010;120:934.

[17]Remarks of the President. February 4, 2009.

[18]Bill Gates, founder of Microsoft, succinctly summarized the trade-offs in a February 9, 2011 interview with Charlie Rose: "Politicians must address tough questions about limited resources. How many teachers are you willing to fire in order to have 78-year olds have a procedure which will be invented 5 years from now that adds 4 months to their life? That sounds terrible, but infinitely choosing those things will shift you away from education for the young, and towards infinite invention of such [medical] procedures."

[19]Hogan C, Lunney J, Gabel J, Lynn J. Medicare beneficiaries' costs of care in the last year of life. Health Aff. 2001;20:188–95.

Nearly a third of elderly Americans undergo a surgical procedure in the last year of life, and most of these within the last month of life. Those patients spend an average of 27 days in hospital, including 5.5 days in critical care. [20]

Signs of Change

There are signs that the growth in USA health care expenditures is slowing. The fraction of the gross domestic product devoted to health care in 2010 was 17.9, unchanged from 2009.[21] More recently, third party payors are taking matters into their own hands by purchasing health care delivery systems—essentially "cutting out the middleman"—making the clinician-employees directly responsive to management.[22] Perhaps more important, they are restructuring health care delivery models. Health care is no longer a matter of illness care, but rather a combination of wellness promotion, illness prevention, and illness care. For example, some integrated insurance delivery systems are investing in physical fitness and balance training to increase strength and prevent falls among the elderly. Such programs appear not only to work but also to be cost-effective.

On March 23, 2010, USA president Barack Obama signed the Patient Protection and Affordable Care Act into law. That Act, perhaps the most far-reaching health care legislation in the USA since the inception of Medicare, acknowledged for the first time that all Americans were likely to need health care; and that all Americans needed to pay into a health care funding mechanism or face a penalty unless exempted for religious reasons. The Act created guarantees that Americans could not be excluded from health care insurance on the basis of a preexisting medical condition and furthermore created structures to promote public health such as screening programs with no co-payment by the patient.

Within days, lawsuits were filed by many states and organizations to block implementation of key provisions of the Act, claiming that the Act violate the Constitution. Lower courts were divided in their rulings, and in 2012 the Supreme Court of the USA heard arguments on key aspects of the Act. On June 28, 2012, the Supreme Court of the USA rendered its decision: the individual mandate requiring national participation in health care financing so that every American could have access to health care was found to be in accordance with the Constitution.

---

[20] Kwok AC, Semel ME, Lipsitz SR, Bader AM, Barnato AE, Gawande AA, Jha AK. The intensity and variation of surgical care at the end of life: a retrospective cohort study. Lancet. 2011;378(9800):1408–13.

[21] Martin AB, Lassman D, Washington B, Catlin A; National Health Expenditure Accounts Team. Growth in US health spending remained slow in 2010; health share of gross domestic product was unchanged from 2009. Health Aff. 2012;31(1):208–19.

[22] Mathews AW. The future of U.S. health care. Wall Street Journal. 12 Dec 2011.

For the first time in American history, health care emerges as a right, and health care financing is established as an individual responsibility. The ruling of the Supreme Court necessarily launches a national conversation about a minimum standard for health care and about finding ways to pay for that standard. The significance for critical care in the USA cannot be overstated: the value of that care will soon be weighed against the value of other types of health care in supporting whatever national standard emerges as a basic American right.

Winston Churchill remarked in the midst of World War II as he saw the tide of battle begin to turn, "Now this is not the end. It is not even the beginning of the end. But it is, perhaps, the end of the beginning."[23]

---

[23] Winston Churchill, November 10, 1942, The Lord Mayor's Luncheon.

Chapter 27
First Critique of Buchman and Chalfin's Analysis

Leslie P. Scheunemann and Douglas B. White

Introduction

Buchman and Chalfin note heterogeneity in how physicians view their obligations to individual patients and society. They correctly note that the traditional Hippocratic view is that physicians should do all they can to advance the interests of their individual patient. However, it is not clear that physicians should or actually do adhere to the Hippocratic vision in modern, publicly funded health care systems. Healthcare rationing involves withholding potentially beneficial treatments from individuals on the grounds of scarcity [1]. Because the population need for healthcare goods is virtually limitless, rationing is unavoidable [2]. Moreover, in most developed countries, healthcare draws its funding from the same public pool as other social goods such as education, infrastructure, environmental protection, and defense. Therefore, healthcare's share of resources often grows by proportional cuts in other social programs. As a consequence, when healthcare is at least partially publicly funded—as it is in the United States and most other industrialized nations—the public good becomes an ethically relevant consideration when assessing physicians' obligations to individual patients and to populations of patients. This essay is limited to such contexts.

Note: All work on this manuscript was performed at the University of Pittsburgh.
Acknowledgment: Dr. White was supported by a Greenwall Foundation Faculty Scholars award and a Paul Beeson Award in Aging Research from the National Institute on Aging.

L.P. Scheunemann, M.D., M.P.H. (✉)
Division of Pulmonary, Allergy, and Critical Care Medicine, University of Pittsburgh Medical Center, NW 628 MUH, 3459 Fifth Avenue, Pittsburgh, PA, USA
e-mail: scheunemannlp@upmc.edu

D.B. White, M.D., M.A.S.
Program on Ethics and Decision Making in Critical Illness, Department of Critical Care Medicine, University of Pittsburgh Medical Center, Pittsburgh, PA, USA

Center for Bioethics and Health Law, University of Pittsburgh, Pittsburgh, PA, USA

D.W. Crippen (ed.), *ICU Resource Allocation in the New Millennium: Will We Say "No"?*, 209
DOI 10.1007/978-1-4614-3866-3_27, © Springer Science+Business Media New York 2013

Rationing raises challenging questions for physicians about their professional ethical obligations. Physicians have generally been acculturated to focus on their obligations to individual patients rather than populations of patients—this is the ethos of the Hippocratic tradition. Whether physicians have any role in rationing hinges on whether physicians have obligations to society in addition to their patient centered obligations. This essay discusses the tension between individuals and society from the perspectives of (1) the traditional view of the physician's role, (2) a compartmentalized view, in which physicians have duties to society and to individuals, which they fulfill through separate administrative and bedside roles, (3) a middle view, in which physicians in all roles dynamically consider the interests of both individuals and society.

The Traditional View of the Physician's Role

The traditional view is usually identified with the Hippocratic Oath [3]. It holds that physicians are obligated to benefit their patients to the best of their ability and judgment, without considering the interests of others. When applied to rationing, it forbids physicians to consider how any treatment they offer an individual patient will affect social resources [4]. From the traditional view, physicians would be required to offer annual screening mammograms to women in their 40s because screening increases their average life expectancy by 5 days, despite the staggering cost of $680,000 per quality-adjusted life year [5]. By comparison, the United Kingdom's National Institute for Health and Clinical Excellence usually does not fund interventions that cost substantially more than $35,000 per quality-adjusted life year [6]. Furthermore, the physician could not respond to a critical care bed shortage by transferring out a recovering patient who might derive even a little benefit from continued intensive monitoring; instead, she would have to deny admission to the patient who had acutely decompensated because of her commitment to the patient already under her care.

The merit of the traditional view is its insistence on loyalty to the patient. Loyalty to the patient is an essential part of medical practice because patients are vulnerable due to illness. Loyalty is the basis of the trust by which patients disclose intimate details necessary for diagnosis and treatment, and trust is central to healing relationships [7].

Is It Ever Justifiable for Physicians to Act against a Patient's Best Interests?

However, the Hippocratic ethic encounters serious problems in actual medical practice. One of its most outspoken critics, Robert Veatch, has argued strongly that the moral problems of modern medicine have shifted from an individual to a social model [8]. By focusing exclusively on the benefit of the individual patient, the Hippocratic tradition has collapsed under the weight of the empirical realities of modern medicine. Moreover, because medicine—including the education of physicians—is publicly funded, there arises a reciprocal obligation on the profession to advance the public good [9, 10].

There are numerous ways that medical practice has broken with the Hippocratic tradition. Providers who claim to adhere to it might be surprised that one informal study found 27 generally accepted rationales for overriding the patient's best interests [11]. Among them were 23 at least partially justified by appeal to social interests, such as the welfare of other patients or the common good. These situations include mandatory reporting of communicable disease (violation of confidentiality), forced treatment for conditions like tuberculosis (violation of right to self-determination), and allowing medical trainees to participate in direct patient care, which creates some risk for patients despite the oversight provided by senior physicians. Importantly for a discussion of rationing, eight of the situations were justified by appeal to the scarcity of some resource, such as money, effective antibiotics, physician time, or patient equipment. For example, physicians sometimes prescribe a slightly less effective medication to a patient in order to achieve a significant cost savings, which is a common practice with bisphosphonates, statins, and antihypertensives [12–14]. Additionally, physicians necessarily ration their time—among patients as well as between patient care and other professional or personal commitments [11, 15, 16].

In sum, it is no longer credible to claim that medical practice is guided solely by the Hippocratic ethic of acting for the best interest of the individual patient. Moreover, in nations with public funding of healthcare, the medical profession has important ethical obligations to society, which require rationing of at least some medical treatments.

Conflicting Statements on Rationing from Professional Medical Societies

Medical societies provide professional leadership and guidance for how to consider challenging ethical issues. Yet their statements on the physician's role in rationing fail to provide clear guidance and reflect ambivalence about how rationing fits in the professional ethic. For example, the American Medical Association's (AMA) Code of Ethics statement on the Allocation of Limited Resources endorses a vision of the physician's duty "to do all he or she can for the benefit of the individual patient" that conforms closely to the Hippocratic ideal [17]. It also warns that policies can constrain physicians' ability to fulfill the duty, so that physicians have policy-level duties to "safeguard the interests of patients." However, the same section of the policy provides "ethically appropriate criteria" by which the patient's wishes may be overridden. It tacitly acknowledges the conflict between patient-centeredness and social goals by reminding clinicians that, "[a]llocation decisions should respect the individuality of patients and the particulars of individual cases as much as possible." It concludes with a definitive statement that "[t]he treating physician must remain a patient advocate and not make allocation decisions." This stands in stark contrast to the empirical reality that physicians make rationing decisions every day.

In contrast, the American College of Physicians' Charter on Medical Professionalism takes a more balanced view. Patient welfare and autonomy come first, but are never given complete reign: the Charter articulates a population ethic, social justice, under which it charges physicians with the fair distribution of resources [18]. It reminds physicians that they are required to practice "wise and cost-effective management of limited clinical resources." The Charter distinctly turns away from the traditional view by insisting on facing both the inevitability of rationing medicine and medicine's social obligations to approach this task in an ethical manner.

A Compartmentalized View of the Physician's Role

Another conceptualization of the physician's role in society acknowledges physicians' duties to both patients and society, but compartmentalizes them. In this view, physicians participate at an administrative level to set rationing policies, which may take the form of guidelines, authorization for testing or treatment, or ethical groups that help to mediate conflicts between individuals and society [7, 11, 19]. Bedside clinicians are responsible for adhering to the limits set by the administrative groups, but are obligated to advocate fully for their individual patients within those limits. The administrative portion of the middle view would work much like the United Kingdom's National Institute for Health and Clinical Excellence, which sets guidelines based on the cost effectiveness of healthcare [6]. Therefore, a physician treating a patient with severe, newly dosed rheumatoid arthritis would not be permitted to prescribe TNF-α inhibitors as part of first-line treatment; they must try at least two disease-modifying anti-rheumatic drugs including methotrexate first [20].

Ethically, the separation is usually justified by appeals to the special role of fidelity within the physician–patient relationship on the one hand, and to the rigor of ethical training and protection from conflicts of interest needed when weighing issues of justice on the other [7, 11, 19]. Maintaining and honoring fidelity within the physician–patient relationship invokes special duties to the patient from the bedside physician. In theory, separation allows bedside physicians to meet all their social obligations by following socially constructed rules, but also focusing within those constraints on advancing the interests of their patients. Within the compartmentalized view, there is no pressure on bedside physicians to ration because they are asked to simply follow the rules put forth by administrative physicians.

Cooperation of Bedside Physicians is Necessary for Rationing to Succeed

A strength of the compartmentalized view is that it recognizes that physicians have legitimate obligations to both patients and society. However, the compartmentalized view fails to recognize that obligations to patients and to society are tightly interconnected at the bedside. This approach to rationing will fail in the short term because strictly enforceable rules do not currently exist [4]. Absent such rules, physicians in the compartmentalized model should do all they can to even marginally increase patients well being and health care costs are unlikely to be controlled. Moreover, any successful compartmentalization would have to disallow any exceptions or appeals, because physicians would in theory be required to pursue such appeals if they held any chance of benefit for their patient.

This raises a second flaw with the compartmentalized view: disallowing exemptions would be insupportably rigid for fair treatment of individuals. One lesson from Oregon's health plan, which paid for a stratified list of about 700 services until the budget was exhausted, was that physicians could often identify individuals who would have benefitted from a (denied) treatment significantly more than predicted by the population-level estimate [4]. Furthermore, one study in the United Kingdom found that proposed guidelines for allocating critical care resources during an influenza pandemic were insufficiently sensitive or specific (AUC 0.74) in predicting survival [21]. They also failed to ration enough beds to meet projected pandemic-level demands, which means that bedside judgment would still be required.

Perhaps the most damning problem for the compartmentalized view is that, although guidelines for cost-effective care exist for a few conditions, they do not exist for the vast majority. The compartmentalized view is unworkable for the foreseeable future unless physicians implicitly ration at the bedside [4].

A Middle View of the Physician's Role

The circumstances noted above suggest that, for the foreseeable future, bedside physicians cannot be guided solely by the goal of promoting the individual patient's interests [4]. Another formulation of the bedside physician's role in rationing is needed. One possibility is a middle view which acknowledges that, because of practical realities, the conflict between the public good and the individual good cannot be removed from the bedside. A middle view would acknowledge that physicians have appropriate roles at all levels of healthcare planning and cost containment: both in the formulation and administration of health policy and at the bedside [22]. Although bedside physicians would understandably give more weight to the patient's interests, they would still be trained to be able to dynamically consider those interests in the context of society [23]. From the other end, administrative physicians would give more weight to upholding society's interests, but would also be accountable for the effects of policies on individuals [19].

This view would require leadership and communication between bedside and administrative physicians. In contrast to the Hippocratic view and the compartmentalized view, the middle view acknowledges the inevitability of bedside rationing and aims to give clinicians the necessary ethical skills and clinical guidance for enacting it. This guidance would have two parts: a clear social agreement that they have a dual role, and ongoing assistance in how to balance obligations to individuals and to the greater good. Of the three current options, the middle view has the greatest potential to effectively satisfy both patient and population level obligations in a transparent way.

A Charge of Responsibility

Unfortunately, medicine is no closer to reaching its goals of cost containment, or of a clear role for physicians in rationing, than it was 15 years ago when Ubel and Arnold urged the importance of accepting bedside rationing [4]. Considerations of the public good are ethically relevant to the practice of medicine and likely to increase in coming years. At the very least, physicians must honestly acknowledge this ethical obligation and commit to developing a more accurate and sustainable conceptualization of their relationship to individual patients and populations of patients.

References

1. Truog R, Brock D, Cook D, et al. Rationing in the intensive care unit. Crit Care Med. 2006; 34(4):958–63.
2. Calabresi G, Bobbitt P. Tragic choices: the conflicts society confronts in the allocation of tragically scarce resources. New York: Norton; 1978.
3. The Hippocratic Oath: Classical Version. *NOVA Oline*. http://www.pbs.org/wgbh/nova/doctors/ oath_classical.html (2009). Accessed 11 Feb 2011.
4. Ubel P, Arnold R. The unbearable rightness of bedside rationing: physician duties in a climate of cost containment. Arch Intern Med. 1995;155(17):1837–42.
5. Truog RD. Screening mammography and the "r" word. N Engl J Med. 2009;361(26):2501–3.
6. Measuring effectiveness and cost effectiveness: the QALY. http://www.nice.org.uk/newsroom/ features/measuringeffectivenessandcosteffectivenesstheqaly.jsp (2010). Accessed 25 Jan 2010.
7. Bloche MG. Clinical loyalties and the social purposes of medicine. J Am Med Assoc. 1999; 281(3):268–74.
8. Veatch RM. The basics of bioethics. Upper Saddle River, NJ: Prentice Hall; 2003.
9. Schneiderman L, Jecker N. Should a criminal receive a heart transplant? Medical justice vs. societal justice. Theor Med Bioethics. 1996;17(1):33–44.
10. Schneiderman LJ, Capron AM. How can hospital futility policies contribute to establishing standards of practice? Camb Q Healthc Ethics. 2000;9(4):524–31.
11. Wendler D. Are physicians obligated always to act in the patient's best interests? J Med Ethics. 2010;36(2):66–70.
12. Stevenson M, Lloyd Jones M, De Nigris E, Brewer N, Davis S, Oakley J. A systematic review and economic evaluation of alendronate, etidronate, risedronate, raloxifene and teriparatide for the prevention and treatment of postmenopausal osteoporosis. Health Technol Assess. 2005;9:1–160.
13. McLain R, Koren M, Bakker-Arkema R, et al. The cost of reaching National Cholesterol Education Program (NCEP) goals in hypercholesterolaemic patients: a comparison of atorvastatin, simvastatin, lovastatin and fluvastatin. Pharmacoeconomics. 1998;14(1):59–70.
14. Hsu J, Price M, Huang J, et al. Unintended consequences of caps on Medicare drug benefits. N Engl J Med. 2006;354(22):2349–59.

15. Strech D, Synofzik M, Marckmann G. How physicians allocate scarce resources at the bedside: a systematic review of qualitative studies. J Med Philos. 2008;33(1):80–99.
16. Konrad TR, Link CL, Shackelton RJ, et al. It's about time: physicians' perceptions of time constraints in primary care medical practice in three national healthcare systems. Med Care. 2010;48(2):95–100.
17. Opinion 2.03—Allocation of Limited Medical Resources. AMA Code of Medical Ethics. http://www.ama-assn.org/ama/pub/physician-resources/medical-ethics/code-medical-ethics/opinion203.shtml (1994). Accessed 5 February 2011.
18. Sox H. Medical professionalism in the new millennium: a physician charter. Ann Intern Med. 2002;136(3):243–6.
19. Benatar SR, Upshur REG. Dual loyalty of physicians in the military and in civilian life. Am J Public Health. 2008;98(12):2161–7.
20. Technology appraisal full table of recommendations. http://www.nice.org.uk/media/F96/CD/NICETADecisionSummaryJuly2011.pdf (2011). Accessed 15 August 2011.
21. Guest T, Tantam G, Donlin N, Tantam K, McMillan H, Tillyard A. An observational cohort study of triage for critical care provision during pandemic influenza:'clipboard physicians' or 'evidenced based medicine'? Anaesthesia. 2009;64(11):1199–206.
22. Reuben DB, Cassel CK. Physician stewardship of health care in an era of finite resources. J Am Med Assoc. 2011;306(4):430–1.
23. Pearson SD. Caring and cost: the challenge for physician advocacy. Ann Intern Med. 2000;133(2):148–53.

Chapter 28
Second Critique of Buchman and Chalfin's Analysis

Lynn Barkley Burnett

Doctors Buchman and Chalfin are to be commended for their emphasis on health education as a means of encouraging adoption of healthy behaviors, and thereby reducing the toll—physical, emotional, and economic—of coronary artery disease, cancer, chronic obstructive pulmonary disease, and other similar chronic illnesses. As they state "To the extent that prevention or even delay of the onset of a chronic illness corresponds to sustained health, prevention of disease is desirable from both personal and societal perspectives." However, even were the populous to modify its health behaviors, thereby achieving what has been termed "compressed morbidity" [1], people *will* still get sick and be injured.

Therefore the concentration of this chapter is the bedside of the individual patient, first via an ethics-focused examination of the stages of decision-making shared among the patient or surrogate and the treating physician(s), with a view to achieving a consensus as to the Plan of Care, and thereby preemptive avoidance of disagreement and confrontation (distasteful to everyone). This is followed by an approach to those situations wherein agreement cannot be achieved, insofar as what interventions are, or are not, appropriate to a given patient in the critical care environment.

This chapter does not address the withholding of medically appropriate interventions that may be made necessary by economic factors (as is the current practice in Oregon). If we get to the stage of debating such an approach on the state or national level, I would strongly suggest we ask ourselves, and one another, if we want our health access policy to be that recently demonstrated on an HBO Special; a situation in which a 54-year-old patient with prostate cancer was informed by the State of Oregon that, given the stage of his illness, further publically financed care was not available, but that the State would pay for Physician-Aid-in-Dying.

L.B. Burnett, M.D., Ed.D. (✉)
Department of Medical Ethics, Community Medical Centers, Fresno, CA, USA

Departments of Emergency Medicine & Forensic Pathology, Touro University,
California College of Osteopathic Medicine, Vallejo, CA, USA
e-mail: drlbburnett@sbcglobal.net

D.W. Crippen (ed.), *ICU Resource Allocation in the New Millennium: Will We Say "No"?*, 217
DOI 10.1007/978-1-4614-3866-3_28, © Springer Science+Business Media New York 2013

Preventive Ethics in Medical Decision-Making

After appropriate work-up and establishment of a diagnosis, discussion between doctor and patient (or surrogates) may address Medical Problems, Patient Preferences, Quality of Life, and Contextual Features [2]. Some of the elements discussed below are essential in every encounter with a critically ill patient; others may become important if there is a problem in shared decision-making. It should be noted that *knowledge of those elements may have prevented the problem had they been known prior to presentation of the disagreement.* Obtaining this information and having these discussions takes time—something in short supply for the physician in the critical care unit. While the physician must be directly involved in some of these elements, Advanced Clinical Practitioners, social workers, chaplains and nurses may be superb sources of information and insight, given their interaction with patient, surrogate, family, and friends.

With regard to the Patient's Medical Problems: what is the diagnosis or diagnoses, how certain is the diagnosis, what is the stage of illness (acute, chronic, acute exacerbation of a chronic illness)? What is the prognosis, what is the likely clinical course, what is the likely trajectory and time frame for such? What is the patient's past medical history? What medically appropriate treatment options exist, of what benefit to the patient is the proposed treatment, what is the probability for treatment success, what are the risks and burdens of treatment, what are the plans in case of treatment failure?

Insofar as Patient Preferences are concerned: Does the patient have Decision-Making Capacity, has the patient previously had Decision-Making Capacity, if the patient cannot communicate is there an Advance Directive or other clear indication as to the patient's values (e.g., prior clear statement of patient, Values Inventory in the Medical Record at a hospital or primary care provider)? Is there a real understanding on the part of the patient or surrogate as to the medical problems and treatment options? What does the patient or surrogate want done now; during later stages of the illness? What are the patient's Goals (Medical and Nonmedical), what modalities and interventions are acceptable to meet those Goals? Has the patient or surrogate had time to reflect on their decision (something that may not always be possible, given exigencies of the patient's clinical status)? If a surrogate is asked to make decisions, does the surrogate know the patient's values and can thereby make a Substituted Judgment, or does s/he not know the patient's values well, and is therefore asked to make a decision that is in the Best Interests of the Patient?

Quality of Life assessment asks "who is the patient": age, life situation and lifestyle, education, profession or occupation? What are the major relationships in the patient's life, the needs of the patient and others, the responsibilities of each in meeting the other's needs? What is the baseline state of the patient's health (physical and mental); is the patient satisfied with the present Quality of Life (is life worth living), what is the likelihood of restoration to baseline with treatment, would the patient want aggressive intervention to return him to his present Quality of Life, what deficits will the patient experience if treatment is "successful" (physical, mental, social)? What are the patient's values insofar as world view, religious beliefs, after death beliefs, and the role of Western medicine in treatment?

Contextual Features involve time constraints, power or personality issues that may influence fairness to all parties; family issues that might influence decisions, such as imposition of values/religious beliefs on the patient, conflicts of interest, abuse/neglect? Disagreement among care givers (if present, is it reasonable), professional integrity versus imposition of the physician's values on the patient? Financial or economic factors, resource allocation? Religious or cultural barriers to communication, understanding, or decision-making; organizational or institutional issues; legal implications?

The Goals of Medical Care

Identification and discussion of these elements enables the physician to understand what is important to the patient: *goals*, the things that patients are striving for in their decisions; *constraints*, various factors limiting the patient in his/her decision; and *protected values*, typically moral and religious issues [3], examples of which could include blood transfusion to a Jehovah's Witness, abortion in cases of maternal-fetal conflict when the mother is Catholic, and similar situations. When there is a conflict among the patient's goals (e.g., "I want to live as long as I can"; "I don't want to be in pain"), the patient/surrogate and physician seek to develop a plan, based on the prioritized goals of greatest importance to the patient, that will best meet as many of those goals as possible.

These goals, constraints, and protected values of the patient inform and guide us as we address this question: What is it we (patient/surrogate, health professionals) are trying to accomplish? What is our overarching Goal of Medical Care?

Table 28.1 is adopted from the EPEC Curriculum [4]. (The EPEC Program [Education in Palliative and End-of-Life Care] was originated by the American

Table 28.1 Primary Goal of Care

	Curative	Supportive	Palliative
Approach	"Win"	"Fight"	"Accept"
Impact on disease	Eradicate	Arrest progression	Avoid complications
Limitations	No limitations of treatment options that are within the standard of practice, and consistent with the patient's values	Some limitations of burdensome treatments	Limitation of treatment, with focus on relief of pain and suffering
Acceptable adverse events	Major	Major-moderate	Minor-none
CPR	Probably yes	Probably yes	Probably not[a]
Symptom prevention/relief	Secondary	Balanced	Primary

[a]CPR is not consistent with Palliative Care. Under very unusual circumstances (e.g., "my Sister is flying in from Europe, and if possible I would like her to at least be able to see and touch a warm person, even if I'm unconscious") it could be attempted, even when the Goal of Care is Palliation

Medical Association, and is now a program of the Feinberg School of Medicine at Northwestern University.) The table has been modified by inclusion of some criteria from the Community Medical Centers Policy on Medically Ineffective Interventions (Futility) [5]. Reflected above are the primary emphases of different goals related to life-threatening illness or injury.

When Patient and Physician Do Not Agree

When there is an absence of agreement between patient (directly or via surrogate) and physician, the cause of the disagreement is often Medically Nonbeneficial Care, also referred to in the literature as Medical Futility, Futile Care, Clinical Futility, Medically Ineffective Care, and Medically Nonindicated Care. It is instructive to view the evolution of this issue of Medically Nonbeneficial Care.

Let's begin with The Father of Medicine, who said [6]

First I will define what I conceive medicine to be.
  In general terms, it is to do away with the sufferings of the sick, to lessen the violence of their diseases, and to refuse to treat those who are overmastered by their disease, realizing that in such cases medicine is powerless.

Hippocrates

Since the days of Hippocrates, however, medicine is no longer powerless when patients are "overmastered by their disease." Indeed, as Drs. Buchman and Chalfin state, "the function of nearly every organ can be manipulated and improved and, if necessary, replaced (mechanically, pharmacologically or via transplantation) indefinitely."

Such tremendous scientific advancements have resulted in the emergence of "The Technological Imperative" [7]. The Technological Imperative, summarized as it pertains to clinical care, is that "If we have the technologic means to do x, we must use it to do x." This imperative, that we *must* use technology, has been played out on both sides of the physician–patient relationship.

The relationship between physician and patient was, for millennia, one best described as "paternalism." After all, it was doctors who had studied disease and knew what treatments were available; accordingly, it was they who made patient care decisions. With the advent of ventilators, renal dialysis, artificial nutrition, and similar interventions, patients survived who previously would have died as a result of their illness. With patients lying in Persistent Vegetative and Other States, seemingly without end, families began to say "Stop." The response of many physicians—motivated sometimes because such technological support was believed to be the "Standard of Care," sometimes out of a perceived ethical obligation to support life by all means available, sometimes out of a fear of potential criminal charges—was "If we have this technology to keep the patient alive, we must use it."

The result of this, in addition to well-known law suits concerning Karen Ann Quinlan and Nancy Cruzan, among other factors, was the development of a "Patient's

Rights Movement" [8]. The unifying principle of the Patient's Rights Movement was "It's my body, and I have a right to determine what will or will not be done to it."

Over time, and in part because of Supreme Court rulings upholding the rights of patients to die, coupled with the effect of such legal decisions easing physician concerns about criminal liability for discontinuing artificial ventilation and other life-support technologies, the pendulum swung. Physicians began to tell families "Just because we can use technology, does not mean that we must, or should, in all cases." In response thereto, the previous posture of patients and their families of "It's my body, and I have a right to determine what will or will not be done to it," was often expanded to "It's my body, and I have a right to whatever medical treatment I want."

Is it true that patients have a right to "whatever medical treatment" they want? No. While patients obviously have the right to consent to or refuse tests, interventions or procedures that are medically indicated, patients *do not* have a right to receive *whatever* treatment they (or their surrogates) demand [9]. For evidence of that ethical wisdom, one need only look at the example of Michael Jackson making demands on his physician for provision of Propofol—whose use was appropriate only in a monitored environment, to induce nightly unconsciousness.

Consistent with the weight of ethics literature, the Code of Medical Ethics of the American Medical Association [10] states, in relevant part

> Physicians are not ethically obligated to deliver care that, in their best professional judgment, will not have a reasonable chance of benefiting their patients. Patients should not be given treatments simply because they demand them. Denial of treatment should be justified by reliance on openly stated ethical principles and acceptable standards of care.

How is "inappropriate" treatment defined, specifically the subtype that is often called futile care or futile treatment? Ronald Cranford MD, the neurologist closely associated with elucidation of the Persistent Vegetative State, said "Whatever futility means, it seems obvious that this is not a discrete clinical concept with a sharp demarcation between futile and non-futile treatment" [8].

The foregoing notwithstanding, a framework that may prove helpful is provided by the categorization of types of futility [8]:

- Physiologic futility (quantitative futility)

  - Based on knowledge of outcomes of specific interventions for categories of illness states, there have been No survivors out of the last 100 cases (5% or less survival, and other survival figures, have also been proposed as criterion).

- Imminent demise futility

  - Despite the proposed intervention, the patient will die in the near future.

- Lethal condition futility

  - The patient has a terminal illness that the intervention does not affect, and that will result in death in the not-too-distant future (weeks to months) even if the intervention is employed (e.g., Coronary Artery Bypass Graft in a patient with severe Coronary Artery Disease, who also has end-stage metastatic cancer).

- Qualitative futility

    - The intervention fails to lead to an acceptable Quality of Life for the patient
      (as viewed from the patient's perspective).

Drane and Coulehan, as quoted by Clark [8], present a variation on a theme when
they state

Medically futile treatment is an action, intervention, or procedure that might be
physiologically effective in a given case, but cannot benefit the patient, no matter how
often it is repeated. A futile treatment is not necessarily ineffective, but it is worthless,
either because the medical action itself is futile (no matter what the patient's condition)
or the condition of the patient makes it futile.

Leaving aside the term "worthless," a word I would not have employed, the key
element is the definition of that all-important term "Benefit"? First, a general guideline, of which we should always be mindful, is that "the goals of medical treatment
are not merely to cause an effect on a portion of the patient's anatomy, physiology,
or chemistry, but to benefit the patient as a whole person" [11].

Treatment Benefits increase the well-being of the patient [9], and may be

(a) *Health benefits* (positive and empirically measurable effects in curing, arresting,
    or relieving the patient's disease, condition, symptoms and pain), or
(b) *Quality-of-life benefits* (improvements in mental status or added days or months
    of life that are mutually rewarding to the patient and others)

If Benefits can be achieved, attention next turns to the Burdens imposed to attain
same. Treatment Burdens diminish the well-being of the patient [9]

(a) When treatment provides no measurable health benefits, and increases pain,
    suffering, or debilitation, or
(b) When treatment reduces the patient's Quality of Life

Analysis of Benefits-to-Burdens, also known as the Test of Proportionality [12],
may well assist in resolution of disparate views as to appropriateness or nonappro-
priateness of interventions.

What if Such Resolution Is Not to be Had

The AMA Code of Medical Ethics [13] states that

When further intervention to prolong the life of a patient becomes futile, physicians have an
obligation to shift the intent of care toward comfort and closure. However, there are necessary value judgments involved in coming to the assessment of futility. These judgments
must give consideration to patient or proxy assessments of worthwhile outcome. They
should also take into account the physician or other provider's perception of intent in treatment, which should not be to prolong the dying process without benefit to the patient or to
others with legitimate interests. They may also take into account community and institutional standards, which in turn may have used physiological or functional outcome measures.
Nevertheless, conflicts between the parties may persist in determining what is futility in the

particular instance. This may interrupt satisfactory decision-making and adversely affect patient care, family satisfaction, and physician–clinical team functioning.

What steps should be taken to prevent or respond to disagreements over medically futile care? The EPEC Curriculum [4] recommends a Process-Based Approach to the Resolution of Issues of Clinical Futility, also called a "Due Process-Based Approach" by the American Medical Association in its Code of Medical Ethics, which states in relevant part [13].

> To assist in fair and satisfactory decision-making about what constitutes futile intervention: (1) All health care institutions, whether large or small, should adopt a policy on medical futility; and (2) Policies on medical futility should follow a due process approach. The following seven steps should be included in such a due process approach to declaring futility in specific cases.

(a) Earnest attempts should be made in advance to deliberate over and negotiate prior understandings between patient, proxy, and physician on what constitutes futile care for the patient, and what falls within acceptable limits for the physician, family, and possibly also the institution.

(b) Joint decision-making should occur between patient or proxy and physician to the maximum extent possible.

(c) Attempts should be made to negotiate disagreements if they arise, and to reach resolution within all parties' acceptable limits, with the assistance of consultants as appropriate.

(d) Involvement of an institutional committee such as the ethics committee should be requested if disagreements are irresolvable.

(e) If the institutional review supports the patient's position and the physician remains unpersuaded, transfer of care to another physician within the institution may be arranged.

(f) If the process supports the physician's position and the patient/proxy remains unpersuaded, transfer to another institution may be sought and, if done, should be supported by the transferring and receiving institution.

(g) If transfer is not possible, the intervention need not be offered.

The rationale underlying this Process-Based Approach [8] is that such an approach

- Is securely rooted in the moral tradition of promoting and defending human dignity
- Protects the patient's right to self-determination (another physician or institution may be sought)
- Protects the physician's right of professional integrity (the patient cannot force a physician to do something s/he believes is medically inappropriate)

Even though it is a fundamental tenet of medical ethics that physicians do not have an ethical obligation to offer or provide treatment that is medically inappropriate [9, 10], often physicians do provide such care, out of fear of litigation. To address such concerns, as well as to provide protections to patients, ten states (Alaska, California, Delaware, Hawaii, Maine, Mississippi, New Jersey, New Mexico, Tennessee, Texas and Wyoming) have codified the right of physicians to refuse to provide "medically ineffective" or "medically inappropriate care," as well as the procedures to be followed in such situations [14].

In California, for example, state law reflects the principles of clinical ethics heretofore discussed. Sections of the Probate Code [15] address issues of surrogate decision-making; the professional integrity of the physician; and those situations wherein there is lack of agreement between patient/surrogate and physician because of perceived futile care, thereby providing guidance to, and legal protection of, physicians, hospitals, and patients. Representative examples include:

Probate Code Section 4714
A surrogate, including a person acting as a surrogate, shall make a health care decision in accordance with the patient's individual health care instructions, if any, and other wishes to the extent known to the surrogate.

Probate Code Section 4735
A health care provider or health care institution may decline to comply with an individual health care instruction or health care decision that requires medically ineffective health care or health care contrary to generally accepted health care standards applicable to the health care provider or institution.

Probate Code Section 4736 states, in relevant part

1. A health care provider or health care institution that declines to comply with an individual health care instruction or health care decision shall do all of the following:

   (a) Promptly so inform the patient, if possible, and any person then authorized to make health care decisions for the patient.
   (b) Unless the patient or person then authorized to make health care decisions for the patient refuses assistance, immediately make all reasonable efforts to assist in the transfer of the patient to another health care provider or institution that is willing to comply with the instruction or decision.
   (c) Provide continuing care to the patient until a transfer can be accomplished or until it appears that a transfer cannot be accomplished. In all cases, appropriate pain relief and other palliative care shall be continued.

Other states may well wish to consider such codification so that the ethical precepts discussed herein are reinforced by the weight of law.

Some Thoughts in Closing

It is hoped that the approaches presented in this chapter will be helpful to clinicians in reducing the potential for confrontations concerning medically futile care, as well providing the ethical (and in some jurisdictions, legal) foundation on which doctors can "Say No."

Each situation must be addressed based on the unique facts that are present, looking at the totality thereof. Honest communication with patient, family or other surrogate is absolutely essential in providing the necessary information, understanding the patients values, and developing the trust that is necessary in life and death situations.

Throughout our discussions we need to be humble. As we make recommendations and decisions, we need to constantly remember that, as human beings, we are not perfect. While such fallibility may seem absolutely antithetical to overriding the request of patient or surrogate, based on medical futility, what other realistic option exists? We do the best we can, with humility.

Furthermore, as we speak with patients and families, they need to know that we care. As the famous physician Peabody said years ago [16].

The secret of the care of the patient is in caring for the patient.

And finally, we should be ever mindful of what we are about, as reflected in the (modern English version of the) Hippocratic aphorism [17], that medicine seeks

To cure some, help many, comfort all.

References

1. Old J. Hospice care and end-of-life issues: paradigm shifts. American Academy of Family Physicians Scientific Assembly. [Online] September 2010 [Cited: 4 June 2011]. http://www. aafp.org/online/etc/medialib/aafp_org/documents/cme/courses/conf/ assembly/2010handouts/136.Par.0001.File.tmp/136-137.pdf.
2. Jonsen AR, Siegler M, Winslade WJ. Clinical ethics 5th edition. New York: McGraw-Hill; 2002.
3. Schwartz A, Bergus G. Medical decision making: a physician's guide. Cambridge: Cambridge University Press; 2008.
4. Emanuel LL, von Gunten CF, Ferris FD, Hauser JM, editors. The education in palliative and end-of-life care (EPEC) curriculum. Chicago: The EPEC Project; 2003.
5. Community Medical Centers. Policy on medically ineffective interventions (futility). Fresno: s.n.; 2010.
6. Shaner DM. Up in the air - suspending ethical medical practice. N Eng J Med. [Online] 30 Nov 2010. [Cited: 5 June 2011.] http://www.nejm.org/doi/full/10.1056/NEJMp1006331#t=articl.
7. Short S. Too much for that: a social critique of medical technology in late modernity. *Australian Review of Public Affairs*. [Online] November 2010. [Cited: 5 June 2011.] http://www.austra-lianreview.net/digest/2010/11/short.html.
8. Clark PA. Medical futility: legal and ethical analysis. Am Med Assoc J Ethics: virtualmentor. org. [Online] May 2007. http://virtualmentor.ama-assn.org/2007/05/msoc1-0705.html.
9. Fletcher JC, Lombardo PA, Marshall MF, Miller FG, editors. Introduction to clinical ethics 2nd edition. Frederick, MD: University Publishing Group; 1997.
10. Council on Judicial and Ethical Affairs. Opinion 2.035 - futile care. Am Med Assoc. [Online] June 1994. [Cited: 5 June 2011.] http://www.ama-assn.org/ama/pub/physician-resources/ medical-ethics/code-medical-ethics/opinion2035.page?
11. Brigham and Women's Hospital. Policy on medical futility and end-of-life care. Boston : s.n.; 2003.
12. Knapp van Bogaert D, Ogunbanjo GA. The principle of proportionality: foregoing/withdrawing life support. SA Fam Pract. [Online] 2005. [Cited: 5 June 2011.] www.safpj.co.za/index.php/ safpj/article/download/295/295.
13. Council on Judicial and Ethical Affairs. Opinion 2.037 - medical futility in end-of-life care. Am Med Association. [Online] June 1997. [Cited: 5 June 2011.] http://www.ama-assn.org/ ama/pub/physician-resources/medical-ethics/code-medical-ethics/opinion2037.page?

14. Anderson JS. Managing patients or families who demand medically futile care. [Online] [Cited: 5 June 2011.] http://www.fammed.ouhsc.edu/Palliative-Care/documents/Medically%20 Futile%20Care.ppt.
15. State of California. California probate code. http://www.leginfo.ca.gov/calaw.html. [Online].
16. Peabody, FW. The care of the patient. JAMA Classics. [Online] 19 March 1927. [Cited: 5 June 2011.] http://jama.ama-assn.org/content/suppl/2009/04/16/301.16.1710.DC1/ JAMAclassics042209.pdf.
17. Cure sometimes, treat often, comfort always. A Country Doctor Writes. [Online] 14 September 2010. [Cited: 5 June 2011.] http://acountrydoctorwrites.wordpress.com/2010/09/14/cure-sometimes-treat-often-comfort-always/.

Chapter 29
Third Critique of Buchman and Chalfin's Analysis

Charles L. Sprung

Drs. Buchman and Chalfin's chapter "Changing normative beliefs and expectations in critical care for the McDonnell Norms Group" identifies several important issues related to critical care medicine and health care in general worldwide. In formulating a workable plan to care for sick people wordwide in the future, several important issues which have been identified in this and previous chapters must be addressed. These include issues related to individual rights vs. collective good, the physician's duty to the individual patient or to society, the patient's right to die evolving into a duty to die, who decides, what is benefit who defines it and the process of intensive care triage.

Is the physician's Duty to the Individual Patient or to Society?

Patient autonomy enables patients to refuse life-prolonging treatments. Since the decision in Quinlan [1], legal decisions have reaffirmed this right repeatedly. The patient has a right to choose or reject a specific treatment which leads to an obligation on the physician to provide it or not. Over the past few years, there have been more reports of physicians unilaterally refusing patient or family requests for medical therapies, including life-prolonging interventions.

The physician's primary responsibility is to his/her individual patient [2]. Doctors are more frequently being pressured to consider society's requirements in addition to their individual patient's needs [2]. The principles of patient autonomy and beneficence are giving way to the principles of distributive justice and proportional advocacy. Physicians cannot discharge their responsibility to their individual patients if they at the same time attempt to conserve societal resources [2]. If physicians

C.L. Sprung, M.D. (✉)
Department of Anesthesiology and Critical Care Medicine, General Intensive Care Unit,
Hadassah Hebrew University Medical Center, P.O. Box 12000, Jerusalem 91120, Israel
e-mail: charles.sprung@ekmd.huji.ac.il

D.W. Crippen (ed.), *ICU Resource Allocation in the New Millennium: Will We Say "No"?*, 227
DOI 10.1007/978-1-4614-3866-3_29, © Springer Science+Business Media New York 2013

become more interested in societal needs than their patient's needs, the impetus for discovering the correct diagnosis, reversible illnesses or new potentially life-saving but expensive drugs will disappear.

Society and institutions, not the individual physician, must create guidelines limiting the availability of therapies [2]. Doctors have a duty to assist in developing societal policies as experts [3] but not at the expense of a patient while they are functioning as that patient's primary care physician. Physicians must continue to be their patient's chief advocate and do everything they believe may benefit their patient unless society or institutions establish guidelines that limit treatments they use.

Patient's Right to Die has Evolved into a Duty to Die [4]

Different opinions as to what is beneficial for a patient especially arise when a patient or family request continued life-prolonging therapies conflicting with doctor's recommendations. In the past, patient or family requests were usually accepted based on the principle of autonomy or the physician's primary responsibility being for the individual patient and beneficence as defined by the patient's values.

Changes in health care have led to alterations in the classical patient–doctor model and concerns for societal needs as noted above have increasingly overruled requirements for individual patients. Our ability to maintain patients with little chance of recovery and with great health costs has caused doctors to look more to the needs of society than individual patients.

Doctor's value judgments may be confused at times with medical indications for treatment. Doctors have defined treatments as futile or nonbeneficial because of their own values. They have even withdrawn life-prolonging treatments without patient or family input or against patient/family wishes. For some of their patients, the right to die has led to a duty to die.

Who Should Decide?

Many believe that the doctor should make the medical decisions and define what is medically reasonable [5]. In fact, it is the physician who decides when the medical situation is such that limiting life-sustaining therapies should be discussed. Doctor's recommendations to continue or forgo life-sustaining treatments are based not only on their medical expertise but often also on their own value systems, religion, and culture [6, 7]. Many patients, families, and even some doctors still believe patients or their surrogates should decide what is best for each individual patient based on autonomy and informed consent.

The most workable approach and the one that has become accepted on both sides of the Atlantic is shared decision making [8]. Shared decision making occurs when patients or their surrogates are involved with their providers in making health care decisions. Three professional societies in the United States (American Thoracic Society, Society of Critical Care Medicine, and the American College of Chest Physicians) strongly support the shared decision-making model and none advocated that the ultimate decision rest with doctors [8]. Currently, a project to develop Consensus Guidelines for *W*orldwide *E*nd of *L*ife *P*ractice for Patients in *I*ntensive *C*are *U*nits; WELPICUS is in progress.

It will be impossible to develop true societal guidelines and rules until there is a professional and societal consensus on the controversial issues. There have been consensus statements [9] that have proved helpful in offering some solutions. They have been used by professional organizations in developing guidelines and helpful for determining medical standards in court decisions.

Who Should Define Benefit?

Although there may be some agreement that doctors do not have to provide "futile," [10, 11] "useless," or "inadvisable" [11] treatments requested by patients or families, it is difficult to come up with an exact or objective definition of these terms. Factors that are important in determining benefit for ICU triage decisions include: likelihood of a successful outcome, patient's life expectancy due to disease, anticipated quality of life of the patient, wishes of the patient and/or surrogate, burdens for those affected and missed opportunities to treat other patients [12]. Presently there is no societal or even professional consensus on the definitions of futility or benefit or who should define benefit—the doctor or patient/surrogate. The patient's "best interests" may also be based on subjective factors including economic issues, quality of life as observed by the doctor, what may be best for the family, caregivers and society or even pressure caused by the needs of other patients. Experienced physicians may have disagreements as to what is beneficial or not. Interestingly, current practices include admitting patients to intensive care unit with little chance of survival or "benefit" [13, 14]. It is important in making assessments of benefit that value judgments not be confused with medical indications.

Process of Intensive Care Triage

The demand for intensive care unit (ICU) beds often exceeds supply and rationing of this scarce resource is common [15–17]. Even in the United States, where there are more beds than in the rest of the world, there are bed shortages and critically ill patients must sometimes be treated in medical or surgical departments [15, 18, 19]. With less available ICU beds, less patients are admitted and those admitted are usually more severely ill.

Unfortunately, ICU beds are not always utilized appropriately even in countries with adequate ICU beds. The majority of patients and families choose to be admitted to ICUs for a one month survival [20] and ICU physicians admit patients with little hope of surviving [13, 14]. The present criteria for ICU admission and refusal are not based on scientific facts. The criteria vary from institution to institution or even within the same institution. In addition, decisions can be arbitrary. Identifying the patient who may benefit from ICU care is extremely difficult. In general, patients in the middle illness severity ranges (APACHE II 11-20) admitted to the ICU rather than a ward had the greatest mortality benefit [21]. Patients at the two extremes of the risk of death spectrum, relatively low risk of death and high risk of death are respectively "too well to benefit" and "too sick to benefit" from critical care services [22].

Several medical societies have published recommendations for the method to be used to triage patients to the ICU. The Society of Critical Care Medicine recommended giving priority to patients likely to benefit most [12, 22] whereas the American Thoracic Society recommended using a "first come, first served" basis [23]. Unfortunately, doctors do not follow the intensive care triage recommendations [24].

Is There a Consensus?

There probably is consensus in most countries that people should receive a certain minimum level of health care. It is probably also agreed that health care is extremely expensive, inefficient, not equitably distributed and inadequate for all those that require it. Most individuals also believe that as bad as the problem of scarce health resources is today it will probably worsen in the years to come with increasing costs and an aging population with many chronic disorders. Most people around the world do theoretically understand that with the rising costs of health care and the lack of sufficient resources, many patients who qualify for certain treatments, interventions or care (especially intensive care) will not always be able to receive it. Unfortunately, they typically do not accept this when it involves themselves or their family members.

There is still no consensus related to the doctor's duty to the individual patient vs. society, the patient's right to die vs. duty to die, who decides, what is benefit who defines it and the process of intensive care triage.

As far as the Buchman and Chalfin strategies,

Strategy 1: Focus on health and health benefit, not on financing
It is clear that we should focus on health and not financing but what is health especially in terms of critical care? How much do ICUs really improve health except for patients with an acute life-threatening illness? Medical uncertainty is a big problem in terms of restoring health and benefits. It is hard to know when to stop interventions and therapies as the end approaches. How do we balance the high cost of ICU, medications and procedures at the end of life when our prognostication may be incorrect and the patient who we think will only live a few days or weeks survives for a few years?

People can choose to pay more to receive more which includes expensive therapies but the state should not have to pay for these choices. How do we decide more intelligently what should and should not be included in state sponsored healthcare? More importantly, who makes the decision?

Strategy 2: Emphasize shared responsibility and collaboration between patient and provider
This is very nice theoretically but what does a physician do when the patient is unwilling to take the responsibility for his/her own heath? Shall we not take care of the chronic lung disease patient who smoked for 30 years and who now has severe COPD or lung cancer? What about the previous alcoholic who now needs a liver transplant?

Strategy 3: Highlight health education throughout general and health professional curricula
Again these are nice words in theory but what about practical realities? It is unclear how education will actually affect and change physician and patient behavior.

Strategy 4: Establish and enforce standard presentations of available estimates of effectiveness, risks, benefits and value using measures that are easily understood by patients, providers and payers
With so much medical uncertainty how can information on optimal therapies and expected outcomes for treatments be given that is accurate? Even if the information is accurate, how will doctors or society force patients and families to use these data to come to "reasonable" decisions? Many families will often still wait for the miracle or because of their religious or cultural background find the doctor's "reasonable" approach unacceptable. As seen by the SUPPORT study, even if information is given and patient–physician communication improves, these changes do not help improve medical decisions or care that impacts on important endpoints [25].

A GLOBAL Health Care Delivery System for the Future

Can a GLOBAL health care delivery system be fashioned? My answer is an unequivocal NO. As noted by the diverse practices around the world and the lack of consensus for the important issues noted above, it is unlikely that a GLOBAL health care delivery system can be developed that would be acceptable and practical for all countries and cultures. Each country will most likely insist on its way of providing health care and it is extremely unlikely that they would be willing to give up their independence for making decisions for their health care. Even if a system were developed and accepted, policing compliance with the new system would be difficult if not impossible in most countries. The minimal level of health care provided by a country for its citizens will also vary widely based on the country, its economic situation and its culture. Some countries such as England have a tradition of accepting rationing while others do not. Several decades ago elderly patients in England with end stage renal failure were told by their doctors that there was nothing more to do for them (as the elderly were not routinely dialyzed) and they accepted their verdict and died. In the United States or Israel, patients or their families would not accept such a decree and do whatever was required to have the patient dialyzed. It appears that a global system for financing and administrating care would not be workable due to different parochial interests and priorities. Clearly, many third world countries will not be able to afford certain minimal levels of health care that will be given in Western countries.

What Then is Proposed to Improve Global Health Care Delivery?

The physician's primary responsibility should always be his/her individual patient. Doctors should help develop societal policies as experts but not at the expense of a patient while they are functioning as that patient's primary care physician. Although from a strict ethical perspective ICU beds are best triaged using a "first come, first served" system, it is most likely that triage will continue to occur giving priority to patients likely to benefit most. Patients with middle illness severity should be the highest priority for admission. All patients meeting ICU admission criteria will not be able to be admitted to the ICU in most countries. Without an adequate number of ICU beds, "saying no" to some patients is ethically permissible [4].

Doctors should work with patients and their families to develop a mutually acceptable course of treatment [9] through shared decision making [8]. This course might include providing medical treatments on a limited basis which even if believed to be inadvisable to the physician, are meaningful to the patient [9]. Patients, families and some health care workers must be taught and will hopefully come to recognize that "doing everything" is not always best for the patient.

For the majority of patients and families, agreement occurs, if not immediately then certainly over a period of time. Most families who do not accept physician recommendations do so in time when they see the around the clock devotion and caring of the ICU medical and nursing staff and the continued deterioration in the patient's condition. There are, however, a minority of patients and families who continue to refuse doctor recommendations for patients with little or no likelihood of improving. These typically include patients with severe brain injuries or chronic intensive care for organ system failure. The families are waiting for a miracle and are unwilling to limit therapy in any way.

If these patients require ICU care with little likelihood of "benefit" and other patients are continually denied life-saving ICU care, what should a physician do? This will obviously depend on the medical realities of each country. For instance, in Israel one half of ventilated patients (many of whom could benefit from ICU care) are treated in regular medical and surgical wards [17]. In most countries, however, ventilated patients can only be cared for in the ICU and cannot be discharged from the ICU. Therefore, in Israel these patients can be discharged to a ward which is impossible in most countries. In situations where the family opposes discharging a patient with little or no likelihood of improving from the ICU to a ward or other type of facility, physicians should ask for an ethics consultation or seek advice from an Ethics Committee. Many institutions have Ethics Committees which hear the facts of the case from the health care providers and the patient and/or the family. Although the committee decisions are usually not binding and only advisory they can sometimes be helpful in resolving difficult issues and conflicts. If this does not help, the final solution may have to be decided in a court of law. Unilateral actions by physicians to withdraw life-sustaining treatments over family objections as occurs in certain Canadian provinces should be avoided. Doctor's actions not honoring patient or family requests especially if based more on society's needs than those of the individual patient will undermine trust in the medical profession.

In summary, there are many difficult issues for which consensus has not yet occurred. Until this does occur, it will be difficult if not impossible to develop a workable plan to care for sick people worldwide.

References

1. In re Quinlan, 70 N.J. 10, 355 A.2d 647, 1976.
2. Levinsky NG. The doctor's master. N Engl J Med. 1984;311:1573–5.
3. Sprung CL, Raphaely RC. Responsibilities of critical care professionals in setting medical policies for foregoing life-sustaining treatments. Crit Care Med. 1990;18:787.
4. Sprung CL, Eidelman LA, Steinberg A. Is the patient's right to die evolving into a duty to die?: Medical decision making and ethical evaluations in health care. J Eval Clin Pract. 1997;3: 69–75.
5. Tomlinson T, Brody H. Futility and the ethics of resuscitation. JAMA. 1990;264:1276–80.
6. Sprung CL, Cohen SL, Sjokvist P, et al. End of life practices in European intensive care units – The Ethicus Study. JAMA. 2003;290:790–7.
7. Sprung CL, Maia P, Bulow HH, et al.; the Ethicus Study Group. The importance of religious affiliation and culture on end-of-life decisions in European intensive care units. Intensive Care Med. 2007;33:1732–9.
8. Carlet J, Thijs LG, Antonelli M, et al. Challenges in end-of-life care in the ICU. Statement of the 5th International Consensus Conference in Critical Care: Brussels, Belgium, April 2003. Intensive Care Med. 2004;30:770–84.
9. Task Force on Ethics of the Society of Critical Care Medicine. Consensus report on the ethics of foregoing life-sustaining treatments in the critically ill. Crit Care Med. 1990;18:1435–9.
10. Truog RD, Brett AS, Frader J. The problem with futility. N Engl J Med. 1992;326:1560–4.
11. Society of Critical Care Medicine. Consensus statement of the Society of Critical Care Medicine's Ethics Committee regarding futile and other possibly inadvisable treatments. Crit Care Med. 1997;25:887–91.
12. Society of Critical Care Medicine Ethics Committee. Consensus statement on the Triage of Critically Ill Patients. JAMA. 1994;271:1200–3.
13. The Society of Critical Care Medicine Ethics Committee. Attitudes of critical care medicine professionals concerning distribution of intensive care resources. Crit Care Med. 1994;22: 358–62.
14. Vincent JL. European attitudes towards ethical problems in intensive care medicine: results of an ethical questionnaire. Intensive Care Med. 1990;16:256–64.
15. Strauss MJ, LoGerfo JP, Yeltatzie JA, Temkin N, Hudson LD. Rationing of intensive care unit services. An everyday occurance. JAMA. 1986;255:1143–6.
16. Lyons RA, Wareham K, Hutchings HA, Major E, Ferguson B. Population requirement for adult critical-care beds:a prospective quantitative and qualitative study. Lancet. 2000;355:595–8.

17. Simchen E, Sprung CL, Galai N, et al. Survival of critically ill patients hospitalized in and out of intensive care units under paucity of intensive care unit beds. Crit Care Med. 2004;32: 1654–61.

18. Singer DE, Carr PL, Mulley AG, et al. Rationing intensive care—physician responses to a resource shortage. N Engl J Med. 1983;309:1155–60.

19. Franklin C, Rackow EC, Mandoni B, et al. Triage considerations in medical intensive care. Arch Intern Med. 1990;150:1455–9.

20. Danis M, Patrick DL, Southerland LI, et al. Patients' and families' preferences for medical intensive care. JAMA. 1988;260:797–802.

21. Sprung CL, Geber D, Eidelman LA, et al. Evaluation of triage decisions for intensive care admission. Crit Care Med. 1999;27:1073–9.

22. Task Force of the American College of Critical Care Medicine, Society of Critical Care Medicine. Guidelines for intensive care unit admission, discharge and triage. Crit Care Med. 1999;27:633–8.

23. ATS Board of Directors. Fair allocation of intensive care unit resources. Am J Respir Crit Care Med. 1997;156:1282–301.

24. Azoulay E, Pochard F, Chevret S. Compliance with triage to intensive care recommendations. Crit Care Med. 2001;29:2132–216.

25. The SUPPORT Principal Investigators. A controlled trial to improve care for seriously ill hospitalized patients. The study to understand prognoses and preferences for outcomes and risks of treatments (SUPPORT). JAMA. 1995;274:1591–8.

Chapter 30
Fourth Critique of Buchman and Chalfin's Analysis

Richard Burrows

Science demands precise measurements with defined tolerances. Flight requires measurements of lift, drag, thrust and gravity without which one leaps off a tower with nothing more than blind faith and a more certain result than that afforded the Wright brothers. The social sciences also need measurements without which there can be no rational programmes to improve the lot of society. But these measurements cannot be made without a definition as to what one is measuring and while some definitions such as infant mortality, maternal mortality and longevity are precise, the social sciences also use abstract terms as "quality", "equality", "fairness" and "justice" that are often highly subjective and measured by "reasonable" consensus that makes the social sciences more Art than Science as illustrated in Dunstan's words: ... It is to be measured by the quality of lives preserved or restored; and by the quality of the dying of those in whose interest it is to die; and by the quality of human relationships involved in each death [1].

Reading the various reports from around the world one is struck by the similarities whereby all jurisdictions insist on for "health for all" measured by these abstractions. As Buckman and Chalfin point out, the World Health Organisation definition of health is "a state of complete physical, mental and social well-being and not merely the absence of disease or infirmity". Ways and means are investigated as to how such a highly subjective "health for all" can be justly, equally and fairly achieved and delivered as a quality service. Small wonder all attempts quickly fall between the stools of lofty policy on the one hand and harsh economics on the other.

In order to try to see where we are going we have to try to understand where we have come from. Modern science, in particular medical science, has been an extraordinarily recent phenomenon with early dramatic gains in the twentieth century that are seldom seen in the new millennium. Prior to the last century medicine was little

R. Burrows, M.B., B.Ch., F.C.A. (SA) (Critical Care) (SA) (✉)
Private Practice - Bon Secours Hospital, Galway, Ireland
e-mail: sworrub@gmail.com

D.W. Crippen (ed.), *ICU Resource Allocation in the New Millennium: Will We Say "No"?*, 235
DOI 10.1007/978-1-4614-3866-3_30, © Springer Science+Business Media New York 2013

more than a belief system based on the observations of Galen. Death was, and still is, the enemy undefeated [2] and disease was visited on man by God. "Influenza" was visited on a sinful humanity and furthermore if life was not lived according to the dictates of those in authority then the "afterlife" was consignment to "fire and brimstone". Physicians, it seems, were not slow in adopting an authoritative, paternalistic attitude either. GB Shaw's Dubedat: "It's always the same with the inartistic professions: when they're beaten in argument they fall back on intimidation. Never knew a lawyer yet who didn't threaten to put me in prison sooner or later. I never knew a parson who didn't threaten me with damnation. And now you threaten me with death".

In recent years education, technology and ethical discourse have subtly secularised society such that the influence of these aforesaid authorities, to a greater or lesser extent, has been thrown over in favour of the authority of the individual who has become master of his or her own destiny with his or her own autonomy viewed as an unassailable right entrenched in law particularly in the First World. In his book "The Logic of Failure" Dietrich Dörner [3] states: "Thought is embedded in a context of feeling and affect; Thought influences, and is in turn influenced by, that context. Thought is also always rooted in values and motivations. We ordinarily think not for the sake of thinking but to achieve certain goals based on our system of values". The autonomous individual is hardly immune from this and quickly creates a context of authority wherein what he or she desires must be delivered unto him or her as a right. It necessarily follows that where an individual considers himself or herself to have a right or entitlement to something then someone else has a duty to deliver that right to him or her.

Of course society can insist on reasonable duties and rights but it is not easy, if even possible, to define reasonable in this context such that a belief system that has conditioned the individual to treat disease and death as something to be defeated or avoided at all costs be set aside. Also, from the clinician's point of view, "There is a point at which the most energetic policeman or doctor, when called upon to deal with an apparently drowned person, gives up artificial respiration, although it is never possible to declare with certainty, at any point short of decomposition, that another five minutes of the exercise would not effect resuscitation" [4]. Shaw's statement illustrates the basic point of the difficulties in stopping resuscitation or treatment—a decision to stop is a decision based on probability such that stopping will result in an unfortunate outcome at some moment, especially in view of the fact that prognostic systems do not apply to the individual case.

Thus the ultimate certainty of death for the individual is quickly measured against the uncertainty of the timing and errors in the diagnosis of that death. Any attempt to limit health care on the basis of diminishing returns is invariably countered by anecdotal cases of recovery against all the odds with criticism of any decision to limit care. Lack of certainty and a perceived duty to persist means costs must increase dramatically as technological imperatives are activated and increased (stacked) as death approaches. The individual's right to life and health is observed and the scene is set for a "Tragedy of the Commons" [5] whereby the resources demanded for the individual trump any limitations that might be considered and the collective good suffers, eventually to such a degree as to finally affect the individual.

The Seductive Promise of Medicine

In spite of much advancement in medicine the most effective interventions are overwhelmingly of the cheap and simple kind [6]. Antibiotics on their introduction were regarded as miracle drugs dramatically reducing mortality in infectious diseases but have been increasingly used to the point of severe bacterial resistance reducing their benefits and perhaps leading to harm. Likewise in Intensive Care the introduction of ventilation in the neonatal tetanus unit of King Edward VIII hospital in Durban reduced the mortality of neonatal tetanus by 50% [7]. Such simple, effective treatment has evolved into sophisticated technology with smaller benefits that require bigger trials in an attempt to show marginal benefits. Simple fluid resuscitation soon falls prey to arguments concerning the nature and type of fluid.

Drugs such as steroids in sepsis have their detractors and supporters wherein arguments develop near religious adherents. Benefits are justified by comparison with the downside of disease and death and can be manipulated by vested interests of profit motive and personal benefit to the point of outright fraud in fields as disparate as Stem Cells, Pain, Obesity, T-Cells, Anaesthesia, Vaccination, etc. Even when modern medicine fails the individual may still reject the inevitable preferring to indulge in the use of unproven medications that, at best, do no harm but also can be frankly dangerous. With the public credulity of success through technology allied with fallible medical decisions it is not difficult to understand why some individuals will insist that resuscitation be continued no matter how vanishingly small the probability of success might be.

The Business of Medicine

Medicine has become a business replete with jargon, plans, customer satisfaction, budgeting, objectives, targets, shareholders and so on. But unlike business in other spheres this business now has an open-ended customer base built on fear and a customer empowered to insist on his or her wants being serviced—and he or she will be serviced. Shaw again "it is not the fault of our doctors that the medical service of the community, as at present provided for, is a murderous absurdity". That any sane nation, having observed that you could provide for the supply of bread by giving bakers a pecuniary interest in baking for you, should go on to give a surgeon a pecuniary interest in cutting off your leg, is enough to make one despair of political humanity. But that is precisely what we have done [4].

Those words were written in 1906 decades before the formation of the NHS at a time when professional remuneration was essentially fee-per-item in the UK. That highlights the problem of a business model wherein the medical decision of the doctor can be influenced by a pecuniary interest [8] and whatever the system that is in operation costs are also increased by a pharmaceutical industry that all too often appears to profiteer out of illness [9, 10]. The cost of technology has forced a rapid evolution from a simple doctor/patient relationship to a more complex relationship as third parties such as the insurance industry and the taxpayer that have assumed control over the financial issues. The doctor has a duty not to waste resources and thus falls between the two stools of duty to patient on the one hand and economic responsibilities on the other [11]. The situation is difficult enough for a clinician of probity and high integrity but doctors, according to Shaw, "are no different from other Englishmen" and Chaucer noted "For gold in physic is a cordial: Therefore he loved gold in special" [12]. Braithwaite observed University scientists to be more than willing pawns as "an FDA survey of safety testing violations that have shown that university laboratories had the worst record for violations than all other laboratories in the survey".

Issues such as bribery, safety testing of drugs, antitrust and drug pushing were common prior to the 1980s [13] but at least in the 1980s there was a corporate attitude that direct consumer advertising was wrong in that it would likely lead to the very real possibility of causing harm by a target audience pressuring physician into prescribing unnecessary medications. This attitude, however, was set aside with the introduction of Acts such as Bayh-Dole and Hatch-Waxman [14] in the USA that allowed fundamental changes in drug research and licensing eventually forcing the Federal Drugs Agency (FDA) to loosen their requirements for direct advertising [15]. The upshot has been an industry that advertises its wares to all and sundry with an enormous take that saw, in 2002, the top ten drug company profits equal the rest of the Fortune 500 companies combined.

And the problem is not confined to drug prescriptions as society indulges itself in gizmo idolatry [16] along with sometimes questionable preventative and prophylactic medicine in colonoscopies, gastroscopies, PSA tests, cholesterol levels, mammograms, bone density scans, carotid Doppler, light speed VCT scans and so on. "It is incident to physicians, I am afraid, beyond all other men, to mistake subsequence for consequence" (Samuel Johnson). In fact the real prize for industry is a credulous medical community seduced into the belief that their heroic efforts are legitimised by the occasional survivor attributed as a consequence of their actions. The upshot of this is a public perception that medical advances are there to benefit the medical industry, reflected in a loss of trust in decisions contrary to one's own perceptions. It would seem that modern medicine has become a sophisticated exercise in marketing, servicing a fear of death and disease well illustrated by out-of-pocket expenses, responsible for 60% of bankruptcies in the USA [17].

Medical Organisation and Costs

It has to be accepted that any service carries a financial burden and, fundamentally, there are only three ways that these financial costs can be addressed. Firstly there is out-of-pocket expenses, secondly taxation and thirdly insurance. Various combinations of these can be employed such as the Beveridge or Bismark models but they all have one indispensable prerequisite—the money must come from somewhere. To put it another way: in order to support a service through taxation or insurance the bulk of the workforce must be gainfully employed. In the extreme a plague such as might be caused by the flu virus has already shown its ability to overwhelm the system should it strike with full effect. In South Africa the HIV virus has already damaged society with little chance of bringing treatment to all who are in need. But such extremes are not really the issue. Depending on what is meant by "basic", "reasonable" or "quality" the issue is what sort of reasonable, basic system can be delivered as a quality service within financial constraints of a particular society— particularly when those finances fail. There has to be a point where the system is stretched to breaking whereby expenditure exceeds income leading to a situation where even deserving cases cannot be treated. That point will be different for different societies and furthermore, in any particular society, that point will be complicated by debilitating arguments of unfair distribution of resources.

In South Africa the government admits to unemployment as high as 40% yet still insists on a degree of universal health care that is little more than a pipe dream and particularly in the AIDS epidemic where health care has been hindered by political denial [18] and political instructions to deliver service beyond the limits of the system. While it is probably unlikely that other jurisdictions will be affected to the same extent the global credit crunch has placed all systems throughout the world under pressure. In fact they have always been under pressure as even the founder of the National Health Service (NHS), Aneurin Bevan, recognised that there would always be short-falls in what could be delivered to the population as a whole and it has undergone change too in spite of Mrs. Thatcher or perhaps even because of her as her policies did not disappear with her resignation. The purchaser/provider scheme introduced in 1991 whereby the general practitioner (GP) "purchases" services for his or her patient from hospital and other "providers" often seems to serve as nothing more than a mechanism to pass the buck while denying treatment [19]. Even in 2011 the standard of care in the NHS has often been found to be severely wanting as care for elderly patients, unable to cope, falls way below that required due in part to cutbacks [20].

Where to Now?

No society is immune from the fear of disease and death. From the USA through the Middle East and Far East the problem is essentially the same. Where clinician decision-making does preclude prolonged dying in ICU it seems clear that the clinicians in charge of those units have learnt to make decisions to the best of their ability, under-standing that their decisions are fallible but are broadly trusted and supported by soci-ety as a whole. Should the culture of that society change in such a way as to refuse to recognise the fallibility of decision-making preferring instead to censure decisions that in hindsight might be less than perfect and thus blameworthy then the only recourse will be to force protracted, futile treatment with protracted legal discussion as appears to be the case in the USA as demonstrated, inter alia, by the Schiavo case [21].

Against this long evolved background Crippen gives us his Fair and Equitable Healthcare Act (FEHA). Inappropriate profit motives will be censured; good care will be rewarded; dishonesty is an unpardonable transgression; those with excessive complications and bad outcomes will be punished; the reasonable will be tolerated and funded. FEHA "will pay for ICU care delivered by fellowship trained and boarded critical care physicians but not for ICU care entirely provided by internists and general surgeons"—this in spite of some evidence that such management by critical care physicians might just have a worse outcome [22]! It will divert the "relatively insensitive tort system to provide our patients with 'no fault' compensation for damages sustained under our care". Furthermore this is only for patients FEHA agrees to insure as other jurisdictions must accept responsibility for those in penury. Put like this FEHA rather grates on sensibilities.

The practise of medicine throughout the ages has been according to the ethos that the physician will help his fellow-man no matter his circumstances and, uniquely, mankind has evolved among species to have the ability to control his environment through the use of technology. Wisdom of control is another matter. In the 1960s Wiesner and York [23] expressed their opinion in the dilemma of deteriorating military security that technological solutions would only serve to worsen the situation. A solution can only lie in a cultural solution. Similarly the use of medical technology requires a cultural solution in the way society deals with the technology. As it is, a decision-making hierarchical, culture has evolved that has given precedence to the patient's choice for good or bad only to be curtailed by the outrageous. To this end an advance directive would, prima facie, appear to solve most problems and failing that the patient's wishes as elucidated from relatives will be next.

The clinician's decision to stop only becomes valid where the demands are clearly outrageous or where there is a degree of certainty agreed by all in respect of the presumptive outcome. Ingelfinger [24] was of the opinion that the clinician's place was not merely that of a "vendor of vegetables inviting individuals to choose" but rather an active advisor, bringing to the table all his or her experience, probity and integrity to decide what is best. Clearly this has importance in helping patients or their relatives avoid bankruptcy through chasing improbable outcomes but it is of no use in the face of belligerent insistence to continue. As it is stated FEHA does not appear to allow for such discussion as patients and clinicians are expected à priory to accept the rules of FEHA and does not address the à posteriori problem of how to deal with the patient who has been rendered to penury through exhaustion of funds with legal action should they refuse to abide by the rules.

In any event some aspects of FEHA probably already exist. Managed Care, HMOs and the National Institute for Health and Clinical Excellence (NICE) already exist as mechanisms to curtail costs but are simply not stated with the same bluntness as FEHA. Where there is a shortfall of resources the problem is compounded. Medical oaths are silent on the topic. Ethicists do little more than simply point out the ambiguity of being advocate for the patient, treating all equally while not wasting resources. Legal threats hang about any decision as a post hoc ergo propter hoc system of redress.

Nobody will ever need more than 40 megabytes—Bill Gates

Predicting the future is the way of madness, the only certainties being birth, taxes and death.

The Future

"The theory that every individual alive is of infinite value is legislatively impracticable. No doubt the higher the life we secure to the individual by wise social organization, the greater his value is to the community, and the more pains we shall take to pull him through any temporary danger or disablement. But the man who costs more than he is worth is doomed by sound hygiene as inexorably as by sound economics." (Shaw) The thought that the economic situation might get so bad that those who could, or should, be treated will not get treated is anathema—both for the clinician and society but it happens in various parts of the world and threatens to happen in wealthier parts of the world. Speaking from personal experience it is relatively easy to rationalise a hierarchy of decisions that patients who are bedridden and terminally ill or who are suffering from a disease such as rabies should not be admitted or that resuscitation should stop in a patient who is no longer responding. The hierarchy might even develop along lines whereby those with chronic but likely treatable disease might be refused admission.

The pressure is quite different when there is no bed for a young mother with eclampsia or a child with acute severe asthma or any other patient who could rationally be expected to benefit from technological medicine. It simply goes against the grain of what it is to be a doctor or a nurse to be unable to treat a patient for economic reasons but that is what happens when resources fail to the extent they have in some parts of the world. Where resources do fail, either because of unrealistic expectations on the part of the public or because of a failed fiscus, then those with perceived accountability will quickly attempt to get any blame deflected. It is far easier for the clinician to use the excuse of "no beds" instead of accepting responsibility for the decision not to admit or to stop because survival is limited in the present circumstances. Administrators will be quick to point out that clinicians have not "managed" the resource, thus deflecting blame. Where there is a dispute between the clinicians and the administrators both sides will be quick to broadcast that "no harm will come to patients" as they seek public approval.

The clinician who stands up to justify a decision to stop because of economic reasons is more likely to be vilified as a general in the pay of the enemy than a doctor doing his or her best in difficult times. The German word Zeitgeist loosely translates into English as "the spirit/culture of the times" and is an expression that applies to society at a particular time and place. Where health is concerned it places a long evolved, albeit imperfect, autonomy above other interests. It is a spirit that also directs individuals to look after others who have fallen on hard times and it does not tolerate abuse but has extreme difficulty defining abuse but, paradoxically, easily apportions blame [25] as a punitive, corrective mechanism for events considered to be adverse outcomes [26]. It measures a nebulous definition of health against the ultimate certainty of death but fails in any decision that attempts to limit treatment where there is an aggressive push to continue in the face of the inevitable.

FEHA is just another symptom of an unsatisfactory politically expedient system developed in the face of limited resources. Unless the Zeitgeist changes in such a way that individuals are more comfortable with the inevitability of death and medical fallibility then the situation will remain as it is—resources as are available will be used until they are exhausted wherein waiting lists will lengthen, priorities of treatment will be assigned as best as possible and when those resources recover, which history teaches us is likely, then the situation will simply revert to what it previously was—waiting lists will shorten and pressures to assign priorities will lessen—but at all times death will be something that is to be avoided at all costs. It can be viewed as a paradigm [27] that allows "the system" to be blamed instead of any one individual or group of individuals, so everyone remains satisfied yet dissatisfied.

References

1. Dunstan GR. Hard questions in intensive care. A moralist answers questions put to him at a meeting of the Intensive Care Society, Autumn, 1984. Anaesthesia. 1985;40:479–82.
2. Illich I. Death undefeated: from medicine to medicalisation to systematisation. BMJ. 1995; 311(23–30):1652–3.
3. Dietrich Dörner. The logic of failure: recognizing and avoiding error in complex situations. New York: Metropolitan Books; 1996. p. 8.
4. George Bernard Shaw. Opening remarks - Prologue to Doctor's Dilemma.
5. Hardin G. The tragedy of the commons. Science. 1968;162:1243–8.
6. Illich I. Medical nemesis: the expropriation of health. New York: Random House; 1976.
7. Wright R, Sykes MK. Intermittent positive pressure respiration in tetanus neonatorum 278. Lancet. 1961;278(7204):678–80.
8. Price MR Broomberg J. The impact of the fee-for-service reimbursement system in the utilisation of health services: part III a comparison of caesarean section rates in White Nulliparous Women in the private and public sectors. SAMJ 1990;78:130–2.
9. Sade RM. Introduction dangerous liaisons? Industry relations with health professionals. J Law Med Ethics. 2009;37(3):398–400.
10. Fugh-Berman A, Ahari S. Following the script: how drug reps make friends and influence doctors. Plos Med. 2007;4(4):e150.
11. Levinsky NG Sounding Board "The Doctors Master" NEngl J Med 1984;13:1573–75.
12. Pertinax. BMJ 1967.
13. Brownlee S. Overtreated: why too much medicine is making us sicker and poorer. New York: Bloomsbury; 2007.
14. Angell M. The truth about drug companies. New york: Random House; 2004.
15. Brownlee S. Overtreated. New York: Bloomsbury; 2008.
16. Leff B, Finucane TE. Gizmo idolatry. JAMA. 2008;299(15):1830–2.
17. Himmelstein DU, Thorne D, et al. Medical bankruptcy in the United States, 2007: results of a national study. Am J Med. 2009;122(8):741–6.
18. Johnson RW. South Africa's brave new world. London: Penguin Books; 2010. pp. 186–93.
19. Paton C. Present dangers and future threats: some perverse incentives in the NHS reforms. BMJ. 1995;310:1245.
20. Quality Care Commision 2011 at http://www.cqc.org.uk/public/reports-surveys-and-reviews/reports/state-care-report-2010/11/access-care-and-services.
21. Jay W. A report to Governor Jeb Bush and the Sixth Judicial Circuit in the matter of Theresa Marie Schiavo. (2003) abstractappeal.com. Retrieved 2006-01-06. pp. 2, 8, 10.
22. Levy MM, Rappoport J, Lemeshow S, Chalfin DB, Phillips G, Danis M. Association between critical care physician management and patient mortality in the intensive care unit. Ann Intern Med. 2008;148(11):801–9.

23. Wiesner JB, York HF. Sci Am. 1964;211(4):27.
24. Ingelfinger JF. Arrogance. N Engl J Med. 1980;303:1507–11.
25. Marx D. Whack-a-mole: the price we pay for expecting perfection your side studios. ISBN-10: 0-615-28307-1.
26. Hobbs RJ. A managing maintenance error: a practical guide. Avalon: Ashgate Publishing; 2003.
27. Thomas S Kuhn. The Structure of Scientific Revolutions. University of Chicago Press Chicago - Library of Congress ISBN 0-226-45808-3 pp 10–22. Third edition; 1996.

Part II
The Fair and Equitable Health Care Act

Chapter 31
The Fair and Equitable Health Care Act

David W. Crippen

The Fair and Equitable Healthcare Act (FEHCA) is a health care insurance vehicle offered to persons with sufficient income to make the premium payments. The FEHCA is not available to poor persons, who are covered by government Medicaid and medical assistance programs. The FEHCA is a not-for-profit organization with a board of directors elected from the community at large. It is committed to providing the most effective health care possible. FEHCA premiums are kept low through aggressive risk management and prioritization of services.

Insurance, by its nature, cannot and does not totally cover every possible need of the patient pool. Such a policy would be unaffordable for any employer or individual. The FEHCA interprets prioritization as (1) the avoidance of disproportionate expense benefiting relatively few patients and (2) the maximization of availability of services that support the most patients in the most need. The FEHCA covers care, at the most affordable price possible, for injuries or illnesses most likely to befall most of its insured members.

What follows is an explanation of how the FEHCA responds to the needs of its insured.

Health care delivery. The FEHCA subscribes to the maxim that fee for service induces a profit motive. Physician-business people have an incentive to create demand and then to supply at a monolithic cost. Accordingly, the FEHCA employs providers who have a greater incentive to ensure quality care while maintaining a commitment to cost-effectiveness. Providers are on salary with the FEHCA, and it is a fair salary. They are financially comfortable, enjoy good personal benefits, and are not overworked. If they maintain a reasonable record, everyone is happy. If their

D.W. Crippen, M.D., F.C.C.M. (✉)
Departments of Critical Care Medicine and Neurological Surgery,
University of Pittsburgh School of Medicine, Pittsburgh, PA, USA
e-mail: crippen@pitt.edu

D.W. Crippen (ed.), *ICU Resource Allocation in the New Millennium: Will We Say "No"?*,   247
DOI 10.1007/978-1-4614-3866-3_31, © Springer Science+Business Media New York 2013

patients tend to have complications and bad outcomes, the physicians will no longer work for the FEHCA.

Physicians who sign on to provide services for the FEHCA will have their FEHCA activities tracked in an extensive database. Information entered includes how much time physicians spend in patient care, what resources they use, and the results of use of these resources. Physicians working the system for profit will be found out because of the data recorded, and they will be warned once and then dismissed if they do not conform. Anyone minimizing resource use to the detriment of patient care, either out of laziness or for profit, will receive the same treatment.

Paying for good care. The FEHCA does not believe that pay for performance is effective and instead pays for consistent good care. FEHCA statisticians and actuaries know what the outcome should be for any disease process because of the existence of large databases containing records of millions of similar patients. FEHCA physicians are expected to stay within a reasonable range of proficiency, established by the FEHCA.

Family practice/internal medicine. The FEHCA pays a reasonable amount for office workups of medical patients, including time spent considering diagnosis and treatment, laboratory tests, films, electrocardiograms, and the like. The FEHCA also pays for follow-up office visits, for which patients have reasonable and affordable co-pays. The FEHCA does not believe that hospitalists should take over the care of hospitalized patients. The FEHCA pays internists and family physicians to admit and care for hospitalized patients, and will pay for reasonable specialty consultations.

Emergency care. The FEHCA understands that emergency departments (EDs) offer 24/7 care for urgencies and must have adequate staffing around the clock. To cover the expense of maintaining readiness for the relatively few emergencies, EDs therefore also offer treatment for nonemergent care on demand. ED care is exceptionally expensive compared to the same care rendered in a family practitioner's office.

The FEHCA understands that the general public does not reliably discern what is an emergency and what is not. The FEHCA maintains a list of approved EDs that provide good care at a reasonable price and charge reasonable rates for nonemergent care. That list is provided to the FEHCA patient population. If a patient visits an approved facility, the FEHCA will pay for the care. If a patient visits an unapproved facility (unless he or she has an authentic emergency and is out of range of an approved facility), the patient will pay a hefty co-pay. In addition, the FEHCA will review all such admissions retrospectively. FEHCA medical directors will review ED admissions, looking for cases in which the ED was used for nonemergent care. These patients will be retroactively billed for some or all of the cost.

Surgery. The FEHCA pays for surgical procedures that its stable of trusted surgical advisors agree are needed. The FEHCA pays for follow-up and an amount of rehabilitation deemed reasonable by its physical medicine advisors. The FEHCA requires two opinions that surgery is indicated and has a fair chance of success. The FEHCA does not pay for cosmetic procedures of any kind, and its decision on what constitutes a cosmetic procedure is final. The FEHCA does not pay for chiropractic.

Intensive care. The FEHCA subscribes to "leapfrog"[1] criteria in dealing with the care of hospitalized critically ill patients. Accordingly, the FEHCA will pay for ICU care delivered by fellowship-trained and board-certified or board-eligible critical care physicians. The FEHCA will not pay for ICU care provided entirely by internists and general surgeons.

The FEHCA understands that adequate intensive care takes time. The FEHCA maintains a chart of how much ICU time is usual and customary for particular disease processes normally seen in ICUs. This list is compiled from large outcome databases. As soon as the FEHCA receives a report from an ICU physician that the patient has no significant chance of recovery, an FEHCA social worker will explain the available option (which the FEHCA will pay for) to the family: an immediate tracheostomy, a percutaneous gastrostomy tube, and transfer to a skilled nursing facility (where care will not be financed by the FEHCA). If the family refuses this option, the FEHCA will stop payment to the hospital and the family can begin meeting with hospital representatives to determine how continued ICU care will be paid for.

Skilled nursing facilities. The FEHCA will not pay for transfers to EDs or ICUs of skilled nursing facility patients with baseline dementia or permanent multiple organ system failure. If the families of these patients desire this care, they can arrange for financing it.

Other medical services. The FEHCA does not pay for any drug or procedure that its panel of experts considers experimental. The panel's decision is final. The FEHCA does not cover outpatient ophthalmological care or dental services, as these are usually not cost-effective to indemnify.

Complications. The FEHCA understands that complications occur and believes that patients and their families deserve some compensation when they do. The current tort system of compensation for medical negligence generates only a relatively marginal and unpredictable benefit at great expense and labor. Many patients with real injuries cannot get an attorney to finance the case because of the low potential for a large monetary award. Other patients might hire an attorney who will bully a malleable jury into inappropriately large judgments by evoking emotional responses. The FEHCA is sensitive to this issue and therefore diverts the relatively insensitive tort system to provide no-fault compensation for injuries sustained by patients receiving care from FEHCA's medical professionals.

The FEHCA maintains a schedule of compensation for injuries that is not negotiable. If an FEHCA surgeon leaves a clamp in an abdomen or amputates the wrong foot, or if a patient dies unexpectedly after being discharged from one of FEHCA's approved EDs, FEHCA will compensate the patient or the patient's family according to the schedule of compensation created by FEHCA's expert physicians, attorneys, and actuaries. Payments will be made in 30 days and will come from FEHCA's liability insurance account. More esoteric claims for injuries due to alleged negligence may require more review.

---

[1] http://www.leapfroggroup.org/for_hospitals.

As a stipulation of being insured by FEHCA, patients must agree to engage in arbitration before filing a lawsuit. If an allegedly injured patient refuses FEHCA's fair and equitable offer of compensation, and rejects the arbitration opinion that the compensation is indeed fair and equitable, the patient is at liberty to hire a private attorney and sue for a larger judgment. Should that occur, the patient will be referred to the agreement he or she signed with the FEHCA concerning compensation for injury, as follows:

"The FEHCA will make every effort to fairly and equitably compensate patients for harm resulting from complications while under the care of our health professionals. We will negotiate this monetary amount in good faith. If you, the patient, desire to negotiate through an attorney for monetary damages in excess of those tendered by us, we will then consider the bargain process not to be in good faith. We will then litigate your claim, and we will not at any point entertain further settlement. We will take every case to a jury. In most cases, physicians win jury verdicts for malpractice. If necessary, we will countersue you in order to be compensated for attorney expenses and time spent litigating your case, which we will consider capricious and without foundation. You will be required to hire private counsel for such a countersuit, as it does not fall under the contingency system of litigation."

Chapter 32
First Critique of the Fair and Equitable Health Care Act

Michael A. Rie and W. Andrew Kofke

Dr. Crippen has assembled this volume to explore how critical care medicine (CCM) professionals approach the care of patients when they or their proxies request marginal or no-benefit care consuming increasingly limited resources in the global critical care village. Most chapters in this volume attest to how CCM professionals operationalize the dilemma of supply versus demand imbalance in their national systems and cultures under variable numbers of ICU beds, technology infrastructure, professional manpower, and historical and cultural precedents in the context of worldwide bioethical pluralism [1].

The Fair and Equitable Health Care Act (FEHCA) legislative proposal seeks to address cost efficiency in individual patient care by imposing a professional behavioral norm of quasi evidence-based standards of care on the preexisting autonomous choices of individuals. FEHCA creates a code of professional conduct driven by efficiency (cost containment) and limitation of patient preferences and priorities *in the absence of national or regional differences in the rule of law.*

We assume that the FEHCA is intended for democratic nation states with some defined antecedent rule of law. This would not apply at present to countries like mainland China with one-party autocratic rule and limited personal freedoms, or possibly other countries as a manifestation of the phenomenon of bioethical diversity [1]. In democratic nations all citizens, regardless of wealth or penury, are entitled to vote for their elected governmental officers and legislators unless the society creates various classes of citizenship. In the United States, one person, one vote has been the rule since the birth of the republic though large numbers were historically

M.A. Rie, M.D.
Department of Anesthesiology, University of Kentucky College of Medicine,
Lexington, KY, USA
e-mail: marie00@email.uky.edu

W.A. Kofke, M.D., M.B.A., F.C.C.M. (✉)
Department of Anesthesiology and Critical Care, University of Pennsylvania,
7 Dulles, 3400Spruce Street, Philadelphia, PA 19104-4283, USA
e-mail: kofkea@uphs.upenn.edu

D.W. Crippen (ed.), *ICU Resource Allocation in the New Millennium: Will We Say "No"?*, 251
DOI 10.1007/978-1-4614-3866-3_32, © Springer Science+Business Media New York 2013

denied citizenship and voting rights. For American physicians the health care professions were lawfully created by the Common Law of America that originated in the English Common Law of centuries past. This system predates the American Constitution and sets forth the underlying fiduciary trust relationships to which all healthcare professionals are lawfully held. We have previously reviewed the framework for CCM under cost containment [2, 3] and explored in some depth how "benefits definitions" for medical service limitation of either patient or professionally driven demand within the fiduciary trust relationship could be crafted in health insurance contractual policy to achieve the mesoallocatory *(mid level) policy ends* sought in the FEHCA [4, 5]. However, a FEHCA proposal in legislation by itself would face multiple objections because it would change the entire doctrinal foundation (*Parens Patria*) of the American law regarding patient care [3, 4, 6]. Furthermore, CCM professionals of all accredited licensures in hospitals treating FEHCA-insured patients would have to contend with legally robust complaints of unequal standards of treatment for these patients compared to patients in other less generous health insurance plans.

It should be pointed out that Englehardt [7] very nicely articulated the ethical and moral developments in the United States which underlaid the current US health system. The buoyant expectations of the 1960s and 1970s produced an ideology that led Americans to expect:

1. The best of care
2. Equal care, and
3. Physician/patient choice, without
4. Runaway costs

Engelhardt eloquently argued, as does Crippen embodied in his FEHCA proposal, that these expectations were neither possible nor plausible.

The Oregonian ICU is a conceptual creation to address an entire healthcare program enacted in Oregon in the early 1990s for the indigent population and other patients who would be covered in the combined Federal and state Medicaid program under American law. Its creator, Dr. John Kitzhaber, who was and is again the Governor of Oregon, has indicated in a recorded presentation at the University of Kentucky public forum [4] the moral principle that when resources are insufficient, Oregon democratically chose to allocate resources and not patients [8]. In so doing Oregon chose morally prioritized (rationed) benefits [8–12]. Oregon became the first lawfully created explicit system for "saying no."

Moral, political, and legal critiques from outside Oregon were globally recorded in the early 1990s and are likely to reappear in 2012 when Governor Kitzhaber's administration submits the details of the new Oregon allocation system that will explicitly reformat health care funding priorities to the "medical home" [13] while decreasing resources to high-technology medical procedures and CCM consumption rights of Medicaid patients.

Table 32.1, based on the Oregon system, records the conceptual moral answers to the FEHCA's concerns about professional legal liability for saying no while providing the ability to raise or lower the quantity of benefits for the poor to the

Table 32.1 Fundamentals of a Pure Oregonian Health Care Policy

- When resources are insufficient, allocate resources not patients.
- In secular pluralist democracies, moral authority is acquired by negotiation between providers and consumers to prioritize prospectively the creation of a list of health insurance benefits. These policies define the macroeconomic moral response to illness and related financial burdens.
- Health policy that lacks moral prioritization of beneficial care when resources are insufficient mandates declining standards of care with measurable increases in the morbidity and mortality of patient populations.
- In democracies, the creation of universal access to health care in a population of unequal personal wealth requires transfer payments to the poor to be limited by the extent to which the majority consent to taxation.
- People and businesses create wealth. Governments requisition wealth.
- The liberal maxim that moral conflict requires moral neutrality while resources are consumed by fiat is abandoned.
- Religious and moral diversity of patients requires public toleration of these expressions of health care desires and financial support and accountability from private sources of wealth.
- Futile medical care is not a disability under the Americans with Disabilities Act. Futility is a publicly derived, morally contentful definition of unfortunate human circumstances defining the limits of human life and our communal resources.
- Resource allocation fuses responsibility and authority in a contractually limited, medically directed managerial structure.
- The fiduciary trust relationship between doctor and patient is preserved, but the right to requisition the resources of a basic health insurance plan is limited.
- Health insurance morality that fuses responsibility and authority to ration is regionally created.
- Oregon morally prioritizes liberty, equality, access, and prosperity in health care.
- Resource allocation policy legislatively limits individual entitlement claims in tort law.
- Unequal health care is acknowledged, and the basic package of care is guaranteed to all. Freedom to purchase more health care is preserved.
- The best of care for all is not attainable. The moral dimensions of medical prognostic probability and clinical impression must be monetarized in individual entitlement allocations at the mesoallocatory policy level.

Adapted from Rie [4]

extent that the majority of society is willing to be taxed. This was previously defined in moral terms as the "egalitarianism of altruism" in contrast to the present implicit resource allocation system of "egalitarianism of envy" [14, 15] widely employed in First World countries, especially with reference to CCM.

The Oregon Health Plan of 1989 permits the existence of multitiered health insurance systems in which the private ownership of money to purchase more healthcare benefits is lawful [8–12]. The malpractice protections for providers in more generous packages of insurance (like FEHCA) were and remain available in law for nongovernmental packages choosing to create coherent explicit benefit limitations (e.g., limitation of malpractice action) in their insurance contracts with new purchasers for prospective patients. The recently enacted changes in the Oregon Health Plan (HB 3650, 2011) [16] prioritizing preventive medical care and the "medical home" for the Medicaid population will rekindle attention in the existent malpractice tort exemptions for "saying no" in somewhat more generous insurance packages like FEHCA within the Oregon hospital system and the domain of procedural and technologically advanced care specialties of medicine.

The Oregonian ICUs seem necessary to invent as neither Oregon nor the Netherlands, which underwent a major health system reform adopting a universal basic health benefit package including critical care, failed to prioritize explicit benefit limitations within the CCM component of the basic health insurance. Whether one is a critical care professional in Amsterdam, Auckland, New Zealand; Albany, New York; or Ashdod, Israel, we are all needing some unifying principles of healthcare policy distribution to preserve our foundational ethical integrity of professionalism in critical care. Should we fail at this task, all patients will stop trusting us as their individual caregivers. Any hope of developing an American or more global FEHCA will have to be grounded in democratic policy creation by the responsible authorities creating health insurance in all nations with the understanding that a basic package of insurance for the poor must underlie the ability of a morally coherent society to adopt more generous packages of insurance like the FEHCA.

Specific Critique of FEHCA

While Dr. Crippen is quick to point to FEHCA as a legislative proposal creating a nonprofit insurance agency for the non-impoverished, why would anyone who is not poor purchase such a seemingly heartless insurance policy? If one is not impoverished and has funds, then nonprofit Blue Cross offers full care while assuring each patient full autonomy. Of course anyone buying such insurance is not "leeching as a taker" from communal largesse. In order for FEHCA to work, it has to be a replacement for government-sponsored insurance with a needs-based mandate to pay the basic premium. As such it is imposing a tax to support a rational Oregonian system [4, 9–12, 16] where everyone has a base FEHCA program and those who choose to have full autonomy pay for it either in cash for service or cash for more generous insurance benefits. As such the program has to include impoverished patients who are supported by the taxpayer. Notably, such a system has been tried in Oregon and is being further financially implemented in public health policy expenditures and primary care expansion in Oregon at this time [16]. This change will explicitly mandate letting some people die without the benefit of highly expensive unproven, yet rational, therapies and resisting the public outrage that will ensue. The revised Oregon Plan will provide comfort, palliative, and hospice care in the home most probably decompressing these services in ICU and hospital settings.

A program such as this will insist on strong evidentiary support for any expensive therapy. It means a profound change in the mores of mainstream America and perhaps other nations. It necessarily means a period of harsh societal debate with no assurance that FEHCA as envisioned would survive.

FEHCA will only pay for modalities that have strong evidence. Unfortunately sensible as it sounds, the concept of good evidence is increasingly problematic [17]. Will the assumptions of monetary value, the so-called Quality-Adjusted Life Years

(QALY) [18, 19], of 2007 be those of 2012? Or will the definitional criteria themselves be subject to change by financial imperative? This is a complex topic and its difficulty is addressed in the series of articles defining how to read and decide on the evidentiary value of a research publication [20]. Some of the issues are the following: How does one decide about a therapy that might induce harm [21]? How does one design a study to evaluate a therapy designed to treat a disease with multiple pathophysiologic pathways and significant physiologic and health system noise [22]? How does one confirm that an obviously effective modality is really evidence based, e.g., parachutes [23]? What if there is abundant evidence in focused preclinical studies but there is not yet formal finding of efficacy in noisy clinical trials, especially for a disease with a severe and expensive bad outcome like stroke or spinal cord injury? It is not straightforward deciding what evidence is good, what is bad, and what is plausible enough to support use in the context of a certain expensive life changing negative outcome without therapy. So any FEHCA program that is implemented will need to have a serious and open-minded consideration of these system structural determinants.

FEHCA Prioritizes

This means an operational strategic change avoiding great financial commitments for the few with reallocation of an acceptable somewhat less for the many with measurably greater overall health status for the many. Vaccinations do more good than ventilators. This means medical care for catastrophic illness may not be funded by a debt-ridden society that has to support infrastructure, defense, and debt service. It means that assessments will be needed of the specific thresholds for the adjusted life quality index [18, 19] versus the cost of sustaining life without consideration of resulting quality of life. It means assessing the probability of returning to a long-lived maker status versus the probability of becoming short-lived or a permanent long-lived taker of communal funds. Such a drastic and draconian imposition of explicit rationing will only happen if there is a priori debate and societal agreement. Given the strident position of a right to life of a large proportion of the US population, this will only occur when the United States faces the financial abyss.

To paraphrase the nineteenth-century philosopher Georg Hegel [24] when we cannot express our morality in money it is of a meager and unsatisfactory kind. Lord Kelvin [25] (nineteenth-century British physicist) said:

> "When you can measure what you are speaking about, and express it in numbers, you know something about it. But when you cannot measure it, when you cannot express it in numbers, your knowledge is of a meager and unsatisfactory kind: it may be the beginning of knowledge, but you have scarcely, in your thoughts, advanced to the stage of science, whatever the matter may be."

Since the earliest days of the Oregon Health Plan, religious and conservative elements of the US society have promoted the sanctity of a right to life but have not expressed what happens when this results in nonaccountable transfer payments for such rights (through taxation) due to insufficient resources and when the consequence of pursuing that right for individuals will result in a decrement in the quality of care to other patients under our care. The reality of these circumstances violates the very notion of unbridled freedom due to constraints by the human rights of others. This conflict of increasing secularization versus the Christian roots of the United States was discussed by Englehardt [7] who pointed out that even the Supreme court at one time ruled that the United States is a Christian country [26].

One of the virtues of improved outcome data systems via the electronic medical record (EMR) and integrated healthcare information systems will be the ability of resource constraints to document when the vector of cost versus quality achieves a negative slope as opposed to a positive one on a cost–benefit curve. We have previously identified this as Continuous Quality Decrement (CQD) based on application of Pareto economic principles [2, 3]. The ability to elucidate it epidemiologically will become more evident as information systems in real time improve that are both detailed in their data and user friendly for CCM professionals to benefit from them under production pressure stress in the clinical environment. The combination of resource stringency and resulting decrements in quality with denial of resources must clash with religious conservative insistence on a right to life. Negotiated reasonable outcomes achievable by critical care investments will make the FEHCA's strengths compelling to resource-limited societies.

FEHCA will alter current incentive structures in health care in multiple ways as can be seen to be developing presently in Oregon's new legislation [16]. First, in FEHCA fee for service is abolished as wasteful and inefficient. This will remove many incentives to work on the part of physicians and produce waiting lists for care. Given that many on the waiting list will die before they make it to the top of the list, this becomes an excellent form of implicit rationing that will save money for FEHCA in addition to stopping most unneeded expensive procedures. Dr. Crippen does not suggest this as a possible outcome or strategy for cost containment though it was a well-described and documented failing of managed care in the 1990s. Indeed, he seems to have a utopian notion of unlimited physician supply. Perhaps the FEHCA legislation will include a long-range plan to increase the number of physicians in the United States such that all can work reasonable hours and earn less money. In the meantime, Oregon's new initiative seems likely to make high-technology consumptive medicine, surgical and other procedures, and ICU utilization scheduled for reapportionment of the public health budget likely to spread to nonpublic health insurance-funded care.

What if a physician wants to work hard? The FEHCA seems to have no opportunities for such individuals except to undertake lawful medical practices outside the FEHCA system. This seems an inevitable consequence of the FEHCA as occurs in the UK. Any suggestion that such extra system work will not be allowed will undoubtedly not survive challenges to constitutionality. Therefore, either a multitiered market must be tolerated explicitly or a black market will develop as occurs in most totalitarian societies.

The FEHCA will have a database recording costs and complications of medical interventions. This will act to further decrease costly interventions as doing no intervention will guarantee no expense and no complications and thus continuous employment for involved physicians with their described comfortable lifestyle. Indeed, the work ethic may subtly shift from working to improve patient health to working to keep one's job ... given that there may be cost-driven overseers monitoring and incentivizing the physicians. This mission change to a goal of job maintenance is a well-known aspect of large bureaucracies.

Dr. Crippen counters this scenario by assuring the reader that monitoring mechanisms will be in place to punish resource nonusers who may be creating complications of omission. This seems naïve given that the whole structure of FEHCA is based on saving money. Such complications due to omission of care will be difficult to document even with the best of IT systems. One should not underestimate the cunning of bureaucrats, which is what physicians are in this system, whose jobs will be to remain employed.

Dr. Crippen states that pay for performance does not work. Why? It is because incentives to do things that evidence-meisters call good tend to get done if there is financial incentive. As noted previously the problems defining acceptable evidence underscore issues with pay for performance. Research in this area underscores the heterogenous and sometimes unintended consequences of pay for performance [27–34] with some suggesting that it induces a decrement in medical professionalism [35].

Similarly, and notwithstanding Dr. Crippen's thoughts about pay for performance, physicians in FEHCA will use resources to as minimal an extent as possible to achieve a modicum of acceptable outcomes, and be incentivized for their behavior. Is this not a form of pay for performance? Dr. Crippen seems insistent that stupid medical care will not be countenanced. Physicians are smart and truly stupid stuff will not be done if loss of employment is the result. However, given the unending uncertainty in our knowledge of medicine in the gray zone and unremittingly undulant evidence that guides medical care [17], it should be relatively easy for physicians or their overseers to create the cheapest possible modes of practice. It will be their incentive, either through more income or simply through employment retention dependent on the cultural environment set by the FEHCA leadership.

The FEHCA finds that most emergency care is not emergent and will set the system so that patients are strongly discouraged to use EDs. This is precisely the goal of the "medical home" approach just enacted into Oregon Statutes of law [13, 16]. This will result in the patient flow being insufficient to support the number of EDs currently open and may result in many hospitals electing to close EDs entirely or decrease their capacity. The obvious result will be fewer functioning high-quality EDs and longer travel times during true emergencies, and thus less expensive care provided as the consequent moribund on arrival cases will undergo mandated withdrawal of support. Telemedicine and telemedicine ICU leveraging of CCM may be needed here to both mitigate delayed deterioration of necessary insured care as well as prevent transfer of end-of-life care patients who cannot benefit from CCM resources in FEHCA. Though some will call this implicit health policy rationing, with proper tort exemptions in place as in Oregon, there will be measureable overall system cost containment with improved overall health of the living population. Those asserting this to be a heathen Malthusian state of affairs will have to contend with resource stringency producing CQD and possibly increased mortality for the medically salvageable. This is the mandatory price resource-limited democracies must bear to preserve fiduciary integrity under law of the health care professionals and their patients.

Oversight of surgical procedures under FEHCA will minimize unnecessary surgery and, arguably, some necessary surgery. Given the overall lack of evidence on efficacy of most surgical procedures this should be a very effective cost containment system and may not produce an overall health decrement. Notably, however, the very notion of the cost-effectiveness of an aggressive approach to management embodied in many surgical procedures is unsettled. Gawande [36] and others [37, 38] suggest that the aggressive approach is overall ineffective. Silber et al. [39, 40] challenge this.

The critical care component of FEHCA is designed to put an end to prolonged fruitless ICU stays with ongoing oversight of each patient by FEHCA. Decision making is removed from families and patients and resides with the doctors and FEHCA. Given the incentives built-in of FEHCA to calculate benefits versus costs it will mean withdrawal of support or withdrawal of financing of support of individual patients based on population data without necessarily consideration for individual issues. Several authors suggest that use of mortality prediction models from the literature is not appropriate when used to make decisions about resource allocation or limitation of life support [41–43]. Conflict will undoubtedly accompany such cultural changes and impose an adversarial element to the patient–physician relationship. In order for such a shift in attitudes to work there will be needed a prolonged educational campaign directed to the public supporting the notion of the ICU physician as really authoritative and indeed suitable to take on the oft-derided practice of acting like a god. There was a time when this was acceptable but American culture in recent decades has shifted to one that encourages patients and families to question the judgment of their physicians. Thus, given the large number of freedoms that are taken for granted by Americans, a large shift in American cultural attitudes back towards physicians would have to occur before FEHCA could ever be successfully implemented. Simply mandating acceptance of such overlording behavior on the part of physicians will never work in the United States.

FEHCA makes essentially no provision for patients in need of special nursing facility (SNF) (i.e., nursing home) care. Given that primary care has created many patients who now slowly physically deteriorate to states of sentient feebleness, this may create huge problems for families and societies in general as invalid and enfeebled elderly with continued cardiopulmonary failure are not financially supported. Poverty will follow with medical bankruptcy, the very consequence FEHCA is supposed to prevent. In fact, it seems that FEHCA's abandonment of patients who have been lifelong insurance premium payers during their years as makers is an egregious breach of obligation to its customers who have undergone the inevitable slow transition to takers. Conceivably, the FEHCA policies might be structured such that the money spent on a patient's last 6 months of life is transferred to that patient, in the form of lower premiums. Of course the result of such a Faustian bargain will be terminal sedation and withholding of nutrition once a threshold of enfeebled dependency has developed. Professional integrity will erode for all of us if FEHCA is seen as the CCM professions acting as enablers of "A Duty to Die." This latter scenario seems to be what the FEHCA will develop into although Crippen does not articulate it quite this harshly.

References

1. Engelhardt Jr HT. Critical care: why there is no global bioethics. Curr Opin Crit Care. 2005;11(6):605–9.
2. Kofke WA, Rie MA. Research ethics and law of healthcare system quality improvement: the conflict of cost containment and quality. Crit Care Med. 2003;31(3 Suppl):S143–52.
3. Rie MA, Kofke WA. Nontherapeutic quality improvement: the conflict of organizational ethics and societal rule of law. Crit Care Med. 2007;35(2 Suppl):S66–84.
4. Rie MA. The Oregonian ICU: multitiered monetarized morality in health insurance law. J Law Med Ethics. 1995;23(2):149–66.
5. Lohr KN, Yordy K, Harrison PF, et al. Health care systems: lessons from international comparisons. Health Aff. 1992;11(4):239–41.
6. Prosser W. Handbook on the law of torts. St. Paul, MN: West Publishing Company; 1941.
7. Engelhardt Jr HT. Critical care: why there is no global bioethics. J Med Philos. 1998;23(6):643–51.
8. Fein IA, Wiener JM, Strosberg MA, Baker R. Rationing America's medical care. The Oregon plan and beyond. Washington, DC: Brookings Institution Press; 1992.
9. Kitzhaber JA, Kemmy AM. Access to care, controlling costs, and the Oregon health plan. Acad Emerg Med. 1994;1(1):3–6.
10. Kitzhaber JA. The Oregon Health Plan: a process for reform. Ann Emerg Med. 1994;23(2):330–3.
11. Kitzhaber JA. Prioritising health services in an era of limits: the Oregon experience. BMJ. 1993;307(6900):373–7.
12. Kitzhaber JA. Oregon Act to allocate resources more efficiently, The proposal would guarantee healthcare access to all. Health Prog. 1990;71(9):20.
13. "Institute_for_Healthcare_Improvement". Pursuing the Triple Aim: CareOregon. http://www.careoregon.org/carenews/2008/Winter-2008-9/documents/CareOregon_Case_Study.pdf (2011). Accessed 6 July 2011.
14. Engelhardt HT. The foundations of bioethics second edition. Oxford: Oxford University Press; 1995.
15. Engelhardt Jr HT, Rie MA. Intensive care units, scarce resources, and conflicting principles of justice. JAMA. 1986;255(9):1159–64.
16. JOINT_SPECIAL_COMMITTEE_ON_HEALTH_CARE_TRANSFORMATION.   House Bill 3650. http://www.leg.state.or.us/11reg/measpdf/hb3600.dir/hb3650.intro.pdf (2011). Accessed 6 July 2011.
17. Ioannidis JP, Ioannidis JPA. Why most published research findings are false. PLoS Med Public Library Sci. 2005;2(8):e124.
18. Cleemput I, Neyt M, Thiry N, et al. Using threshold values for cost per quality-adjusted life-year gained in healthcare decisions. Int J Technol Assess Health Care. 2011;27(1):71–6.
19. Zhao FL, Yue M, Yang H, et al. Willingness to pay per quality-adjusted life year: is one threshold enough for decision-making?: results from a study in patients with chronic prostatitis. Med Care. 2011;49(3):267–2.
20. Guyatt GH, Haynes RB, Jaeschke RZ, et al. Users' guides to the medical literature: XXV. Evidence-based medicine: principles for applying the users' guides to patient care. Evidence-based medicine working group. JAMA. 2000;284(10):1290–6.

21. Levine M, Walter S, Lee H, et al. Users' guides to the medical literature. IV. How to use an article about harm. Evidence-based medicine working group. JAMA. 1994;271(20):1615–9.
22. Kofke WA. Incrementally applied multifaceted therapeutic bundles in neuroprotection clinical trials...time for change. Neurocrit Care. 2010;12(3):438–44.
23. Smith GC, Pell JP. Parachute use to prevent death and major trauma related to gravitational challenge: systematic review of randomised controlled trials. BMJ. 2003;327(7429):1459–61.
24. Dyde SW. Philosophy of right. http://srzbiz.com/intellectualism/philosophy/hegel/Hegel,_ G.W.F._-_Philosophy_Of_Right.pdf (2011). Accessed 18 July 2011.
25. Thomson (1st Baron Kelvin) W. Some Favorite Quotes. http://nowandfutures.com/quotes.html (2011). Accessed 18 July 2011.
26. US_SupremeCourt. United States v. Macintosh, 283 US 605; 1931.
27. Serumaga B, Ross-Degnan D, Avery AJ, et al. Effect of pay for performance on the management and outcomes of hypertension in the United Kingdom: interrupted time series study. BMJ. 2011;342:d108.
28. Nicholas LH, Dimick JB, Iwashyna TJ. Do hospitals alter patient care effort allocations under pay-for-performance? Health Serv Res. 2011;46(1 Pt 1):61–81.
29. Chen TT, Chung KP, Lin IC, et al. The unintended consequence of diabetes mellitus pay-for-performance (P4P) program in Taiwan: are patients with more comorbidities or more severe conditions likely to be excluded from the P4P program? Health Serv Res. 2011;46(1 Pt 1):47–60.
30. Augustine S, Lawrence RH, Raghavendra P, et al. Benefits and costs of pay for performance as perceived by residents: a qualitative study. Acad Med. 2010;85(12):1888–96.
31. Van Herck P, De Smedt D, Annemans L, et al. Systematic review: effects, design choices, and context of pay-for-performance in health care. BMC Health Serv Res. 2010;10:247.
32. Alshamsan R, Majeed A, Ashworth M, et al. Impact of pay for performance on inequalities in health care: systematic review. J Health Serv Res Pol. 2010;15(3):178–84.
33. Kahn JM, Scales DC, Au DH, et al. An official American Thoracic Society policy statement: pay-for-performance in pulmonary, critical care, and sleep medicine. Am J Respir Crit Care Med. 2010;181(7):752–61.
34. Tanenbaum SJ. Pay for performance in Medicare: evidentiary irony and the politics of value. J Health Polit Policy Law. 2009;34(5):717–46.
35. Qaseem A, Snow V, Gosfield A, et al. Pay for performance through the lens of medical professionalism. Ann Intern Med. 2010;152(6):366–9.
36. Gawande A. The cost conundrum: what a Texas town can teach US about Health Care. The New Yorker; 2009:36–44.
37. Fisher ES, Wennberg DE, Stukel TA, et al. The implications of regional variations in Medicare spending. Part 2: health outcomes and satisfaction with care [Summary for patients in Ann Intern Med. 2003 Feb 18;138(4):I49; PMID: 12585852]. Ann Intern Med. 2003;138(4):288–98.
38. Fisher ES, Wennberg DE, Stukel TA, et al. The implications of regional variations in Medicare spending. Part 1: the content, quality, and accessibility of care [Summary for patients in Ann Intern Med. 2003 Feb 18;138(4):I36; PMID: 12585853]. Ann Intern Med. 2003;138(4): 273–87.
39. Kaestner R, Silber JH. Evidence on the efficacy of inpatient spending on medicare patients. Milbank Quart. 2010;88(4):560–94.
40. Silber JH, Kaestner R, Even-Shoshan O, et al. Aggressive treatment style and surgical outcomes. Health Serv Res. 2010;45(6 Pt 2):1872–2.
41. Lemeshow S, Klar J, Teres D. Outcome prediction for individual intensive care patients: useful, misused, or abused? Intensive Care Med. 1995;21(9):770–6.
42. Keegan MT, Gajic O, Afessa B. Severity of illness scoring systems in the intensive care unit. Crit Care Med. 2011;39(1):163–9.
43. Chawda MN, Hildebrand F, Pape HC, et al. Predicting outcome after multiple trauma: which scoring system? Injury. 2004;35(4):347–58.

Chapter 33
Fixing the Foundation of Critical Care at the End-stage of Life: Second Critique of the Fair and Equitable Health Care Act

Jack K. Kilcullen

Introduction

The Fair and Equitable Health Care Act (The FEHCA) is one intensivist's vision of how health care spending might be reduced even as demand continues to surge. Unlike the federal government stymied by political opponents, the FEHCA as a private insurance plan faces no such obstacles. It will cut spending by refusing to finance prolonged intensive care unit (ICU) stays. It will further deny payment to individuals with dementia and multiorgan failure seeking hospital admission from nursing facilities. It will not finance unproven therapies. The FEHCA will employ its own physicians whom it can corral into following guidelines that would spur greater efficiency and improved outcomes.

Why we spend so much, however, cuts far more deeply, well into our view of medicine itself. This chapter examines how modern medicine's goal of saving lives poorly serves the millions nearing the end of life. These patients endure increasingly ineffective attempts to keep them alive at both great emotional cost to their families and economic cost to society. Ironically, all this suffering and waste lies well within the prevailing standard of care. As such, physicians drive this care even as they believe they are helpless before a demanding, unrealistic, and litigious public.

Yet, there are physicians already proving to their colleagues that as a person's health naturally declines, the medical response can be moderated to bring greater comfort at less cost. More fundamentally, there are those at work helping patients and families approach death without fear. As bodies decline, there can be alternatives to the emergency department, where treatments can prolong freedom from pain and time at home with family. When physicians begin to offer the means for a good death rather than just more of life, they will potentially reduce spending while providing what their patients ultimately want. In examining the impact the FEHCA might have on health care costs, this should be our standard.

J.K. Kilcullen, M.D., J.D., M.P.H. (✉)
Department of Critical Care Medicine, Virginia Hospital Center, Arlington, VA, USA
e-mail: jkkilcullen@gmail.com

D.W. Crippen (ed.), *ICU Resource Allocation in the New Millennium: Will We Say "No"?*, 261
DOI 10.1007/978-1-4614-3866-3_33, © Springer Science+Business Media New York 2013

Our Current Health Care Is Unaffordable

For many, the alarming state of health care spending is already painfully apparent. While average households spend 6% of income on health care [1], median out-of-pocket health spending for those on Medicare increased from 12% in 1997 to 16% in 2006. By 2020, it will exceed 25%. For one in four households, it already exceeds 30% and for one in ten, 50% [2]. For families with health insurance at work, premiums have risen 114% over the past 10 years, while their own contributions have risen faster, to 147% of levels in 2000 [3]. At the extreme is bankruptcy, where unpaid hospital and doctor bill weigh significantly for half the people filing for relief [4].

However, households themselves pay only part of the cost. Employers pay the far larger share of most families' coverage, with plans averaging $13,770 in 2010. This sum nearly matches an annual minimum wage salary. For a family with a median income of $52,000, their plan equals over 25%. However even this understates the problem, as much of health care is also paid for by government, charities, and a myriad of other sources. National health expenditures totaled $2.5 trillion in 2009, or $8,086 per person. For a family of four, their share would come to $32,000 or 62% of what they earn [5].

Ironically, health insurance companies are on a roll. The New York Times reported that "the nation's major health insurers are barreling into a third year of record profits, enriched in recent months by a lingering recessionary mind-set among Americans who are postponing or forgoing medical care" [6]. For the government, there is no such windfall. Medicare costs are projected to grow substantially from approximately 3.6% of GDP in 2010 to 5.5% by 2035, and to increase thereafter to about 6.2% by 2085. Medicare's Hospital Insurance Trust Fund will be exhausted in 2024, 5 years earlier than estimated in last year's report [7]. In short, Medicare is costing more than the taxes collected to pay for it. With trust funds to meet the shortfall running out in 13 years, we will need to borrow to make up the difference.

What the Fair and Equitable Health Care Act Offers

The FEHCA is a proposal for a not-for-profit organization insuring those able to afford its premiums. Unlike major insurers, the FEHCA will not deny coverage to individuals with chronic medical problems, i.e., people with preexisting conditions. In fact, it promises financial protection to anyone in the working-age population able to meet its monthly fee. There are three major sources of savings the FEHCA cites to finance its ambition.

The first is "aggressive risk-management," which means in part, restricting reimbursements to major medical expenses. The FEHCA does not offer coverage for everything, but rather insurance against the unusual and unpredictable illnesses and injuries that can devastate a family financially. Like similar catastrophic coverage plans that operate with high deductibles, customers will be expected to pay out of pocket for most routine medical expenses.

The FEHCA's second strategy is the "prioritization of services," that is, the refusal to pay for high-end treatments "benefitting only a relative few patients." For instance, it will not pay for any drug or procedure that its panel of experts believes is "even remotely experimental. Their decision is final." The FEHCA will not pay for patients in skilled nursing facilities and presumably all long-term care facilities "with baselines of dementia or permanent multiple organ system failure to be transferred to EDs or ICUs. If their families desire this care, they can arrange for financing it."

As for all others admitted to the ICU:

The FEHCA understands that time is needed to assure adequate intensive care. The FEHCA maintains a chart of how much ICU time is usual and customary for a list of disease processes normally seen in ICUs. This list is compiled from large outcome databases. As soon as the FEHCA gets a report from an ICU physician that the patient is stalled with no meaningful chance of recovery, a FEHCA social worker will explain to the family that their options are then tracheostomy, PEG (which the FEHCA will pay for) and transfer out to a SNF (financed by someone other than the FEHCA). If they refuse, the FEHCA will cut off payment to the hospital and they can start having meetings with hospital representatives to determine how this care will be accomplished.

That is, the FEHCA reserves the right to determine the limit of an ICU stay and will stop paying once it has performed procedures that would permit a patient's transfer to another facility. If the patient is too ill to transfer, the FEHCA will pay for hospice. If the family or the patient insists on continuing aggressive ICU care, they will have to begin assuming the costs.

Third, the FEHCA will hold down medical malpractice costs that arise from employing its own physicians by requiring customers agree in advance to a set schedule of compensation. If an injured patient or the surviving family considers the compensation inadequate, the FEHCA promises blistering litigation, including counterclaims for breach of contract, attorneys' fees, and trial expenses.

The FEHCA has other strategies to hold down costs. Central is its function as a staff model managed care plan like Kaiser Permanente. Employed physicians would not have the personal incentive to overtreat as is the case under the current fee-for-service system. Instead, they will receive a "fair salary" leaving them "comfortable [and able to] enjoy good personal benefits and not [be] overworked."

In addition, the FEHCA envisions a significant investment in information technology able to monitor physician behavior, in particular any possible tendency to undertreat:

Any physician that signs on to provide services for the FEHCA will be part of an extensive database of their activities, how much time they spend in patient care, what resources they use, what the results of these activities are and so on, adjusted for the usual statistical realities ...

Anyone inappropriately working the system for profit will show up on the screen quickly and will be warned once then fired if it doesn't get back into line. Anyone inappropriately minimizing resource use to the detriment of patient care, either out of laziness or profit will get the same treatment.

The FEHCA rejects the financial penalties and bonuses available through existing pay-for-performance schemes. At the same time, physicians' outcomes will be routinely analyzed:

> The FEHCA's statisticians and actuarials have a very good idea what the outcome should be for any disease process, based on millions of similar patients treated elsewhere in large databases. There will be a reasonable range of proficiency established and our physicians will be expected to stay in it. Anyone that gets out of it for some reason will be asked to explain why. There should be a good answer. If not, we don't want doctors that consistently get out of a range of reasonableness. We'll find doctors that will stay in that range.

Finally, the FEHCA will attempt to reduce inappropriate use of emergency departments, specifically those outside its network of facilities. It will require that requests for elective surgery be approved by two surgeons. The FEHCA will not pay for what it considers cosmetic surgery.

Challenges for the FEHCA Model

As a new arrival to a mature market with high barriers of entry, the FEHCA will face fierce competition. This will be true especially if, as it assumes, the Affordable Care Act is nullified by either the Congress or the Supreme Court. In order to keep its promise to provide coverage to anyone able to afford its premium, it will need a large pool of customers in good health. It will compete against existing national insurers who will also be targeting the healthiest population while they exclude those with preexisting conditions, which the Affordable Care Act would have outlawed.

For the FEHCA, its potential pool of customers is narrower than it would hope. While its price will no doubt be attractive, many of its potential customers would be those not willing to buy insurance from anyone, regardless of the price. Former Governor Mitt Romney faced the same need to expand the pool of contributors when crafting health care reform in Massachusetts in 2004. He found that the market among those without insurance divided into three groups: 20% were eligible for Medicaid but had not enrolled; another 40% were working full time but could not find an affordable plan; the last 40% chose not to seek insurance because they considered themselves too healthy to likely need coverage [8]. To draw in that 40%, he required everyone buy insurance or face a penalty. Absent a federal mandate to buy insurance, the FEHCA will not likely reach this low-risk pool.

Second, though savings from malpractice reform will reduce national health expenditures, it will not help the FEHCA compete with traditional insurance plans that do not employ physicians and thus do not pay for their errors. The CBO estimates that recent changes in state laws would save the nation 0.2% by reducing malpractice premiums and 0.3% by reducing unnecessary tests and procedures [9]. The FEHCA as a managed care plan would be able to reduce needless testing by controlling its own physicians. At the same time, its competitors like United Health Care or Blue Cross/Blue Shield have other means to discourage wasteful procedures without having to pay physician malpractice premiums. Thus, the FEHCA will likely remain at a competitive disadvantage.

Moreover, the FEHCA's plan to fight patients or their survivors who reject their predetermined awards stands in sharp contrast to malpractice reform efforts that seek early settlements that both sides find acceptable. Under a $3 million federal grant, New York City is piloting a program for early judge-directed negotiations in malpractice cases by judges with medical training [10]. Based on models in Michigan and South Carolina, then-Senators Clinton and Obama co-sponsored a federal bill that would promote programs that coupled early disclosures of error with changes in practice aimed at reducing future harm. Anticipated results would be improved safety, lower recoveries, and higher satisfaction because the results are based on mutual agreement, not acquiescence to one-sided demands [11].

Third, what the FEHCA might save by reducing ICU length of stay would be too little, too late. In a study of 1,778 mechanically ventilated ICU patients admitted a minimum of 48 h, Kahn and others calculated how shaving off days at the end of an ICU stay saves only hundreds not thousands of dollars. Two factors explain this. First, most expenses are fixed, incurred by the availability of the bed. A much smaller share of costs are variable, incurred only when the bed is occupied. Secondly, the majority of expenses are in the first 48 h, during which most procedures and consultative time occur [12]. True savings happen when the patient does not enter the ICU in the first place.

In our mixed market of private and public payers, it is no accident that private insurers score winning profits while their government counterparts struggle with mounting losses. Private carriers are required to maintain profitability through the setting of premiums and deductibles and by limiting those they insure to people with little likelihood of needing medical care [3]. The insurance model becomes unworkable as we age, when medical treatment becomes commonplace. A 70-year-old healthy person living independently will predictably spend about $136,000 in services (hospitals, physicians, rehabilitation) with a life expectancy of 14 years. A 70-year-old who is limited in at least one activity of daily living (such as mobility, toileting, feeding) will spend an average of about $145,000 as he or she lives another 12 more years [13]. One-third of lifetime medical expenses occur in just the last year of life alone [14].

Faced with a growing population retiring without health care coverage, the federal government in 1965 stepped in with Medicare. Soon after, Medicaid was enacted to pay health care bills to providers for eligible low-income patients. Medicare uses insurance terminology but it is not truly an insurance system. People are entitled to coverage when they reach a certain age and have made minimal contribution requirements, even though the risk they pose would exclude them from traditional plans. While Medicare charges annual "premiums" and "deductibles," it relies on the non-elderly workforce to provide much of the funding. Not surprisingly, the federal government through Medicare and Medicaid shoulders almost 75% of beneficiaries spending in the last year of life [15, 16].

Albeit not-for-profit, the FEHCA is not a charity and not intended to target such a disadvantaged population. In fact, its success depends on its ability to enroll healthy customers. In so doing, it must market itself to a skeptical public in a post-managed care era, where customers deeply mistrust plans that might seek savings at the expense of their health [17]. Moreover, the FEHCA must make significant investments in information technology to be able to regulate physician behavior. At the same time, it must offer an attractive work environment to recruit enough physicians to serve the large consumer base necessary to sustain itself.

As a social tool to expand coverage, the FEHCA's promise is modest. Since it offers catastrophic coverage with a high deductible, it will be of greatest attraction to those with low out-of-pocket expenses. Thus, while not "cherry-picking" as other insurers might do, it financially excludes much of the same population by requiring those with multiple prescriptions who must make regular visits to different specialists to pay these expenses out of pocket.

But these issues are secondary to whether the FEHCA will have any noticeable impact on reducing the spending that threatens our nation's long-term solvency. If it plans to behave like all other "indemnification plans," the answer is no. The reason is that unlike other countries, our national "system" lacks any authority that would prevent private plans such as the FEHCA from shifting costly patients to public programs. Its own business plan provides three examples.

First, the FEHCA will not serve those with limited income. If these individuals have a major disability or have dependent children, they might turn to Medicaid. Otherwise, they will be like most uninsured people, many of whom are employed full time, forced to wait till their conditions become an emergency before they can see a doctor. Second, the FEHCA will not cover patients from long-term facilities with dementia or multiorgan dysfunction coming to the emergency department or the ICU. These patients will likely be covered by Medicare or again Medicaid for their inpatient care. Third, when the FEHCA decides a patient is not recovering in the ICU, it will thereafter pay only for a tracheostomy and PEG or a transfer to hospice. All remaining hospital expenses will be the family's responsibility. This means, in reality, the hospital itself will be stuck with the bill. Hospitals can offset part of this loss through tax deductions on expenses, which means once again the public pays in the end.

Why We Spend so Much?

So, we are back to where we began, facing medical expenses we cannot afford. The FEHCA concept is born out of frustration over paying for unproven cancer therapies, endless ICU days with no progress, or repeated admissions for patients with neurologic or multiorgan dysfunction. It demands that we collectively confront what the physician who inspired this model sees daily: the endless delivery of intensive care that cannot hope to cure.

But none of this is new. Over 10 years ago, the Institute of Medicine's Committee on Quality of Health Care in America sensed that our efforts were misdirected:

The health care system as currently structured does not, as a whole, makes the best use of its resources … What is perhaps most disturbing is the absence of real progress toward restructuring health care systems to address both quality and cost concerns, or toward applying advances in information technology to improve administrative and clinical processes … For several decades, the needs of the American public have been shifting from predominantly acute, episodic care to care for chronic conditions. Chronic conditions are now the leading cause of illness, disability, and death; they affect almost half of the US population and account for the majority of health care expenditures [18].

The implication is that we are aggressively treating episodic crises when patients require long-term strategic health management which the system is not designed to deliver. Notwithstanding, such unnecessary intensive treatments fall squarely within the prevailing standard of care. Moreover, this excess spending is determined not by the patient's condition but by geography. There are hospitals treating patients at the end stage of life for considerably less than their neighbors. For years, the Dartmouth Atlas Project has documented the geographic variation in hospitalizations, ICU admissions, and specialty consultations for matching populations of chronically ill patients during the last 2 years of life. Not only does utilization vary in different states, but even within the same state. For example, they reviewed Medicare records of patients within different hospital referral regions in California. Compared to the Sacramento region, per capita Medicare spending was 69% higher in Los Angeles. Providers in Los Angeles used 61% more hospital beds, 128% more intensive care beds, and 89% more physician labor [19].

Measures of quality also varied to suggest that more spending was associated with worse outcomes. In Los Angeles, 33% of Medicare deaths in Los Angeles involved an admission to intensive care, compared to 19% in Sacramento. More than half of Los Angeles hospitals were rated below average by patients who had used them, while patients at 13% of Sacramento hospitals rated them below average. The authors calculate that had Los Angeles hospitals provided care at the rate of the Sacramento standard during the 5 years studied, savings to Medicare would have been approximately $1.7 billion.

Were the FEHCA willing to venture into the Medicare population with its monitoring technologies and uniform standards, it is not clear that it would fare much better. In a study where researchers at Kaiser Permanente applied the Dartmouth Atlas Project, they found to its credit that its inpatient utilization rates were lower and hospice use rates higher than the non-Kaiser community standard. However, when comparing its own facilities across different regions there was still a two- to fourfold variation in hospital admission [20].

The Congressional Budget Office analyzed the national implications for Medicare if high-spending areas could be brought down to low-spending areas. They assumed that reduction in payment rates in high-spending areas would be phased in over 5 years and capped at 20%. It would apply to all payments—including those to hospitals, physicians, and providers of post-acute care—made on the basis of a fee schedule. If implemented, this option would reduce Medicare spending by an estimated $12 billion over the 2010–2014 period and by $51 billion over the 2010–2019 period [21].

Why there should be excess utilization rates in the last 2 years of life is the macroscopic side to the question, why should anyone approaching death spend their final days in an ICU. Certainly, few people seek this kind of end, yet preventing it resists well-constructed solutions. In the study to understand prognoses and preferences for outcomes and risks of treatments (SUPPORT), researchers documented profound gaps in physicians' understandings of their patients' wishes for end-of-life care. Among dying patients, 38% spent no less than 10 days in the ICU. Among the many who did not want cardiopulmonary resuscitation (CPR), only half of their physicians were aware of this wish. A 2-year intervention period ensued with trained nurses working with patients, families, and hospital staff. They sought to improve understanding of outcomes, encourage attention to pain control, and facilitate advance care planning. Despite all this, patients experienced no improvement in patient–physician communication or in the targeted outcomes such as increased use of do-not-resuscitate (DNR) orders, reduced days in the ICU, or better management of pain [22].

Why physicians overtreat forcing Medicare to overspend during patients' last years has no easy answer. The FEHCA implies one view that the problem is patients and primarily families who "force" physicians to order aggressive treatments or keep them in the ICU. This may be an oversimplification in only because it is rare. Researchers from the SUPPORT study looked at the 16% of the original 9,105 who had ICU stays of 14 days or longer [23]. Fewer than 40% of patients or their surrogates reported that their physicians had talked with them about their prognoses or preferences for life-sustaining treatment. Only 29% of those seeking palliative care thought that the care they actually received was congruent with their desire. This implies that physicians, not families, were more often in control of the treatment plan.

A similar conclusion was found among patients receiving what has been called "futile" care in Texas. In 1999, the Texas Legislature passed The Texas Advance Directives Act, following recommendations by a coalition of medical providers and "right-to-life" organizations. Along with reforming provisions regarding advance directives, the act provides a mechanism for resolving disputes with patients' families over the continued provision of what physicians deem "medically inappropriate" care. A hospital ethics board is empowered to hear disputes between physicians who seek to withdraw care and families who insist on continuing aggressive measures. If the ethics board sides with the physicians, the family is given 10 days to arrange a transfer to another facility during which time current care would continue. In a report to the Texas legislature in 2005, of 2,922 ethics consults, including an estimated 974 futility consults, only 65 10-day letters were issued. Of those 65 cases, 11 patients were transferred within 10 days, 22 patients died during the 10-day period, 27 patients had the disputed treatment withdrawn, and 5 patients had treatment extended or were transferred later [24].

The hope a family clings to may have been fueled by the medical team itself, who can more easily let go when progress stalls. Imagine a 69-year-old man with a history of hypertension who collapses from a massive bleed in his brain. The neurosurgeon sees signs of soaring pressures in his skull and scrambles to insert a ventriculostomy tube, a sensor placed deep in his brain to measure pressures and drain fluid. The early days are tense as the team fights to control these pressures, pushing the patient into a coma in order to still any activity. Maybe there is a worsening bleed below the skull caused by the trauma during the fall, so the neurosurgeon takes him to the operating room for drainage. He is on a ventilator, his vitals marching along the monitor as his family listens to every precious word from physicians and nurses alike for something hopeful.

After a week, the patient's pressures resolve and the ventriculostomy tube is pulled, yet his eyes remain closed. Now, the family finds it hard to catch the neurosurgeon making rounds. They keep watching the patient and all the care given as part of the ICU routine. He is receiving nutrition, respiratory treatments, and regular skin care. He is breathing spontaneously, maybe his eyes open and stare, but still nothing. A fever is picked up, there is a pneumonia, antibiotics are started, the family is informed, and, in the absence of any word to the contrary, they continue to hope.

Into the third week, they want answers and now, they are told not to expect improvement, given how serious his original injury was. After so many days of drama, the suggestion to withdraw care is made and it comes across as cold and confusing. Why all the treatments, just to end up here? After more family meetings, a tracheostomy and feeding tube become the middle ground, to "give the family more time," and now there is a social worker with forms and brochures about long-term facilities.

Anthropologist Sharon Kaufman spent 2 years in an ICU studying the dying process. She spoke with families, physicians, nurses, and administrative staff. She attended family meetings and morning rounds and sat at the bedside talking with patients able to speak. She describes dynamic pathways directing the course of care, down which physicians, patients, and families follow, where the challenge to break free is daunting.

In particular, the hope families live off of comes from the heroic mission that hospitals embody. This heroic pathway operates with the force of an airport moving walkway—with high sides. Once a patient and a family are placed on one, its logic is more powerful, at least initially, than any individual voice, lay or medical. Everyone is stuck there—doctors, patients, and families.

Though patient autonomy (and its extension, in practical terms, to family members) serves as one important source of an ethic of medical practice, the notion of patient autonomy is actually applied only within a narrow sphere—decision-making about specific medical treatments offered by individual physicians … Patients and families are given choices but only among the options made available by hospital norms and regulations and within the framework of the almost unstoppable march of treatment … And while many physicians in recent years, especially younger ones, tell patients and families in no uncertain terms when the end of life is approaching, they present that information within the hospital environment, a place where many assume miracles can be worked. The message is mixed: "You (or your relative) are dying." That is the fact. "But let's do this procedure." That is the hope.

What's more:

> other than in relation to certain kinds of diseases (e.g., terminal cancer or end-stage AIDS), death is rarely spoken of or foreseen until shortly before it occurs … A waning life is rendered invisible, or nearly so, in the reading and then treating of signs of the body's pathology. Disease is treated until there is no more physiologic response to therapy. Only then is death expected. Only then does it "need" to be acknowledged [25].

This heroic pathway is deeply rooted in the mission of modern medicine that emerged in the last century. In 1900, the top three causes of death were pneumonia, tuberculosis, and enteritis with diarrhea. Its victims were often young wager earners. New medicines along with public sanitation and clean water have transformed the demographics of dying so that by 2000, people died of heart disease, cancer, and strokes. Despite this demographic shift, hospital marketing strategies continue to portray themselves as centers of breathtaking technological cures, an image entrenched in the popular imagination. However, dramatic recovery occurs in healthy patients or those with the early stages of chronic disease. For those with advanced disease, recovery only returns them to a life of daily disability, where the next exacerbation or infection is only a matter of time [26]. This is the population that receives the disproportionate share of health care resources [27].

The vast majority of patients with progressive disease follow one of the three trajectories [22, 28]. Almost a third are those with cancer, whose life expectancy is predictable and who comprise the majority of hospice patients. A smaller percentage are those with chronic organ failure, i.e., heart, lungs, liver, and kidneys, whose life course and period of disability are longer. Because physicians are less confident in predicting death, these patients often have little recourse but the hospital when their final exacerbation occurs. Third and more often are patients with frailty often accompanied by neurological deterioration, who have a long period of mounting requirements in daily care. They suffer frequent infections complicated by septic shock, each of which lends itself to heroic intervention in what Kaufman describes as "the revolving door" pathway [29]. The incurability of their underlying dementia and its causative role in the infection itself are all pushed aside by the drama of saving their lives.

With more than 80% of those dying covered by Medicare, what Medicare finances dominates is how we care for those at the end stage of life. Medicare pays for acute hospitalization followed by, at most, a brief period of targeted rehabilitation. In turn, physician salaries and reimbursement (not to mention career choice) emphasize specialties treating acute conditions with invasive procedures rather than the primary care geriatricians and internists [30]. Therefore, patients seeking medical attention approach a system which for all its sophistication in halting death offers so much less in the months and years that follow.

Crossing the ICU Threshold

For the family, admission to the ICU reflects the heroic intention of the institution. Indeed, admission commits the hospital to a significant commitment of resources, guaranteeing it the smallest return in the face of a fixed reimbursement by Medicare. Yet, criteria for ICU admission are influenced by many factors, often reflecting a lack of alternatives more than an absolute need for critical care itself. Some hospital floors cannot manage an insulin infusion, others cannot manage noninvasive ventilation, and some can handle a ventilator but only if the patient has a tracheostomy in place.

Some telemetry units will continue certain infusions, and others only at certain doses. Some patients are clinically stable but frequently agitated and if a hospital cannot afford a nursing assistant to sit at their bedside, the patients go to the ICU.

What admission often does not depend on are standardized criteria such as those promulgated by the Society of Critical Care Medicine (SCCM) [31]. These guidelines provide models for systematic and transparent decision-making to ensure effective use of limited resources. The fundamental principle is that given "the utilization of expensive resources, ICUs should, in general, be reserved for those patients with reversible medical conditions who have a reasonable prospect of substantial recovery." There is a prioritization model that articulates four levels of need. Highest priority goes to unstable patients demanding intensive treatment including ventilator support or continuous vasoactive drugs. Next are those who are stable but require tight monitoring because of a high risk they may suddenly decompensate, such as with certain high-risk gastrointestinal bleeding. Third are patients who "are critically ill but have a reduced likelihood of recovery" given their "underlying disease or [the] nature of their acute illness." They may receive intensive treatment with established limits of care in place such as no mechanical ventilation or cardiopulmonary resuscitation.

Finally, there are those considered inappropriate for admission, who have terminal and irreversible illness and thus too sick to benefit from ICU care. Examples include severe brain damage, irreversible multiorgan system failure, metastatic cancer unresponsive to therapy, brain dead non-organ donors, and patients in a persistent vegetative state or otherwise permanently unconscious.

How ineffective these guidelines have proven to be was shown in a recent survey of 146 directors of medical ICU directors at academic medical centers [32]. Eighty-eight percent reported having written admission criteria, but only 25% said that these were used on a regular basis. Even during the weekday daylight hours, attendings made triage decisions only 40% of the time, falling to 27% at night. Only 21% had written criteria that would deem a patient ineligible for admission. In almost 30%, not even attendings had authority to refuse a given patient who would not likely benefit from ICU care. Just 35% described themselves as "very familiar" with the SCCM triage guidelines. The guidelines arguably have even less impact in the majority of nonacademic open ICUs, where intensivists play no managing role and where the ICU director is typically a nurse.

But apart from a lack of alternatives and a lack of triage standards, ICU admission results from a more fundamental inability to define dying itself. Everyone who is born will one day die, but whether death is as natural as it is inevitable has been contested with every new medicine that can forestall it. As Kaufman notes, "although decline of bodily functions leads, inevitably, to death, death is not necessarily inevitable, even at the end of life, during any given medical crisis or any given hospitalization" [33]. A physician comes to decide that a patient is dying only after all therapies have been tried and the patient does not recover. Families are expected to grant physicians consent to proceed with testing and treatment; indeed, that is the presumption when the patient arrives in the emergency department.

Medicare reimbursement patterns slant this debate in two ways. Kaufman reports the practice in hospitals before the 1980s to place individuals in the terminal stage of disease on a death watch, where nurses would tend to a patient's comfort needs that families could not provide at home while offering them support. Hospitals were paid for each day the patient was cared for. When prospective payment was introduced in 1983, hospitals lost money when patients stayed longer than that anticipated for the patient's diagnostic related group (DRG). In kind, private insurance carriers hired former nurses to review charts and disallow days where patients did not receive services exclusive to a hospital. Kaufman writes, "Several young doctors told me, 'Dying is not billable. You can't treat it'" [34]. Second, enrolling a patient in a hospice program requires physicians attest to a life expectancy of 6 months. The Office of the Inspector General (OIG) of the Department of Health and Human Services regularly surveys programs' compliance with federal regulations, citing those where patients are living longer than regulations "permit" [35].

Yet, while predicting when one might die is difficult, deciding someone is in the end stage of life is not. Lynn and others have tested the question "Is this person sick enough that it would be no surprise if he or she died within the coming year?" to capture those with serious illness while excluding many who have been able to recover from an acute insult and live a substantial time thereafter [36].

For this population, ICU admission might forestall their death but not improve their lives. Rockwood and others showed that half of those over 65 who survive one trip to an ICU die within a year [37]. While age predicts less than 5% of the variance of absolute risk of death compared with younger patients, age along with previous ICU admission, acuity of illness score, and length of stay predicted morality in 82%. Age may not be a barrier for elective surgery, but when comorbidities create a surgical emergency, the odds of survival fall dramatically. A retrospective cohort study examined outcomes in 578 consecutive patients aged 80 years or more who underwent either elective or emergency surgery. Mean survival for planned surgery patients was 2,104 days. In patients whose surgery was unplanned, median survival was 28 days [38].

Like any diagnosis, deciding a patient is in the end stage of life requires deliberation followed by communication with patient and family. Best made by a patient's primary physician, planning treatments for this condition takes time and resources. The worst time to make it is at two a.m. when the patient arrives in the ED in respiratory failure requiring immediate intubation. When the ICU team thereafter approaches the family, all that lost opportunity to plan the patient's care has now become their nightmare. As Kaufman saw repeatedly, families are gripped with grief, shock, confusion, refusal, and what I observed to be a striking inability to cope with making "choices" about procedures. Many families demonstrated this inability, regardless of education level or any other sociodemographic feature … [F]amilies sometimes resist speaking for patients (or sometimes contradict patients who can speak), and that frustrates physicians. Families rarely want to shoulder medical decision-making responsibility, and they view procedural choices ("Should she be defibrillated?" or "Should she be placed on or removed from a ventilator?") not as options for managing death but as assuming responsibility for "killing" the patient [39].

More often, the decision to admit the patient to the ICU has already been made before the family's arrival. However, ICU admission dramatically raises the emotional stakes. What the family now sees is an institutional determination that the patient deserves aggressive care. They see their 81-year-old mother septic from a urinary tract infection next door to the 51-year-old engineer suffering a heart attack. When she fails to improve, the only way the medical staff can move her from the unit is by withdrawing care, that is, going from everything to nothing. To many families, this may be an abrupt "giving up." Despite all the doctors the family took her to see over the preceding year, likely none of them told them that she was likely to die during this time. Now, after all this aggressive care, her dying is a shock and they must be the ones to let her doctors put an end to her breathing. Not surprisingly, some just say no, not yet.

At least in the hospital, much of this anguish and expense could be avoided if alternative care settings were established. Just because an ED has aggressively stabilized a patient that does not mean a hospital must offer nothing short of an ICU. If a unit existed with fewer nurses and monitors with a plan not to escalate care, families would understand from the outset that aggressive measures are inappropriate. Families would not need to stop anything. Treatment would continue but expectations would be reduced. They would have this period of stability as a time to understand the fact that their loved one is in the end stage of life and no amount of critical care can change that. But few facilities offer palliative care units with attendings on call at all hours to help families make this decision. The lack of alternatives makes the heroic pathway the default choice.

The FEHCA and the Way Out

One night on call during my critical care fellowship, I reviewed the chart of a new admission from the oncology service. She was in her forties with advanced disease. What stunned me was the presence of a DNR and advance directives. When I caught up with the oncologist, the youngest in his group, I blurted out a "Thank you" as I explained my surprise at this brave foresight. He chuckled and said, "Well, I have partners who don't consider death a consequence of this disease."

Models for compassionate care mindful of the reality of life's end exist. At a Kaiser Permanente facility in California, specially trained liaison staffs coordinate with all subspecialty physicians to identify patients with advanced heart and lung disease while they are still stable. Families are trained and supported by regular follow-ups by hospice staff. After only 8 months, emergency transport calls and unplanned hospitalizations were cut in half. Two large hospital systems in La Crosse, Wisconsin, organized and trained community volunteers to help elderly patients understand and complete advance directives. Titled "Respecting Your Choices," training soon encompassed health care providers, who were trained as facilitators [40]. In these and numerous other programs, families and patients have the time and help to think through these difficult issues and share their wishes to all who matter in their care.

No one wants to die in an ICU. For those who might have a choice, they need sound medical advice to help them prepare. Primary care physicians who have been part of a patient's history can use the trust they have established to ask the difficult questions about how to shape care when that inevitable time comes. These discussions must include the family, who at times are the biggest obstacle and who often must be the messenger of a patient's final wishes.

For patients fortunate to survive the ICU, discharge should include documented discussions with families about how to avoid the next admission. With the memory fresh in everyone's minds of all the commonplace suffering patients endure, physicians can start the discussion with how to plan for the next exacerbation so that alternatives to the ED and ICU can be pursued.

Unlike traditional commercial carriers, the FEHCA can demonstrate unique leadership. By employing its own physicians with technology to coordinate their actions, it can be proactive. Patients with progressive disease can be identified and managed by a collaboration between traditional and palliative care providers. Rather than calling all this hospice, the better rubric is an alternative treatment pathway that emphasizes limited interventions with increased symptom management. The family can be educated to react to early signs of instability and plan hospital treatments as needed before they become emergencies. When signs and symptoms progress and hospital treatment becomes unproductive, care can move into the home to ensure that the patient's and family's needs are met during the patient's final days. This whole trajectory can be undertaken early enough so that savings close to the low-spending patterns can be reached by minimizing resort to the hospital setting. Most importantly, the patient would be spared the ICU. The FEHCA would have the technological means to perform the needed data analysis in order to demonstrate that savings can be achieved.

With systems such as the FEHCA, we can give our patients the care they need in their final years right up to their final moments. The result will be a potential way out of the upward spending spiral physicians and their desperate patients keep generating. As health care providers, we can direct spending back downward while meeting our patients' unvoiced demands for a closing of life that is conscious and connected. To pursue this simple goal would be no less than our best calling.

References

1. Pankow D. Taking charge of family finances: how much should we spend? (2009) Retrieved from http://www.ag.ndsu.edu/pubs/yf/fammgmt/fe440w.htm.
2. Kaiser Family Foundation. How much "skin in the game" is enough? The financial burden of health spending for people on medicare (2011). Retrieved from http://www.kff.org/medicare/8170.cfm.
3. Claxton G. How private insurance works: a primer (2002). Kaiser Family Foundation. Retrieved from http://www.kff.org/insurance/2255-index.cfm?RenderForPrint=1.
4. Himmelstein DU, WE, Thorne D, Woolhandler S. Illnes and injury as contributors to bankruptcy. Health Affairs, Online (2005). Retrieved from http://content.healthaffairs.org/content/suppl/2005/01/28/hlthaff.w5.63.DC1.
5. Center for Medicare and Medicaid Services. National Health Expenditure Fact Sheet. (2011) Retrieved from Retrieved from https://www.cms.gov/NationalHealthExpendData/25_NHE_Fact_Sheet.asp.

6. Abelson R. health insurers making record profits as many postpone care. The New York Times (13 May 2011). Retrieved from http://www.nytimes.com/2011/05/14/business/14health. html?e.

7. Social Security and Medicare Boards of Trustees. Status of the Social Security and Medicare Programs. Social Security Actuarial Publications (2011). Retrieved from http://www.ssa.gov/ oact/trsum/index.html.

8. Lizza R. Romney's dilemma. The New Yorker, 6 June 2011.

9. Congressional Budget Office. CBO's Analysis of the Effects of Proposals to Limit Costs Related to Medical Malpractice ("Tort Reform") Letter to Orrin Hatch, 9 October 2009. Retrieved from http://www.cbo.gov/doc.cfm?index=10641

10. Glaberson W. To curb malpractice costs, judges jump in early. The New York Times, 12 June 2011.

11. Clinton HR, Obama B. Making patient safety the centerpiece of medical liability reform. NEJM. 2006;354:2205–8.

12. Kahn JM, Rubenfeld GD, Rohrbach J, Fuchs BD. Cost savings attributable to reductions in intensive care unit length of stay for mechanically ventilated patients. Med Care. 2008;46: 1226–33.

13. Lynn J. Sick to death and not going to take it anymore! Reforming health care for the last years of life. Berkeley, CA: University of California Press; 2004. p. 8.

14. Lubitz JD, Riley GF. Trends in medicare payments in the last year of life. N Engl J Med. 1993;382:1092–6.

15. Hogan C, Lunney J, Gabel J, Lynn J. Medicare beneficiaries cost of care in the last year of life. Health Aff. 2001;20:188–95.

16. Center for Medicare and Medicaid Services. Last year of life expenditures (2003). Retrieved from https://www.cms.gov/mcbs/downloads/issue10.pdf.

17. Blendon RJ, Brodie M, Benson JM, Altman DE, Levitt L, Hoff T, Hugick L. Understanding the managed care backlash. Health Aff. 1998;17:80–94.

18. Committee on Quality of Health Care in America. Crossing the quality chasm: a new health care system for the 21st century. Washington, DC: National Academy Press; 2001. p. 21.

19. Wennberg JE, Fisher ES, Baker L, Sharp SM, Bronner KK. Evaluating the efficiency of california providers in caring for patients with chronic illnesses. Health Affairs Online (2005). Retrieved from http://content.healthaffairs.org/cgi/content/full/hlthaff.w5.526/DC1.

20. Stiefel M, Feigenbaum P, Fisher ES. The Dartmouth Atlas applied to Kaiser Permanente: analysis of variation in care at the end of life. Permanente J. 2008;12:4–9.

21. Congressonal Budget Office. Budget Options Volume I Health Care (2001). Retrieved from http://www.cbo.gov/doc.cfm?index=9925.

22. The SUPPORT Investigators. A controlled trial to improve care for seriously ill hospitalized patients. The study to understand prognoses and preferences for outcomes and risks of treatments. JAMA. 1995;274:1591–8.

23. Teno JM, Fisher E, Hamel MB, Wu AW, Murphy DJ, Wenger NS, Lynn J, Harrell Jr FE. Decision-making and outcomes of prolonged ICU stays in seriously ill patients. J Am Geriatr Soc. 2000;48(5 Suppl):S70–4.

24. Fine RL. Point: the texas advance directives act effectively and ethically resolves disputes about medical futility. Chest. 2009;136:963–7.

25. Kaufman S. And a Time to Die: How American Hospitals Shape the End of Life. New York: Scribner; 2005. p. 29.

26. Lynn J, Adamson DM. Living well at the end of life: adapting health care to serious chronic illness in old age. Rand Corporation. ISBN: 0-8330-3455-3; 2003.

27. Hogan C, Gabel LJ, Lynn J. Medicare beneficiaries cost of care in the last year of life. Health Aff. 2001;20(4):188–95.

28. Lunney JR, Lynn J, Hogan C. Profiles of older medicare decedents. J Am Geriatr Soc. 2002;50:1108–12.

29. Kaufman S. And a Time to Die: How American Hospitals Shape the End of Life. New York: Scribner; 2005. p. 98.

30. Guglielmo W. Medscape physician compensation report: 2011. The specialist/primary care split continues and not just in compensation. Medscape (2011). Retrieved from http://www.medscape.com/viewarticle/740086.
31. Task Force of the American College of Critical Care Medicine, Society of Critical Care Medicine. Guidelines for intensive care unit admission, discharge, and triage. Crit Care Med. 1999;27:633–8.
32. Walter KL, Siegler M, Hall JB. How decisions are made to admit patients to medical intensive care units (MICUs): a survey of MICU directors at academic medical centers across the United States. Crit Care Med. 2008;36:414–20.
33. Kaufman S. And a Time to Die: How American Hospitals Shape the End of Life. New York: Scribner; 2005. p. 78.
34. Kaufman S. And a Time to Die: How American Hospitals Shape the End of Life. New York: Scribner; 2005. p. 91.
35. Office of the Inspector General of the Department of Health and Human Services. Medicare hospice care for beneficiaries in nursing facilities: compliance with medicare coverage requirements (2009). Retrieved from oig.hhs.gov/oei/reports/oei-02-06-00221.pdf. How American hospitals shape the end of life. New York: Scribner; 2005.
36. Lynn J. Sick to death and not going to take it anymore! Reforming health care for the last years nof life. Berkeley, CA: University of California Press; 2004. p. 180.
37. Rockwood K, Noseworthy TW, Gibney RTN, Konopad E, Shustack A, Stollery D, Johnston R, Grace M. One-year outcomes of elderly and young patients admitted to intensive care units. Crit Care Med. 1993;21:687–91.
38. Jonge ED, Rooji SED, Levi MM, Korevaar JC. Long-term survivial and functional and cognitive outcomes after ICU admission in very elderly patients. Crit Care Med. 2005;33:A83.
39. Kaufman S. And a Time to Die: How American Hospitals Shape the End of Life. New York: Scribner; 2005. p. 59.
40. National Coalition for Health Care and Institute for Health Care Improvement. Promises to Keep: Profiles of institutions and organizations that have demonstrated excellence in end-of-life care (2002). Retrieved from http://www.ihi.org/IHI/Topics/Improvement/ImprovementMethods/Literature/PromisestokeepChangingthewayweprovidecareattheendoflife.htm.

Chapter 34
Third Critique of the Fair and Equitable Health Care Act

Leslie M. Whetstine

Evaluating a healthcare system can be done from a number of perspectives including those that focus on legal, economic, and policy concerns. This article, however, analyzes the Fair and Equitable Healthcare Act (FEHCA) from a bioethical approach in order to determine if it would serve as a viable alternative to the current health-care model. To this end, the ethics of rationing, no-fault compensation, and a discussion of what constitutes fair and reasonable treatment as outlined by the FEHCA proposal will be considered.

On Rationing

As a practicing bioethicist specializing in end-of-life care, my experience with ethics consultations at the bedside often focuses on patient and family education. Typical discussions involve explaining what it means to forgo life support, how to make advance directives, and what can be expected as the care plan transitions from curative to palliative. Most patients and their families eventually reach a consensus with the healthcare team regarding a treatment plan. They express no interest in continuing therapies when it is clear that the burdens outweigh the benefits, and a comfort care plan is adequately explained. Occasionally, however, conflicts become intractable.

The vast majority of such cases focus on the type of scenario discussed earlier in this book: a family refuses to stop life support despite the treatment team's judgment that such interventions no longer benefit the patient. As a result, many thousands of dollars are expended, time and resources are occupied, staffs (particularly

Dr. Whetstine would like to thank Walsh University for its generous support of this research through the faculty scholar award.

L.M. Whetstine, Ph.D. (✉)
Division of Humanities, Walsh University, North Canton, OH, USA
e-mail: Lwhetstine@walsh.edu

nurses) feel that they are violating the precept to do no harm, and the patient dies, sometimes as his or her ribs are cracking under yet another round of family-mandated CPR.

It seems difficult to believe that anyone would deliberately abuse a family member with technology at the end of life. Yet for a multitude of reasons (faith, fear, denial, etc.) no amount of education, graphic or otherwise, will persuade some families that moving to a palliative care plan is the most humane and dignified response. Often this sort of situation escalates into a power struggle where families feel that they can demand specific therapies, typically misguided by advice gleaned from the Internet, while the healthcare team feels that their expertise in medicine ought to prevail. The unfortunate reality is that when the relationship between the physicians and family becomes irreparably fractured and egos vie for authority, the patient suffers.

With few exceptions, the current approach to such conflicts is to defer to families. Institutions have a vested interest in avoiding any negative publicity that would suggest that their physicians are not highly rated in customer satisfaction. Furthermore, surrogates have decision-making authority over physicians to prevent a revival of the paternalistic model that has historically dominated medicine. Circumstances that would support overriding a family's wishes are typically threefold: (1) when the requested therapies would cause quantifiable harm to the patient, (2) when the physician ethically objects to the treatment (in which case the physician must arrange to transfer the patient to another institution and the family must approve the transfer), and (3) when the treatment meets the definition of strict medical futility and is therefore contrary to the standard of care. Medical futility is a complex topic now mired in equivocation, but in its strict sense medically futile treatments are those treatments that cannot perform the function intended. That is, if interventions can sustain vital signs in a moribund patient, they do not satisfy the narrow definition of medical futility and may not be unilaterally forgone.

There are serious ethical implications to automatically deferring to families who demand open-ended therapies that do not offer human benefit. Data show that Medicare spends more than $50 billion dollars a year for patients in their final 2 months of life and the United States spends 17.6% of its Gross Domestic Product on healthcare [1]. It is shameful that this disproportionate expenditure occurs in a developed nation when 21% of its children suffer food insecurity [2]. As a society we have an ethical obligation to examine whether it is legitimate to refuse to utilize interventions that can sustain vital signs (and are not technically futile) but are considered medically inappropriate when they will not restore the patient to a functional baseline and will serve only to create a protracted dying process.

The Harvard trained philosopher Daniel Callahan is perhaps the most notable proponent of rationing healthcare resources [3]. In short, his argument espouses a traditional utilitarian calculus—the greatest good for the greatest number of people. Since healthcare resources are finite, the argument would follow that an individual cannot impose limitless demands for them to the detriment of the many. Some critics would claim that saving money on healthcare expenditure would not ensure that the funds would be redirected to other areas of need, thus nullifying the so-called Robin Hood theory. Clearly more data are needed before public policy is crafted, but in theory reducing spending in one area should logically lead to a surplus to be applied to another. The FEHCA appeals to judicious rationing based on the premise that more people will have access to necessary services if less money is spent on expensive therapies that benefit few and yield insignificant benefit at that.

Such a position likely sounds shocking to American readers but, as demonstrated by various authors in this book, other countries have no qualms openly discussing and setting specific parameters on rationing. Americans seem to have a cultural aversion to the notion, however. The American mindset has historically shown itself to be rabidly autonomous, if not a bit schizophrenic.

In the polarizing debate regarding healthcare reform many protesters armed themselves with signs announcing that the government should keep its hands off its Medicare. Perhaps only marginally more ironic are the farmers, or agribusinesses, that decry big government while reaping agricultural subsidies. In truth the public relies on many socialized programs (the armed services, the interstate highway system, and yes, even Medicare), but any explicit attempt to put limitations on treatment has heretofore been met with accusations of socialized medicine, death panels, or threats of lawsuits. The fact that insurance companies by nature work diligently to ration care by only partially covering prohibitively expensive procedures seems to go unnoticed.

A fundamental difference between a for-profit insurance company and the FEHCA is that the former attempts to cut costs by rationing using a series of complex codes, co-payments, deductibles, and paperwork whereas the latter overtly rations by identifying what procedures are covered prospectively. The issue is not whether rationing is taking place—it occurs in this country already, albeit implicitly. If you are one of the 40 million uninsured and cannot afford certain healthcare procedures out of pocket, or are a senior who cannot afford to purchase prescription drugs, you are clearly subject to price rationing [4]. The FEHCA claims to ration fairly and equitably, but this needs to be examined more closely.

On Defining Fairness and Equality

Aristotle advised that we "treat equals equally and unequals unequally." This is merely a starting point, however, as criteria must be established that determine equality and fairness or it simply begs the question. The FEHCA is subject to a similar criticism; it offers a good starting point but it relies on terms that lack specification and are often subject to interpretation such as " Providers will be paid a fair salary …" or "The FEHCA will pay a reasonable amount for the office work-up of medical patients …." Certainly what one might consider a "fair salary" will vary, as will one's understanding of what a "reasonable" amount of office work-ups amounts to. At its core, however, the FEHCA seeks to uphold the principles of justice and equality objectively using metrics similar to the Oregon Health Plan (OHP) [5].

Though the FEHCA would not apply to low-income patients, conceptually it closely resembles the OHP, which prioritizes healthcare services based on clinical effectiveness as determined by clinical experts, which are then assessed by independent actuaries. The benefits to both programs are transparency and a reliance on evidence-based medicine rather than the capricious decision making invoked by insurance companies. Another advantage for patients under these plans is that physicians do not have an incentive on either end of the spectrum found in capitation or fee-for-service models.

Both the FEHCA and the OHP cover services that fall under the scope of the basic minimum of healthcare such that they support continued health and maintenance, what the philosopher Norman Daniels would classify as therapies that promote the "Normal Opportunity Range" for its citizens [6]. Similarly, Christopher Boorse would argue that we have an obligation to help citizens attain "Species Typical Functioning, or biological normality" [7]. Under this theory, there is a baseline where humans perform optimally, thus healthcare services that support such a range would be covered whereas others would be regarded as nonessential.

While this may sound intuitively fair, there is no bright line, however, between what constitutes the basic minimum and what qualifies as enhancement. The OHP covers growth hormone for children only with pituitary dwarfism, Turner's syndrome, Prader–Willi syndrome, Noonan's syndrome, short stature homeobox-containing gene (SHOX), chronic kidney disease (stages 3, 4, 5 or 6), and those with renal transplant. Studies show that tall people derive a greater advantage than their counterparts in terms of employment and mate selection [8]. If being short inhibits one's Normal Opportunity Range and Species Typical Functioning, the argument follows that short stature is a disease and ought to be corrected; thus the OHP covers such treatment as a medical necessity. If a child were short because of the genetic lottery, however, growth hormone treatment would be considered enhancement and not be covered despite the fact that the consequences of short stature are the same regardless of etiology. The FEHCA must articulate how it defines disease versus enhancement in order to cover the same therapies for some but not others, thus calling the principle of justice into question.

On No-Fault Compensation

The FEHCA essentially seeks tort reform by instituting a no-fault compensation package, ostensibly to minimize disproportionate settlements and consequently lower physicians' malpractice insurance. There is no question that the USA is a litigious society, and nearly everyone is familiar with the infamous McDonald's coffee case where a woman was awarded a multimillion dollar judgment from a burn after she spilled a cup of hot coffee on her lap. This typifies what most would consider a frivolous case (after all, coffee is supposed to be hot), and one where juries award damages that far exceed reason.

Perhaps with the intent of ending such baseless cases, in 2003 the state of Texas put a liability cap on noneconomic damages on malpractice lawsuits at

$250,000 [9]. The goal was to reduce defensive medicine, which occurs when doctors overprescribe tests and procedures to avoid being sued, as well as reduce malpractice insurance and overall healthcare costs. It seems that the FEHCA aligns with this goal.

However, data show that the opposite has occurred in Texas. In fact, its Medicare spending has increased over the national average [10]. Moreover, evidence shows that the 10% reduction in malpractice insurance caused by tort reform translates to a mere 0.1% reduction in healthcare spending [11].

Further, it bears revisiting the McDonald's case, arguably the paradigm of trivial cases frequently cited as a reason for tort reform. McDonald's corporate standard was to hold its coffee between 180 and 190°F, and was aware such high temperatures could cause third-degree burns, as evidenced by the 700 customers who reported injuries over a ten-year period from 1982 to 1992. This specific case involved a 79-year-old woman who was scalded while she was seated in the passenger side of an unmoving vehicle when she removed the lid to add sugar. She suffered full thickness burns over her legs and genitalia that required skin grafts and debridement. She attempted to settle for $20,000, which McDonald's refused. Ultimately, she was awarded $160,000 in damages and 2.7 million dollars in punitive damages, which was reduced to $480,000 [12].

The public did not hear these facts, however, and the six-figure judgment for punitive damages amounted to the profits McDonald's earned from coffee sales in 2 days. Similarly, tort reform is a red herring; it simply does not make a dent in the healthcare expenditure any more than this case hurt the fast food giant. Insurance companies make millions annually and even publicized cases where a multimillion-dollar judgment is entered do not affect the system.

The FEHCA's no-fault proviso is not unethical so much as inconsequential when it comes to keeping healthcare costs low. It is questionable, however, to intimidate prospective customers into not pursuing further action and the FEHCA's threat to countersue individuals *even when a jury rules in the patient's favor* seems rather unsavory. Individuals should always have recourse through the justice system without fear of reprisals, but again this will not make a tangible difference in cost control.

One difficulty with any no-fault program is determining a schedule of compensation. Perhaps it is beneficial to look to the Swedish no-fault system as a metric [13]. According to their plan: "The Fund does not attempt to compensate all injuries caused by medical intervention (or lack thereof). From the outset, planners considered and rejected a range of possible considerations that might have been used to establish rights to compensation (for example, provider fault, unsuccessful or unfortunate results, rare occurrences, relative seriousness, or individual need for compensation). Instead, in an attempt to delineate a more objective and rational basis for compensation, the Fund chose to make compensation contingent upon the occurrence of an 'avoidable' medical injury." The FEHCA could adopt a similar approach, although it remains to be seen how a socialized program that is successful in Sweden would translate in an aggressively profit-driven market.

A Model for the Future

The FEHCA is probably best understood as a mission statement rather than a comprehensive health plan at this stage. It offers a laudable foundation from which to build a system that clearly addresses the need for rationing based on objective criteria. Since the program would function primarily as an affordable stop-gap for those who do not have private insurance through their employers, are not poor enough for Medicaid, nor old enough to qualify for Medicare, coverage would be stripped down to offer the basic minimum only. Individuals who desired more coverage than the FEHCA but who could not afford it would simply not receive services. This may sound callous but for-profit insurance companies operate under the same principles by setting limitations on reimbursement; the FEHCA would offer at least a safety net for people who were ineligible for private insurance or public assistance programs.

Some of the most egregious cases that contribute to the budgetary crisis occur when families demand to keep dying nonagenarians on life support. Because Medicare pays only a fraction of the cost, the balance is absorbed by the hospital, forcing it to offset its loss by charging at a premium. The current all or nothing approach is neither sustainable nor ethical; the system is hemorrhaging money at the end of life while creating a vicious cycle of covert rationing.

The FEHCA has merit as a platform for further responsible discussion that does not appeal to political agendas or fear mongering. It reaches a populace that would not have access to any type of equitable coverage and provides a safety net until they can transition to private insurance or public programs. It is, on balance, a viable alternative for the 50 million uninsured who truly are subject to "death panels" due to price rationing.

References

1. http://www.cms.gov/NationalHealthExpendData/02_NationalHealthAccountsHistorical.asp#TopOfPage.
2. http://feedingamerica.org/hunger-in-america/hunger-studies/~/media/Files/research/state-child-hunger-2010.ashx?.pdf.
3. Callahan D. Setting limits: medical goals in an aging society. Washington, DC: Georgetown University Press; 2003.
4. http://www.npr.org/templates/story/story.php?storyId=1444391.
5. http://www.oregon.gov/OHA/OHPR/HSC/docs/L/Apr11List.pdf.
6. Daniels N. Just health: replies and further thoughts. J Med Ethics. 2009;35:36–41.
7. Boorse C. On the distinction between illness and disease. In: Caplan AL, McCartney JJ, Sisti DA, editors. Health, disease and illness: concepts in medicine. Washington, DC: Georgetown University Press; 2004.
8. Judge TA, Cable DM. When it comes to pay, do the thin win? The effect of weight on pay for men and women. J Appl Psychol. 2011;96:95–112.
9. Tex. Civ. Prac. & Rem. Code ch. 74.

10. Carrol, A. (2011, 06 02). Meme-busting: Tort reform = cost control [Web log message]. Retrieved from http://www.washingtonpost.com/blogs/ezraklein/post/meme-busting-tort-reform–cost-control/2011/06/02/AGpb0DHH_blog.html.

11. William Thomas J, Ziller EC, Thayer DA. Low costs of defensive medicine, small savings from Tort reform. Health Aff. 2010;29:91578–1584.

12. 1994 Extra LEXIS 23 (Bernalillo County, N.M. Dist. Ct. 1994), 1995 WL 360309 (Bernalillo County, N.M. Dist. Ct. 1994).

13. Studdert DM, et al. Can the United States afford a"no-fault" system of compensation for medical injury? Law and Contemp Probl. 1997;60(2):1–34.

Part III
Legal and Nursing Viewpoints

Chapter 35
Medical Judgment Versus Capitulation

Gilbert Ross

A patient is admitted to an intensive care unit (ICU) but after a time derives no significant medical benefit from remaining there. The patient's attending physicians begin establishing a plan for transfer (to another section of the hospital, to a skilled nursing establishment, or to a rehabilitation center or other facility). Members of the patient's family refuse to allow the transfer, threatening legal action if the patient is moved from the ICU. A legal nightmare is about to engulf the hospital and the health care professionals—or maybe not.

The funding of health care in the United States is anything but straightforward. A large percentage of Americans are covered by a variety of health insurance programs, with benefits ranging from Rolls-Royce to Yugo. Each of these health insurance plans provides reimbursements to health care providers through contracts. A portion of the population, primarily those over 65, is insured through Medicare, a national plan run by the federal government. In addition, for many Americans receiving public assistance, health care is funded by plans administered by individual states. Finally, there are the underinsured and uninsured, as much as 30% of the nation's population.

The wide spectrum of systems and individuals that provide health care in the United States includes not-for-profit hospitals, for-profit hospitals, university medical centers, physicians employed by universities or medical centers, physicians who practice independently but are permitted to practice within hospital facilities, physicians who work for public entities, and physicians whose approach to the delivery of medical services is entrepreneurial. Physician staffing of ICUs in the United States also varies. At university medical centers, physicians are likely to be employees of the medical center or university. Private hospitals tend to enter into staffing contracts with corporations established by groups of physicians; these physician corporations may be specialty specific or multispecialty.

G. Ross, J.D. (✉)
Sussman, Selig & Ross, One East Wacker Dr., Suite 3650, Chicago, IL 60601, USA
e-mail: gil@gilross.com

D.W. Crippen (ed.), *ICU Resource Allocation in the New Millennium: Will We Say "No"?*,   287
DOI 10.1007/978-1-4614-3866-3_35, © Springer Science+Business Media New York 2013

By contrast, the law in the United States regarding the duties owed by hospitals and physicians to patients in ICUs is somewhat more straightforward. A patient in an ICU has the right to receive treatment in accordance with the standard of care. The standard of care (a legal term) is determined through the testimony of an expert witness (or witnesses)—that is, a health care professional practicing within the same specialty of medicine as that practiced by the patient's health care providers. In other words, the standard of care is the care that a reasonably well-qualified physician practicing within that specialty would provide under the same or similar circumstances. The law recognizes that being admitted to an ICU is not like taking one's car to a mechanic to get the transmission fixed; there are no guarantees of a good result. Furthermore, the standard of care owed to a patient in an ICU in the United States is not determined by the patient's ability to pay.

An ICU, whether medical, surgical, cardiac, pediatric, neurological, or neonatal, is not a long-term-care facility. Patients are admitted to an ICU because they are experiencing sepsis, have multiple organ failure, are hemodynamically unstable, have life-threatening cardiac arrhythmias, have had an acute neurological event, or have respiratory compromise. The intent of admission to the ICU is to provide intensive care support for what is hoped to be a reversible condition.

The condition of a patient in an ICU is dynamic. Medical decisions are made constantly, in response to changes in the patient's condition. When choosing treatments, a physician exercises his or her best judgment and considers the totality of the patient's status and history.

There comes a time when the patient's condition stabilizes and his or her treating physicians determine that further interventions in an ICU would be futile. At such a juncture, the patient should be moved from the ICU. The decision to move the patient to another setting is a medical one and should not be dictated by commands, demands, or threats from the patient's family. However, such threats and demands are part of the reality of critical care in the United States. The demands and posturing of the family are often supported by the threat of legal action. It is important to discern whether such a threat is real or mere bellicosity and bluster.

I suggest that a hospital's legal and/or risk management department be placed on notice upon the first hint that a patient's family will threaten legal action. Discussing the events with an ethicist would also be beneficial. (It could be worthwhile for the hospital to explore enacting a formal "intervention" procedure, possibly coordinated by the social services department. Such a procedure could be implemented when a patient's family threatens to take legal action. A discussion involving the doctor[s], a risk management representative, a medical ethicist, social services, and family members could defuse a volatile situation.).

To stand before the court and make good on the threat of legal action, a patient's family must be able to present a sound theory recognized in law. Most commonly, that results in a malpractice lawsuit. To prevail in a malpractice lawsuit, the plaintiff (the patient's family) must first establish the standard of care owed to the patient. The plaintiff must then establish that those involved in the care and treatment of the patient deviated from the standard of care and, further, that this deviation caused or contributed to the injuries complained of. If the care and treatment rendered to the patient were in fact appropriate—were within the standard of care—it is unlikely that a malpractice claim will be filed. It is even more unlikely that such a lawsuit would be successful.

In the issue under discussion, the real threat is that the patient's family will go to court to demand a temporary restraining order (TRO) directing that the patient not be moved from the ICU. To obtain a TRO, the patient's family must convince a judge that failure to grant the restraining order will result in irreparable harm; that the law provides no alternative remedy; and that, if and when a lawsuit is filed, the plaintiff (the patient's family) will prevail. In short, the patient's family would be asking a judge to substitute the judgment of the court for the medical decisions of the patient's health care providers.

There are several reasons to believe that the threat of a lawsuit and the threat of a demand for a TRO amount to bluster. Attorneys who represent plaintiffs (in this instance, the patient and/or the patient's family) in medical malpractice lawsuits almost always agree to take on the representation based on a contingency fee agreement. This means that the attorney receives a percentage of any monies obtained, whether by settlement or verdict. If the case is unsuccessful, the attorney receives no fees for the legal services rendered. In addition, it is custom and practice for the attorney to advance the costs involved in the litigation. If the case is unsuccessful, it is highly unlikely that the patient or patient's family will be eager to reimburse the attorney for those costs.

In the case we are assessing, it would be necessary for the plaintiff's attorney to retain expert witnesses willing to testify that, based on a reasonable degree of medical certainty, a more favorable result would be realized if the patient were to remain in the ICU. In reality, it is exceedingly difficult to convince a judge or a jury of such a contention, in light of the fact that admission to the ICU in the first place was based on the patient's poor condition. The essence of a malpractice lawsuit would be to second-guess medical decisions made by health care professionals involved in the patient's treatment and assert that had the patient only remained in an ICU setting, the result would have been different. Very few plaintiffs' attorneys are willing to undertake such a case.

As discussed above, one of the other legal options is to try to obtain a TRO. In our case, if the patient's family were successful, the court would issue a restraining order only for a very short period. This would be followed by a full hearing on whether the restraining order should be continued. This hearing would allow the hospital to fully set forth its position before the court. The hospital would be able to explain that the patient was no longer receiving benefit from remaining in the ICU and that the standard of care does not require that a patient remain in the ICU in perpetuity. It is highly unlikely that a court would extend a restraining order prohibiting the patient from being moved from the ICU.

Physicians' apprehension about restraining orders and lawsuits stemming from decisions to move a patient from an ICU, or not to administer futile care, far exceeds any actual threat. Such lawsuits are extremely rare. Unfortunately, the few cases that have been filed have not provided sound legal precedents.

In *Gilgunn v. Massachusetts General Hospital,* No. 92-4820 (Mass. Super. Ct. Civ. Action, Suffolk Co. 1995), a lawsuit was brought against the physicians and hospital by Catherine Gilgunn's surviving daughter. Despite her claims that Catherine had "wanted everything done" that was medically possible, a do-not-resuscitate order had been placed. The daughter alleged negligent infliction of emotional distress. The jury found in favor of the hospital and doctors that further care would have been medically futile.

In another case (*In the Matter of Baby "K,"* 16 F.3d 590, 1994), Baby K was an anencephalic neonate, intubated and placed on ventilator support at the demand of the mother. The infant was eventually weaned from the ventilator and transferred to a nursing home. After several readmissions of the patient for respiratory distress, the hospital initiated legal proceedings, asking that a guardian be appointed for the child and that the hospital be allowed to render only palliative care. Unfortunately, the case was decided under the Emergency Medical Treatment and Labor Act, the court never really addressing the issue of futile care being given at the demand of the mother.

Today, the risk of a medical malpractice lawsuit being filed is lower than it has been in decades. According to the National Practitioner Data Bank, in 2010 malpractice payments made on behalf of physicians fell for the seventh straight year. Yet physicians state that they need to practice defensive medicine to insulate themselves from malpractice claims. The fear of lawsuits results in unnecessary costs and, worse, creates an atmosphere of tension and distrust between health care providers and patients and their families. In reality, medical malpractice claims account for less than 2% of total health care costs. According to the Congressional Budget Office, this figure takes into account the costs of defensive medicine, insurance premiums, lawyers' fees, payment to claimants, and administrative costs, including insurance company profits (see http://www.cbo.gov/ftpdocs/106xx/doc10641/10-09-Tort_Reform.pdf).

And yet the clamor for tort reform continues. Tort reform comes in many guises: panels, affidavits, and reports as to the merit of a claim; caps on attorneys' fees; and caps on damages. Caps on damages may apply only to noneconomic damages (such as disfigurement or pain and suffering) or to the total award. Some jurisdictions have placed an absolute cap on total awards, including proven economic loss. Caps on fees apply only to fees earned by the plaintiff's attorney, not by defense counsel. Meanwhile, the Congressional Budget Office estimates that national implementation of tort reform would result in a reduction in national health care expenditures of only 0.02% (see http://www.cbo.gov/ftpdocs/106xx/doc10641/10-09-Tort_Reform.pdf).

Placing a cap on damages has consequences. Very large damage awards almost always involve cases of catastrophic injury in cases where the liability is proven. The costs of future care can be in the millions of dollars, or even tens of millions. At trial, both sides introduce evidence of these anticipated costs, and the amount in damages awarded by a jury is based on that evidence. The state of Indiana, as an example, has an absolute cap on damages. At trial, the jury is not informed of this limitation. After an award is made by the jury, the judge reduces the amount in accordance with the statutory cap on damages. In a case of catastrophic damages, the jury might award damages sufficient to cover the costs of future care. However, after that amount is reduced by the judge, the monies awarded to the catastrophically injured plaintiff are likely to be quickly depleted and ultimately exhausted. Yet the need for extensive and expensive future care continues, forcing the plaintiff to receive public aid. The result is a shifting of the burden of paying for damages caused by the negligent defendant to the taxpayers—hardly an equitable system.

If the United States adopted a single-payer universal health plan, the mechanism for altering malpractice claims could be implemented under the existing law. It is presumed that the federal government would be that single payer. The Federal Tort Claims Act (FTCA) provides legal protections to physicians practicing at qualified health centers. Under the FTCA, claims against covered physicians and health centers are removed to the federal courts. Cases are decided by judges, the act barring juries. There is a cap on fees earned by the plaintiff's counsel. The FTCA supersedes state law.

Health care professionals often voice their frustration over what they perceive to be frivolous lawsuits. There is a relatively straightforward, tested, workable solution to this problem. Many years ago, Illinois passed a malpractice reform act that radically reduced the number of frivolous cases. When a medical malpractice lawsuit is filed in that state, the attorney must attach to the complaint his or her personal affidavit setting forth that the medical records have been evaluated by a physician practicing within the same specialty of medicine as the defendant and that the consulting expert has determined that there is a meritorious cause for the filing of the lawsuit. A written report from the consulting expert, setting forth the basis for the opinion, must then be attached to the affidavit. A separate affidavit and report must be attached for each named defendant. Adopting this aspect of Illinois' medical malpractice law would address concerns over frivolous cases.

In conclusion, the apprehension about lawsuits is often unfounded and out of all proportion to any actual risk. Tort reform is not a panacea and is generally not indicated. When a patient is admitted to an ICU, the presumption is that appropriate care and treatment will be rendered. When that patient is no longer receiving benefit from critical care treatment and interventions, it is a medical decision as to when it is appropriate to move the patient from the ICU setting. Like decisions about any treatment, this decision should never be mandated by demands and threats from the patient's family. When such threats are made, it is essential that the health care providers seek input from the hospital's legal representatives and/or risk management department, as well as from a medical ethicist. Once these steps have been taken, treating physicians should be able to make medical decisions without being subjected to threats and coercion. If the threats of legal action continue, the hospital's legal staff must support the physicians and refuse to capitulate. Standing firm in support of sound medical judgment is morally, ethically, legally, and economically the best course of action. The alternative enables bluster, bellicosity, and threats to hold proper medical care hostage, to the detriment of us all.

Chapter 36
Nursing Aspects of Inappropriate Patient Care

Melanie S. Smith

The recognition of our own mortality is an inevitable moment in our life. Once we grasp the concept that our time in this world is limited, we begin to contemplate the manner in which our lives may end. Whether it be an unexpected accident or a terminal disease, we prefer to imagine a death that is comfortable or at least natural and surrounded by our family and intimates. It would be difficult to find anyone who has envisioned their own death as a drawn out process with their body kept viable by machines. Enduring a multitude of invasive procedures and tests that accomplish nothing to improve someone's state of life will simply prolong an inevitable dying process.

Robbing patients of otherwise peaceful end-of-life care causes many problems for all levels of care givers, but it is particularly onerous at the level of nursing. Nurses traditionally enter the medical profession with a passion for people and a genuine calling for patient care. Nurses exhibit these traits—full of compassion and strong humanistic virtue, displaying an earnest desire to tend to the sick.

During their training, nurses are taught to deliver a multiplicity of treatment measures in collaboration with physicians with the object of producing the most optimal clinical outcome and a care plan reflective of a quality of life acceptable to the patient. However, when patients and their families demand care that increases their discomfort in an open-ended fashion, with no beneficial goal, nurses are the first line of care givers that suffer the consequences.

There is a strong and willful desire in a nurse to alleviate the anguish of a patient, but the point is to improve the patient's condition. Nurses are the first to struggle with exhaustive and unrealistic heroic measures delivered to a patient with a poor prognosis. These are often measures that will prolong the patients' impending death, extending their suffering. This wears heavily on the shoulders of nurses creating an emotional dilemma difficult for them to deal with.

M.S. Smith, R.N., M.S.N., C.C.R.N. (✉)
Department of Neurovascular Intensive Care Unit, UPMC Presbyterian University Hospital,
Pittsburgh, PA, USA
e-mail: smithms@upmc.edu

D.W. Crippen (ed.), *ICU Resource Allocation in the New Millennium: Will We Say "No"?*,   293
DOI 10.1007/978-1-4614-3866-3_36, © Springer Science+Business Media New York 2013

At what point is it appropriate for nurses as a patient advocate to exercise their professional responsibility to question a physician order or the will of a family? Consider the following case scenario detailing the critical state of a 76-year-old woman admitted to the intensive care unit following a fall at home. This patient was on a medication regimen that included Coumadin (warfarin), plavix (clopidogrel bisulfate), and aspirin. This patient had an extensive past medical history of atherosclerosis, peripheral vascular disease, coronary artery disease with stent placement, left ventricular aneurysm, diabetes mellitus, and hypertension.

She presented to the emergency department with lethargy which eventually progressed to an unresponsive state. A CT scan of the brain revealed a large intracranial and subdural hemorrhage with massive edema and impending herniation. The patient's son demanded the most aggressive care against the advice of the neurological surgeon. Ultimately, she went to the operating room for a hemi-craniectomy and evacuation of a massive clot, following which she was transferred to the neuro-ICU unresponsive except grimacing to pain.

Multiple specialists including the critical care and neurosurgical services continued to counsel the patient's son that there was no meaningful hope for recovery using objective criteria including multiple brain CTs and EEG. Overall her stay in the intensive care unit was complicated by multiple decompensations including pneumonia, septic shock, and ultimately multisystem organ failure.

The medical record revealed numerous documented meetings with the son and the medical team outlining the failing condition of his mother. The hospital ethics committee spoke to him as well. The son continued to maintain that he wanted everything done. Not everything reasonable, everything possible. In due course the hospital attorney was consulted but no one was willing to go against the wishes of the son.

Due to her tenuous status she went into cardiac arrest several times and was afforded aggressive resuscitation including CPR at the behest of her son who continued to believe she would eventually get better if given enough time. The patient in the end expired after several weeks following an extended performance of CPR failed to return circulation.

The nurses were the front line of participants, watching helplessly as their patient suffered continuously, wanting little more than to see it end. Participating in multiple scenarios involving the patient inappropriately being brought back from the brink of death over and over again weighed heavily on the nurses, who accurately perceived that the care they were providing was medically futile. The emotional strain on the nurses caring for this patient was evident. Nurses shied away from accepting this assignment, expressing difficulty coping with what they felt was an unethical situation. Several nurses questioned their calling as medical care givers.

These scenarios are playing out more frequently in ICUs as inappropriately optimistic patients and families depart from reasonable care plans. When nurses fail to see improvement in an already grim situation they feel emotional empathy toward the patient and they feel the emotional impact of a patient being denied comfort measures due to the capricious availability of open-ended technology. This is an unacceptable burden.

Medicine is increasingly a collegial and multidisciplinary endeavor. Nurses are strong and capable patient advocates, the bridge of communication between physician, patient, and family. As frontline providers with a strong emotional investment, nurses feel that it is ethically permissible or even mandatory to appropriately lobby families and physicians regarding these issues while avoiding impositions on their decision-making capability.

Part IV
Conclusions

Chapter 37
Where Is "Universal" Health Care Headed in the Global Village?

Michael A. Kuiper and Steven M. Hollenberg

Introduction

In this chapter, we address the question of how health care in general, and, in particular, the highly technical domain of critical care medicine, can evolve to remain obtainable and accessible around the world. Great differences exist around the world—in resources, in cultures, and in political systems. Those differences play out in how health care systems are organized and financed around the globe as well, as should be clear from previous chapters in this volume. It is also clear that many of these differences have been driven by historical developments, often in response to cultural factors.

These differences are deeply rooted and are not likely to disappear. The differences are not limited to the health care systems, but are also related to differences in legal systems and prevailing ethical norms. Thus, solutions to the problems raised in this book will necessarily differ.

This appears to be particularly true in the critical care environment. Critical care is in many ways the most local of services, delivered in highly specialized units by physicians attending patients at the bedside. Global disparities in critical care resources are vast, varying from almost no critical care medicine in developing countries to an overgrowth of critical care facilities in some Western countries. Critical care services are influenced by broader forces, for example the availability

M.A. Kuiper, M.D., Ph.D., F.C.C.M. (✉)
Department of Intensive Care Medicine, Medical Center Leeuwarden,
Troelstraweg 33, 8916 AB Leeuwarden, The Netherlands
e-mail: mic.kuiper@gmail.com

S.M. Hollenberg, M.D.
Department of Medicine, Cooper Medical School of Rowan University,
Camden, NJ, USA

D.W. Crippen (ed.), *ICU Resource Allocation in the New Millennium: Will We Say "No"?*,   299
DOI 10.1007/978-1-4614-3866-3_37, © Springer Science+Business Media New York 2013

of other resources in the hospital, health care needs in surrounding areas, or considerations of resource utilization on a regional or a national level. The consequent problems can differ widely, and thus to an extent the solutions to those problems will need to be tailored to the local environment.

Nonetheless, there are many commonalities to be found in the health care debate, and this chapter focuses on those commonalities. This will be made a bit easier by focus on the global village; the United States will be considered separately in the next chapter. The United States is an outlier in a number of ways, not the least of which is the disjunction between its health care spending and its quantifiable outcomes. We consider the following areas of commonality, and how they are likely to influence the evolution of health care:

- Societal attitudes and norms and how they affect both the accessibility and delivery of health care, with particular attention to whether health care should be considered a right and how universal coverage might be attained
- The inevitable tension between accessibility to the range of health care services and their affordability
- The influence of technological advances on both the quality and cost of health care
- The interplay among patient autonomy, physician's responsibility to an individual patient, and societal notions of distributive justice
- The need for transparent, comprehensible, and easily accessible measures of health care quality
- The need for education about these health care issues, by clinicians at the bedside but also in broad societal terms

Universal Coverage

Is There a Fundamental Right to Health Care in the Global Village?

This is a societal question, but also a moral one. We agree with the health care economist Uwe Reinhardt when he argues that "Every nation's health care system reflects that nation's basic moral values." (Quoted in *The Healing of America: A Global Quest for Better, Cheaper, and Fairer Health Care, T.R. Reid*) [1]. The Universal Declaration of Human Rights of the United Nations includes access to medical care in Article 25 (http://www.un.org/en/documents/udhr/index.shtml#a25).

"Everyone has the right to a standard of living adequate for the health and well-being of himself and of his family, including food, clothing, housing and *medical care* and necessary social services, and the right to security in the event of unemployment, sickness, disability, widowhood, old age or other lack of livelihood in circumstances beyond his control." [2] This view contrasts with the notion of health care as a commodity that can be bought or sold.

Buchman and Chalfin describe a version of this view when they write: "why is health prized, and why should we Americans believe that we are entitled to good health? Neither church nor state makes such a promise. The Constitution of the United States does not assert a right to health care, and there is no evidence that the founding fathers had any intention of guaranteeing health care, much less unbridled access to highly technical critical care." [3] To be fair, they were discussing the notion that patients are entitled to good health, and close reading of their chapter reveals that they in fact advocate provision of a basic standard of health care, but less nuanced views of this sort have been used to justify strict market approaches in the United States.

The case for universal health care in a society where resources will permit it is thus made on the basis of a moral imperative. Put in terms that will resonate in the American context, people have the right to be maintained at least at a threshold below which equal opportunity is compromised. In other societies, universal coverage is motivated by a strong belief that equitable distribution of health care resources is vital to social cohesiveness.

But there are practical reasons for espousing universal coverage as well. Universal coverage provides a powerful incentive for preventive health measures, as these will decrease overall cost. Anything less than universal health care generates the potential for cost-shifting. As T.R. Reid puts it, "A unified system eliminates the gamesmanship and cost-shifting that permeates American health care. In the United States, whichever entity is asked to pay for the treatment of a particular patient will save money if it can shove that patient off to another system." [1] Perversely, many insurers in the United States have a disincentive to reimburse preventive or screening measures with benefits over the long term, as the average worker spends 6 years with a given insurer, and adults over 65 are cared for by Medicare; they absorb the costs, but potential benefits are reaped by other systems.

Gamesmanship by health insurance companies can be constrained however by (governmental enforced) regulation such as a system that is called "health insurance risk equalization": the "bad-risk" (mainly chronic) patients need to be equally shared among health insurers. This, combined with an obligation of health insurers to accept all patients, takes away the urge to shift patients to another system at the moment they are about to become costly. And it has other consequences as well: it makes health programs such as improved diabetes care, weight loss for obesity, and the like beneficial for the health insurance companies, as these programs have the potential to decrease health care costs later in life [4].

Universal Coverage Is Coverage for Everyone, but not Everything

The cost of health care could increase asymptotically if everyone had the right to demand the full range of testing, medication, and treatment that can be afforded by current high-technology state-of-the-art medicine. Offering all possible treatment to every patient puts the health care system on the road to bankruptcy.

The only sensible way to conceptualize universal health care is to view it as having a floor—but also a ceiling. Both will be guided by available resources, but the notion that the floor should be determined by that level sufficient to ensure that every member of society will have an equal opportunity to succeed seems like a good starting point. A kindred notion is that the amount of "potential" or "opportunity" that remains for a given person should guide the amount of health care he or she receives [5].

Judged by measures of cost-effectiveness, it will be apparent that preventive and public health measures will have the greatest yield, as noted by Drs. Buchman and Chalfin [3]. It is difficult to demonstrate that critical care is cost effective, for a number of reasons: the population is extremely heterogeneous, the costs are hard to evaluate, how best to assess the outcomes is uncertain, and some of the care is simply costly without being effective at all. Nonetheless, some patients will benefit, and there are other reasons for providing critical care.

Even if critical care may not always provide the best value for money in terms of life years gained (although this is arguably true of many other fields of medicine, most notably oncology), critical care medicine provides something that is valuable, even when it is not technically able to save a life. Critical care medicine provides *hope*. Hope in the sense that it relays the message to very sick patients that there is a chance that they can recover and that the necessary measures will be taken to make that happen. This is a feature that critical care has in common with organ donation and transplant medicine: even when the chances are slim, we as society transmit the message that we will do everything we can to help that we deem reasonable. And critical care is an important part of pushing the envelope of what modern medicine can achieve so that we can move forward and avoid stagnation. The degree to which we can do this may depend on available resources. But viewing critical care only in terms of measurable costs per life years gained ignores an important part of its true value.

It seems clear that the floor will include some critical care, if only because having established the units and the expertise, there is no going back. But there must also be a ceiling, and it is equally clear that the ceiling will exclude some critical care. Determining where that ceiling should be, however, is a much greater challenge, particularly in the critical care context. Complex and often emotionally charged decisions must be made at the bedside by physicians, patients, and family, often in a time-sensitive fashion, all within the context of societal norms and expectations. The financial overlay of these decisions has usually been tacit, but we may not be able to afford to do that indefinitely. We first discuss the inevitable tension between accessibility to care and overall costs, and then consider the interplay between physicians, patients, and cultural and normative factors in a later section.

Costs and Coverage

While the concept of what "universal" health care entails is viewed differently around the globe, all societies experience the problem of growing imbalance between supply and demand, between cost and yield. In times of economic crisis such as these, the economy is no longer growing in most countries, and there is increased need to limit expenditure in all domains, including health care.

In the United States, where substantial numbers of patients are uninsured and lack access to health care, the standard thinking has been to attempt to limit costs so as to free up funds to increase coverage. We would argue that this is wrongheaded, for two reasons. In the first place, as we have noted, we believe that there is a moral imperative to provide some level of universal coverage. A substantial portion of the uninsured in the United States are children, and not providing them with health care is shortsighted for any number of reasons. But as a practical matter, extending health care coverage to all is essential to create a political dynamic for managing cost. The alternative is to reinforce perverse incentives to shift costs to other payers.

A thorough consideration of methods to lower health care costs is well beyond the scope of this chapter. But in the critical care context, there are some measures worthy of consideration.

Clustering of ICUs

In 2006 the Dutch Institute for Health Care Improvement (CBO) produced a report regarding the organization of intensive care departments. This report gave detailed recommendations about the number of physicians, nurses, and supportive personnel that should be available at the bedside. Furthermore it gave clear-cut definitions about levels of intensive care departments; for instance, a Level 3 intensive care department is an intensive care that can deliver the most complicated intensive care, has more than 3,000 ventilator days per year, and has most people at the bedside.

A Level 1 intensive care only provides basic-level intensive care. This report has had enormous influence in the Netherlands. It was foreseen that it would take at least 5 years before all recommendations would be implemented, but it is fair to say that they are almost all implemented today. It was becoming increasingly clear that the outcome of the high-volume, Level 3 ICUs, as measured in terms of hospital survival, is significantly better compared to smaller, low-volume ICUs. Treating patients with severe acute conditions in high-volume ICUs leads to better outcomes [6].

In 2012 a new guideline about the treatment of intensive care patients will be published, endorsed by the scientific communities of intensive care, internal medicine, and anesthesiology. The starting point of this new guideline will be integrated care, where the actual problem of the patient will be put in the center and the various departments in and outside the hospital will follow the needs of the patient. This sort of effort has the potential to increase both efficiency and quality of care, as patients are cared for by practitioners with appropriate training and resources.

Outreach Teams

Studies show that outreach teams (medical emergency teams), operating from either the ICU or the ED, often encounter acutely ill patients in whom limitations of medical treatment or even palliative care may be more appropriate than admission to the intensive care unit. One study demonstrated that end-of-life care (EOLC) issues were present in approximately one-third of medical emergency team (MET) reviews [7]. These MET calls involved urgent review of elderly medical patients, often after hours, and involved decisions that resulted in limitation of medical therapy. Patients that had a limitation of medical therapy (LOMT) installed were more likely to be older, to be female, and to be a medical admission, were less likely to be admitted from home, and more likely to die in the hospital. Others have published similar findings, with about 10% of changes in DNR or LOMT orders resulting from an MET call [8–10].

The degree to which LOMT was appropriate in these studies is uncertain. LOMT can be a self-fulfilling prophecy in that patients receiving an LOMT are more likely to die because of treatment limitations. Use of limitations of medical treatment and DNR orders is dependent on the culture and health care system. This study uses data from seven hospitals from Australia, Canada, and Sweden. How these data reflect on the delivery of medical care in other countries is uncertain.

Nonetheless, it is clear that we need better ways to deal with patients who are dying. The MET call, triggered by the onset of death, is a poor substitute for advanced care planning. Specialists from different medical specialties have different views on the likelihood of benefit of critical care treatment and/or resuscitation and may be less capable or competent (or consider themselves as such) to manage end-of-life issues. Critical care physicians and nurses are increasingly the providers with the broadest perspective in present-day hospitals, and as such are very capable of providing assistance when needed, and of giving advice when not to refrain from life-sustaining interventions.

Despite the impulse to shift from acute interventions at the end of life to preventive strategies, it is not clear that fewer intensivists will be required. It seems more likely that physicians with training in how to assess the potential benefits of intensive care and how to navigate end of life will prove invaluable in the future of health care. But greater involvement by primary health care providers and by palliative care teams is also important.

Palliative Care and the End of Life

During the twentieth century we have seen a change; dying was removed from our houses and was brought to the hospital, lured by augmented possibilities of treatment and cure brought to us by medical science. We have forgotten the constant closeness of death known to former generations of our families. As we all know, the increase in medical knowledge and technological and pharmacological advancement gave us many good things, but it did not bring us eternal life. This medical advancement seems to have made us less capable, however, of knowing when death will arrive. Entering the hospital in search of a cure that is not there, we face the possibility of dying there during intensive treatment. Why do we seem to have lost the ability to recognize the onset of death? Why is it so hard to differentiate between dying and a curable condition?

Two medical developments played the most important role in this evolution. One of these was the inception of cardiopulmonary resuscitation (CPR), which was developed in the late 1950s and early 1960s [11–13]. While amazingly effective in the operating theater, the extrapolation to many, if not all, patients experiencing cardiac arrest in the hospital was unfounded, and most of these resuscitation attempts proved futile. Another reason is the development of critical care, which also evolved during the same period, and which created the opportunity to support or substitute organ function. Both developments have led to a less well-defined end of life: as hearts can be restarted and organ function can be supported or even completely taken over, how do we know that life's end has come?

Modern dying takes place in the modern hospital [14]. And in recent years there seems to have been a shift of dying towards acute and critical care: the MET was given a place in many hospitals in order to meet the demands of acutely sick patients and to reduce cardiac arrests, but also to reduce acute admissions to the critical care unit of patients with little chance to benefit from ICU treatment. Being open and honest about diagnosis and prognosis has the potential to prevent prolonged continuation of medical treatment, and may improve acceptance by both patients and families that death is inevitable. Intensivists, with their broad perspective and experience, may often be best suited to such discussions.

Primary Health Care

Primary health care can be effective in reducing costs by reducing hospital admissions and thereby reducing the amount of (expensive) specialist consultation. A recent paper showed an organizational model for after-hours primary care that is safe, efficient, and satisfactory for patients and health care professionals [15]. Around the year 2000, Dutch primary care physicians (PCPs) reorganized their after-hours primary care and shifted from small rotation groups to large-scale PCP cooperatives. Physicians expressed high satisfaction with PCP cooperatives; their workload decreased, and job satisfaction increased. In general, patients were also satisfied, but areas for improvement included telephone consultations, patient education, and distance to a pharmacy.

Telephone triage by nurses had positive effects on care efficiency by increasing the proportion of telephone consultations and decreasing the proportion of clinic consultations and home visits. The after-hours primary care system in the Netherlands might set an example for other countries struggling to find a good solution for the problems they encounter with after-hours primary care. Future developments in the Netherlands include integration and extensive collaboration with the accident and emergency departments of hospitals, in which PCPs take care of self-referring patients.

Technology and the Cost of Health Care

There is no doubt that scientific and technological advances have led to enormous improvements in the quality of health care. Nor is there any doubt that innovation and the scientific enterprise represent the way forward. But it is less certain that technologic advances will decrease costs.

Consider the example of Pompe's disease. Medical science has made it possible to treat patients with Pompe's disease, a glycogen storage disease caused by lack of an enzyme that metabolized glycogen, leading to its accumulation. The unwanted storage of glycogen causes progressive muscle weakness and affects various body tissues, particularly in the heart, skeletal muscles, liver, and nervous system. Discovery of the enzyme responsible led to replacement therapy, an effective treatment that greatly extends the life of these patients. The drug (Myozyme™) costs an average of $300,000 a year, and must be taken for the patient's entire life. Technology is finding cures for more and more diseases, which is wonderful, but is as likely to increase costs as to decrease them. Similar considerations pertain to expensive chemotherapy for cancer, therapy that can be effective, but leads to small increases in life expectancy or quality of life at significant expense. The same challenge is evident in critical care medicine, with its expensive technology and workforce, as well.

One of the current technologies with the most promise is genetic testing. Again, there is little doubt that the advent of personalized medicine will allow for more precise diagnosis and targeted therapy, therapy that promises to be more effective with lower toxicity than currently available options. What is less clear is that there will be a salutary effect on health care costs, although its use may drive systems even more rapidly towards universal coverage. Insurance systems are based on large populations in which risks are pooled. Insurers could use genetic information to exclude high-risk patients, while patients with access to such information not shared with insurers could use it to tailor their level of coverage. Either way, the concept of pooling of risk is defeated. Similarly, targeted therapy is eminently desirable but may not prove inexpensive. The notion that advances in genetic technology will solve the problem of medical costs may prove to be wishful thinking.

Physicians, Patients, and Society

It is clear that physician's primary obligation is to the patient. This has been codified from the time of the Hippocratic Oath, through the Oath of Maimonides, down to the present day. The Declaration of Geneva of the World Medical Association [16] binds the physician with the words, "The health of my patient will be my first consideration," and the International Code of Medical Ethics declares, "A physician shall act in the patient's best interest when providing medical care." [17] Similarly, the Declaration of Helsinki [18] (1964, last revised in 2008 in Seoul) makes a statement of ethical principles for medical research involving human subjects, including research on identifiable human material and data, and is also founded on the belief in personal well-being: "It is the duty of physicians who participate in medical research to protect the life, health, dignity, integrity, right to self-determination, privacy, and confidentiality of personal information of research subjects."

These ethical principles, accepted and supported worldwide, make it difficult to limit therapeutic options in individual patients. Nonetheless, in any medical system, regardless of its structure, there must necessarily be a ceiling of some sort, as resources are inevitably limited at some level. Decisions in regard to individual patients and decisions in medicine as a whole are guided by the four main principles of medicine (beneficence versus non-maleficence, futility, autonomy and distributive justice).

A frequent misunderstanding is that if a patient or a surrogate of the patient demands continuation of treatment in the ICU, this demand needs to be obeyed. This view is described in Dr. Sprung's chapter: [19] "The patient has a right to choose or reject a specific treatment which leads to an obligation on the physician to provide it or not. Over the past few years, there have been more reports of physicians unilaterally refusing patient or family requests for medical therapies, including life-prolonging interventions." In our opinion this notion is untenable and unworkable. While patients or surrogates have a strong right to refuse treatment (even treatment that is obviously medically beneficial) according to the principle of autonomy, it does not follow that patients have a right to receive whatever treatment they (or their surrogates) demand. In the first place, physicians do not have an obligation to provide treatment they do not consider beneficial. Who would force a physician to amputate a perfectly healthy limb at the request of a patient? More importantly, both physicians and their patients need to consider distributive justice, meaning that available means need to be allocated amongst all who need these. This principle can conflict with the principle of autonomy in some cases.

These issues are often most challenging in the intensive care unit. In the ICU the choices seem less clear: it may be uncertain whether continuation of mechanical ventilation is beneficial if the chances of ultimate recovery are minimal. And the stakes are higher. ICU physicians are constrained by the inherent uncertainty of the eventual prognosis of their patients and the consequences of their decisions: if there is any chance left that recovery is possible, stopping treatment would mean a certain death.

There is thus an interplay among physicians, patients, and societal norms in making bedside decisions in the intensive care unit. The most wrenching of those decisions concerns saying "no," the situation in which a request/demand for therapy is not felt by the health care team to be in the best interests of the patient or the society. In these situations, the notion of futility is often invoked, but we agree with others in this volume that this notion is commonly less than useful, and its invocation can at times be counterproductive.

The problem is that futility does not have a universally accepted definition. In some societies the notion of futility has a different meaning than in others. As an example, if inability to recover to a meaningful existence is considered futile, then ventilating a patient who has suffered a catastrophic stroke without a chance of recovery will be considered futile in many settings. If, on the other hand, the goal is merely to keep the patient alive, albeit not recovering, then ventilating this same patient will not be considered futile in other settings.

What makes conflict about the appropriate level of care at the end of life even more challenging is that in addition to social and cultural norms, political and economic and sometimes also religious factors can come into play. And these situations will often be negotiated at the bedside by the physician, patient, and family, each influenced by external forces that may well not be aligned in the same direction.

Does this mean we are abdicating our responsibility to provide useful advice? We hope not. Universal solutions to these difficult problems are too much to ask for. As noted, local circumstances, representing an interplay among historical, cultural, societal, political, and yes, economic forces, will influence the decisions, but unique aspects of the clinical situation, along with the physician–patient relationship, an even more local factor, will prove just as important. What we hope to bring to the table is that notion that these decisions, as local and individual as they are, are not made in a vacuum. Clinicians are not required, nor are they in some sense entirely free, to make these sorts of decisions without consideration of cultural and societal norms. And while those norms are societal rather than individual, physicians are in a unique position to bring information to hand to inform the discussion.

What Role Should Government Play in Health Care Resource Allocation?

Except in settings in which health care is paid for out of pocket (generally underdeveloped countries, although, shamefully, this includes those uninsured Americans who do not qualify for other support), the government necessarily plays a role in resource allocation, either by regulating which interventions and treatments will be funded and which will not, or both. The degree to which this occurs is dependent mainly on national politics, but also on historical factors and broadly accepted societal notions of fairness. But the individual physician is neither free to ignore governmental factors entirely nor ineluctably bound to them.

If the statement by Dr. Sprung ("The physician's primary responsibility is to his/her individual patient. Doctors are more frequently being pressured to consider society's requirements in addition to their individual patient's needs. The principles of patient autonomy and beneficence are giving way to the principles of distributive justice and proportional advocacy. Physicians cannot discharge their responsibility to their individual patients if they at the same time attempt to conserve societal resources.") [19] were strictly true, then only a government could regulate and allocate resources; physicians would by definition be unable to do so. Presently this is not the case. Physicians make choices, for individual patients and in their positions as managers of hospitals and hospital departments.

Still, ICU physicians should do their best to avoid no-win situations in which they are required to enforce external limitations on health care while still retaining personal responsibility for the consequent outcomes. In South Africa the ICU capacity is so limited that patients who in other systems might have been admitted to the ICU will be denied an ICU bed. Physicians in that system are forced to make often wrenching triage decisions due to limitations in medical resources on a day-to-day basis.

How is this to be accomplished? For a start, physicians discussing limitation of care should make it clear that they are acting within the context of widely accepted societal norms. In New Zealand, as described in this volume by Dr. Streat [20], this occurs as a matter of course. In other places, this may not be so easy. In particular, when external cost constraints are in play, physicians should be wary of being put in the position of the solitary enforcer of cost limitations. Those imposing those constraints need to take responsibility in an open and straightforward fashion. If an insurer is denying reimbursement for a procedure, then the expectation should be that the insurer must justify the denial. If a system deems that a given therapy does not meet criteria, then that should be done in a public and transparent fashion. The National Institute of Health Care Education (NICE) process in England, whatever its shortcomings, does have this merit. Physicians may need to campaign to ensure that the process of limitation of care is both open and based on a broad societal and normative consensus.

Another important factor is to have discussions informed by transparent, relevant, and easily accessible data. Finally, the role of intensivists as educator—of other staff, of patients, of policy makers, and of the public at large—is already important and will become even more so with time.

Measures of Health Care Quality

As we have noted, health care in general, and critical care in particular, represents an inevitable trade-off between accessibility to services and affordability. In order to navigate these trade-offs, intensivists will need good data to enable them to weigh costs and benefits. The issue of costs will not be considered in detail; they are beyond the scope of this chapter and likely beyond the scope of the book as well. In this respect, we confine ourselves to noting that focusing only on cutting costs provides too narrow a perspective. We should not forget that health care also gives rise to profit, not only for those who work in health care, but also in a much broader perspective. The gains in health produced by the health care system positively influence productivity. And not only that, health care is an economic good, and provides considerable employment in many countries. Just cutting costs in health care systems will probably have a negative effect on the economy [4].

That said, for critical care there are a number of potential future changes that can be incorporated to keep the systems working and affordable. Critical care will change by two main driving forces: safety management and the related issue of clustering of critical care services. Quality and performance indicators manage this process.

Quality Indicators and Performance Indicators

Society as a whole will increasingly demand clear and reliable data on how health care systems are spending money, whether this is paid by individuals, through taxes, or through a system of health care insurance. As most countries have mixed systems, all of these ways in which health care is financed will influence where health care is headed. Patients, governments, and insurance companies are already asking for transparent data on how money is spent and what outcomes have resulted from that expense. This demand for transparent and accessible data will only increase in the near future, and such data are important for the health care system itself, both to justify itself and to allow for process improvement.

Quality indicators have already become important in all areas of health care and their importance will increase globally. Depending on the local system, governmental or independent agencies that control hospitals and other health care facilities will ultimately demand that facilities meet quality criteria in order to be allowed to continue to provide health care. Quality indicators are put in place to regulate the (minimal) demands for a health care facility: number of physicians and nurses per bed, hygiene aspects, control and maintenance of medical equipment, and the like. Quality indicators are relatively easy to define and to monitor. Health care professionals and managers are familiar with using quality indicators to improve care, and in general they have positive attitudes towards the implementation of quality indicators. It is however necessary to lower barriers against implementing quality indicators by influencing behavioral factors. As these barriers and facilitating factors differ among professions, age groups, and settings, tailored strategies are needed to implement quality indicators in daily practice [21].

Performance indicators, on the other hand, are much more difficult to agree on and are difficult to monitor. Despite that, in the end performance indicators can be used to compare outcome of hospitals and ICUs and ultimately this can lead to changes in health care systems: if consolidation of ICUs produces better outcomes, and if ICUs staffed with critical care specialists lead not only to reduced mortality, but also to better care in other aspects in addition to mortality, these performance indicators can be used as agents to spur productive change. This can lead to clustering of ICUs, which can, within limits, lead to a higher general level of performance of these ICUs.

Measuring Effects of Health Care: Cost-Effectiveness

How should the effects of health care be measured? Can we be satisfied with the notion of providing what we as health care workers think is best care, or do we need to measure effects of health care? The answer to this question is simple: we need to measure outcome; and we most probably need to measure outcome in relation to costs. This is of course a very complicated area. The outcomes that are easiest to measure may not be the ones that are most important. In critical care we most often use mortality at a certain time point (28 days; hospital discharge, 90 or 180 days) as a standard outcome. But is this a useful concept? Do we need to take into account Osler's dictum that the goal of medicine is not so much to add years to life but life to years, and measure added quality of life? These measures are more difficult, less certain, and more costly to obtain.

The cost side of the equation is even more complex. Issues such as fixed and marginal costs, opportunity costs, and cost shifting lend a considerable degree of uncertainty to cost estimates. One measure that is relatively quantifiable is health care costs as a percentage of GDP. In this respect, the results are clear. The United States spends by far the highest percentage of its GDP on health care (17% compared to 7–11% in other Western countries) without a commensurate improvement in health-related outcomes, making the American system one of the least cost-effective health care systems in the world [22]. This is unlikely to be sustainable in the long term.

Health Care Education

Physicians are in general well respected, not in small part because of their role as advocates for their patients. In the evolution of health care, we would contend that physicians need to extend their role as patient advocates by educating the public about the issues that face health care systems as a whole. We agree wholeheartedly with Drs. Buchman and Chalfin in their chapter of this book, in which they state that health care education is of primary importance [3]. On an individual level, this means an emphasis on healthy lifestyle choices and a notion of shared responsibility. On a general population level this entails education in methods of risk assessment, outcomes measurement, and performance measures in preparation for interpretation of transparent reporting of the quality and cost-effectiveness of various therapies. Who better than physicians to take the lead in initiatives to improve health care literacy?

Intensive care physicians are particularly suited to be health care educators. At the bedside of a critically ill patient, intensivists need to engage patients and families in discussions of complex and often emotionally fraught issues about which these patients may or may not have a great deal of preexisting knowledge. The same is true for other physicians in the hospital, particularly subspecialists who may not be used to taking the broad view of the patient and his or her co-morbidities and overall prognosis. What is needed are people who think broadly and take the long view, something that serves as a fairly good description of a good intensivist. Critical care physicians are also necessarily involved in analysis of processes of care, and their expertise may translate into issues involving measurement of outcomes and adjustment for preexisting risk, as well as those involving broad systemic issues. Finally, discussion of limitation of care at the end of life, although a broad societal issue, benefits greatly from the perspective from someone with the experience of standing at the bedside of a dying patient.

In addition, physicians may need to take the lead in initiating broad societal discussions of the trade-off between the amount of health care and its costs. It is all too easy to make the argument that all that is required is greater efficiency, but as we have seen, technology can increase efficiency and quality, but usually increases costs as well, by making new therapies possible. The discussions are political and economic, and thus need to have broad representation with an effort to include disparate viewpoints, but physicians are uniquely positioned to be honest brokers in this sort of discussion.

A last issue, one with particular relevance to limitations of care, concerns the necessity of accepting uncertainty. Physicians are men and women of science and seekers of precise answers, but as Dr. Burrows trenchantly points out in his chapter [23], although death itself is inevitable, its timing cannot be predicted with certainty. Thus limitation of care must be based on something less than full certainty, which introduces the possibility of error. Any workable process for limitation of care (and as we have seen, in the absence of unlimited resources, there must be some such process) must thus function with some uncertainty. Put another way, physicians must be prepared to admit that neither they nor anyone else has all of the answers.

Conclusion

We start with the premise that health care systems should aspire to universal health care coverage, for both moral and practical reasons. But universal coverage is coverage for everyone, not everything. The health care edifice needs a floor, but also a ceiling.

Science and technology are making great advances in improving care, but although efficiency may be increasing, costs are not going down. Part of the reason may be that previously untreatable conditions are now amenable to therapy, increasing the range of services that can be provided.

Critical care uses a great deal of resources, and is a promising target for cost reduction. In this context, clustering of specialized services, use of medical emergency teams, and close attention to processes of care may be beneficial.

Nonetheless, decisions to limit care will need to be made, often at the bedside in the intensive care unit. The critical care physician has an obligation to the patients, but should also feel constrained to act within the bounds of societal values and norms. It is hoped that the impulse to do so will result from notions of fairness rather than financial pressures. These notions should inform the tension between individual rights and the collective good. But since both the particular circumstances and the underlying societal norms will vary from place to place, local factors will predominate.

Finally, decisions about levels of care should be informed by data concerning prognosis, risks, benefits, and outcomes. Clear, comprehensible, and transparent measures need to be developed and disseminated. And education—at the bedside, in the hospital, and of the public at large—concerning these complex yet vitally important issues remains paramount.

References

1. Reid TR. The healing of America: a global quest for better, cheaper, and fairer health care. New York: Penguin; 2009.
2. The Universal Declaration of Human Rights of the United Nations. http://www.un.org/en/documents/udhr/index.shtml – a25.

3. Buchman TG, Chalfin DW. Changing normative beliefs and expectations in critical care. In: Crippen DR, editor. ICU resource allocation in the New Millennium: Will we "Say No"? Berlin: Springer; 2012.
4. Pomp M. Een beter Nederland. De gouden eieren van de gezondheidszorg. [A better Holland. The golden eggs of health care]. Amsterdam: Balans; 2010.
5. Caplan AL. Max Harry Weil Honorary Lecture: doing what is right: the ethics of reforming our broken health care system. In: Society of Critical Care Medicine 38th Annual Scientific Conference. Nashville, TN; 2009.
6. Peelen L, de Keizer NF, Peek N, Scheffer GJ, van der Voort PH, de Jonge E. The influence of volume and intensive care unit organization on hospital mortality in patients admitted with severe sepsis: a retrospective multicentre cohort study. Crit Care. 2007;11:R40.
7. Jones DA, Medical Emergency Team End of Life Care Investigators. The role of the medical emergency team in end of life care: a multi-centre prospective observational study. Crit Care Med. 2012;40(1):98–103.
8. Parr MJ, Hadfield JH, Flabouris A, Bishop G, Hillman K. The medical emergency team: 12 month analysis of reasons for activation, immediate outcome and not-for-resuscitation orders. Resuscitation. 2001;50:39–44.
9. Hillman K, Chen J, Cretikos M, et al. Introduction of the medical emergency team (MET) system: a cluster-randomised controlled trial. Lancet. 2005;365:2091–7.
10. Jones DA, McIntyre T, Baldwin I, Mercer I, Kattula A, Bellomo R. The medical emergency team and end-of-life care: a pilot study. Crit Care Resusc. 2007;9:151–6.
11. Safar P. Ventilatory efficacy of mouth-to-mouth artificial respiration; airway obstruction during manual and mouth-to-mouth artificial respiration. J Am Med Assoc. 1958;167:335–41.
12. Kouwenhoven WB, Jude JR, Knickerbocker GG. Closed-chest cardiac massage. JAMA. 1960;173:1064–7.
13. Eltringham RJ, Baskett PJ. Experience with a hospital resuscitation service including an analysis of 258 calls to patients with cardiorespiratory arrest. Resuscitation. 1973;2:57–68.
14. Nuland SB. How we die. Reflections on Life's Final Chapter. New York: Knopf; 1994.
15. Giesen P, Smits M, Huibers L, Grol R, Wensing M. Quality of after-hours primary care in the Netherlands: a narrative review. Ann Intern Med. 2011;155:108–13.
16. WMA Declaration of Geneva. 1948. http://www.wma.net/en/30publications/10policies/g1/.
17. International Code of Medical Ethics. 1949. http://www.wma.net/en/30publications/10policies/c8/.
18. Declaration of Helsinki. 1949. http://www.wma.net/en/30publications/10policies/b3/.
19. Sprung CL. A global health care plan for the future. In: Crippen DR, editor. ICU resource allocation in the New Millennium: Will we "Say No"? Berlin: Springer; 2012.
20. Streat S. Contrasts in global health care resource allocation Where we've been in critical care in New Zealand. In: Crippen DR, editor. ICU resource allocation in the New Millennium: Will we "Say No"? Berlin: Springer; 2012.
21. de Vos ML, van der Veer SN, Graafmans WC, et al. Implementing quality indicators in intensive care units: exploring barriers to and facilitators of behaviour change. Implement Sci. 2010;5:52.
22. Pritchard C, Wallace MS. Comparing the USA, UK and 17 Western countries' efficiency and effectiveness in reducing mortality. JRSM Short Rep. 2011;2:60.
23. Burrows R. Medicine: art, science and political expediency. In: Crippen DR, editor. ICU resource allocation in the New Millennium: Will we "Say No"? Berlin: Springer; 2012.

Chapter 38
The New Shape of Intensive Care in the USA

Derek C. Angus

Anticipating the future of critical care delivery in the USA is obviously no trivial task. Numerous issues abound, ranging across who will develop critical illness, how critical illness will manifest, where it will be treated, who will provide the care, and how care will be financed. Many earlier chapters in this book have documented some of the elements of an approaching "Perfect Storm" of demand outstripping supply. The first demand element is the rising incidence of critical illnesses, such as severe sepsis, in an ever-aging, increasingly frail population. The second is the continued expectation of a general public that no expense should be spared, even though the burden, and therefore the total cost, is rising. The third is that the expectation for care is actually increasing, stretching to expect staffing levels beyond that of past models. Meanwhile, supply is compromised because of a lack of funds and, consequently, a worsening lack of an adequately trained, willing workforce. Against this rather bleak backdrop, what are the likely realities and opportunities?

Supply Constraints

Ten years ago, we forecast that the increasing demand for critical care services based simply on changing demographics would rapidly begin to outstrip supply unless there was a significant change in supply [1]. Since that time, demand has increased faster than anticipated because of additional desires from hospitals and insurers to have greater intensivist presence [2]. Yet, in 10 years, there has been no increase in the number of Medicare-funded training slots for intensivists. The lack of change in supply is hardly surprising, since training slots are funded under the Medicare hospital trust fund, a fund that faces huge demand from multiple directions,

D.C. Angus, M.D., M.P.H., F.R.C.P. (✉)
Department of Critical Care Medicine, University of Pittsburgh School of Medicine,
Pittsburgh, PA 15261, USA
e-mail: angusdc@upmc.edu

D.W. Crippen (ed.), *ICU Resource Allocation in the New Millennium: Will We Say "No"?*, 315
DOI 10.1007/978-1-4614-3866-3_38, © Springer Science+Business Media New York 2013

not just critical care, and one that is projected by the Congressional Budget Office to be insolvent in just a few years [3].

The existing workforce will thus be asked to work longer hours or will simply fail to meet demand. Certainly, many intensivists are now being asked to work in-house at night and are being asked to see higher numbers of patients. The changing nature of these demands will likely result in some adaptations to the job. First, for many, intensive care will become shift-work. Second, with more off-hours work, if intensivists cannot be found, then compensation will likely rise. Recent data suggest intensivists now work more hours than all but one of the 63 medical specialties [4].

Intensive Care as Shift-Work

It is impractical to work in-house at night in addition to working full normal business hours. Graduate medical training has already come unstuck on this issue, recognizing that such a model drives the working week up to an excess of 100 h. With an 80 h workweek, trainees can either work normal days with occasional nights or have large gaps in their week or they can adopt a shift model similar to that used in emergency medicine. Some facsimile of these choices will increasingly permeate through the ICU attending workforce. It is likely that normal business hours followed by in-house call will be a great source of stress and burn-out over the long-run, and thus shift-work models will become an increasingly attractive alternative, even if "attractive" is a somewhat relative term. Unlike Emergency Medicine, Intensive Care requires sustained patient management, and thus, shift-work could pose greater burdens on continuity of care. Again, Graduate Medical Education programs are already grappling with the need for more effective sign-out because trainees are far less present than in the past. The challenge will be similar for attending physicians.

Spreading Intensivists More Thinly

Of course, moving to shift-work will not solve the supply–demand problem. Fundamentally, we must also grapple with solutions that allow intensivists to be spread more thinly. These include identification of patients that should not be seen by an intensivist and identification of approaches that allow intensivists to manage a higher number of patients. The easiest way to ensure an intensivist does not see a patient is to avoid ICU admission in the first place.

Can We Cut ICU Beds?

The USA has many more ICU beds per 100,000 population than most other industrialized nations [5]. France, a country whose healthcare system is frequently ranked number one in the world, has only a third of the ICU bed supply as the US. Yet, as of 10 years ago, an intensivist managed only one in three ICU beds in the USA, whereas an intensivist manages all French ICU beds [6]. Therefore, perhaps the most obvious solution for the USA is to eliminate or reclassify two-thirds of its ICU beds. If the USA were a single payer system, some type of "base closing commission" might well be able to execute such a task simply by mandate. However, de-escalation of ICU services likely requires a more multifaceted approach with consideration of legislative and financial levers.

The only obvious legislative approach would be through statewide certificate of need programs. However, such programs have generally not been very effective in controlling other healthcare technologies and therefore it seems unlikely that legislation is the correct instrument on its own.

A second approach is to restrict the use of the ICU through financial incentives. Unfortunately, we already have such incentives. The DRG system used by Medicare and many third party payers does not reward hospitals for incurring higher per-patient costs. Therefore, building more ICU beds and admitting more patients to the ICU, to the extent that these actions increase per-patient costs, represent actions that hospitals take despite a financial disincentive. From the hospital's perspective, there must be other gains from having more ICU beds. These gains may be financial or otherwise. A financial gain may be that, by having more critical care facilities, the hospital may garner greater patient volumes or a more favorable case-mix. A nonfinancial gain may be that better critical care facilities yields better patient outcomes, which may also help with further gains in market share. Strengthening financial disincentives in the face of these existing forces is potentially contentious. For example, one would not want to provide a financial disincentive against quality. Thus, it seems likely that any realignment of financial incentives would have to be implemented in concert with strong clinical rationale in support of denied ICU care.

Defending Denial of ICU Care

The high number of ICU beds per capita does not appear to yield particular gains in health. Rapoport showed many years ago when comparing Alberta to western Massachusetts that a two- to threefold difference in ICU availability had no measurable impact on population health [7]. The UK, despite having one seventh of the ICU bed capacity of the USA, enjoys longer life expectancy [8]. But, denying ICU care is a local decision, and a long run weighing of risks and benefits is too abstract to figure prominently at the bedside. Much of this book has wrestled with the ICU admission decision. To deny ICU care is either to deny marginally effective care or to deny ineffective care (since we are not yet at the point of deliberately denying highly effective care). Unfortunately, ineffective care is difficult to define, and individuals will cling to any chance of benefit, even if small, especially if the cost to the individual is small.

A starting point, therefore, would be for professional societies to develop standards for ICU admission. These standards, supported by funding agencies, such as Medicare, may make the decision more protocoled and standardized, and thus remove the decision from the bedside doctor and family. Regrettably, even if professional societies took this step, there is no obvious federal process to support this approach. For example, when passing the recent Affordable Health Care Act [9], Congress specifically prohibited the Patient-Centered Outcomes Research Institute (PCORI) [10] from considering any cost-cutting efforts that involved rationing or restriction of services. Nevertheless, there is the desire from Congress to see the growth of Accountable Care Organizations (ACOs). These ACOs may embrace national policies and guidelines developed by professional societies if they appear to be based on sound clinical evidence and are likely to curb costs.

Creating Different Levels of ICU Care

In addition to trying to avoid ICU admission, some attempt to tier the level of care could likely help manage an under-supply of intensivists. Again, professional societies could play a valuable role in defining the appropriate levels of care for different patient groups. Tiered care already exists for neonatal intensive care and for the care of trauma victims. Although neither system is perfect, both allow a high degree of regionalization of services, with concentration of scarce resources, including specialty workforce. Furthermore, considerable evidence supports the findings that higher levels of care both yield better outcomes for sicker patients and incur higher costs. Thus, restricting the higher levels of care for those patients who benefit most seems to be an efficient system goal.

Work exploring the potential for tiered care and corresponding regionalization and triage of adult nontrauma ICU patients has already begun. Kahn et al. has shown that outcomes are better for patients requiring mechanical ventilation when cared for at high volume centers [11]. In subsequent work, we showed that many patients are in fact cared for at low volume centers [12]. What is less clear is where the lower cut-point in severity of illness lies. Must all ventilated patients be sent to a high volume expert center? In addition, we know less about differences in the costs of care. Can the nonexpert centers provide less expensive care? Can lower risk patients be managed safely by nonintensivists? Obviously, if all ICU patients would benefit from care at a high volume center, then there is little prospect for efficiency gains from the creation of tiered care.

Extending the Intensivist's Reach

If we cannot adequately restrict the number of ICU patients seen by an intensivist to the point that demand matches supply, we must consider alternative mechanisms to extend the reach of the intensivist to a larger number of patients. There are two broad approaches: mid-level ICU providers and tele-ICU.

Mid-level ICU Providers

The potential role for mid-level ICU providers in the ICU was described 12 years ago [13]. A number of aspects of the coordination of complex critical care lend themselves conceptually to a model where intensivists and mid-level providers (e.g., acute care nurse practitioners or physician assistants) interact. Mid-level providers can assess patients, order diagnostic tests, institute therapies, initiate and wean life-support techniques, and perform invasive procedures. Standards governing the degree to which these actions can be conducted independently versus under supervision or in consultation with an intensivist are governed by professional society guidelines, state licensing boards, and hospital credentialing committees. Working with a mid-level practitioner, it is therefore conceivable that an intensivist could see a larger number of patients, triaging his or her time and effort, and that of the mid-level practitioner, across the varying mix of tasks and patients in a given ICU. Yet, mid-level practitioners have played a very small role in US ICU care provision until recently.

There are likely several factors that inhibited the growth of the mid-level practitioner workforce in US critical care. First, for many years, there simply were few training opportunities, and, consequently, there was no workforce. Second, for ICU nurses seeking advanced degrees, a natural choice was to seek training as a nurse anesthetist. The nurse anesthetist workforce in the USA is very large and continues to grow, salary is excellent, and the job has a number of positive attributes, including attractive work hours, considerable independence, and a strong professional identity. In contrast, other nurse practitioner salaries have not traditionally been as high, and ICU nurse practitioner jobs often had varied specifications, less attractive work hours, and less well-defined job expectations.

The tide, however, appears to be changing, and one might reasonably expect that there will now be rapid growth. Faced with decreased resident support in the ICU, a number of large, renowned academic institutions have committed to building a large mid-level practitioner workforce. In so doing, these institutions have had to design jobs attractive to potential recruits. As such, jobs have become well described, and include adequate remuneration, adequate professional independence, attractive work hours, and so on. In addition, these institutions, recognizing the lack of potential applicants, have invested in expanded training programs. Recent data are now emerging suggesting that conversion from resident-based care to care delivered in concert with mid-level practitioners is safe and feasible, overcoming residual concerns about whether non-MD intensivists are safe [14].

As resident work hours continue to drop, and as in-house fellow coverage becomes increasingly scarce, one might reasonably expect that mid-level practitioners will increasingly work alongside MD trainees in ICUs across all academic institutions. As the profile of the mid-level practitioner workforce grows, nonacademic centers may also seek them, especially to help intensivists stretch themselves across a larger number of ICU patients, and possibly also to provide in-house coverage during off-hours in compliance with Leapfrog Group standards for adequate access to intensive care expertise.

One roadblock to the growth of mid-level practitioners will be the funding mechanism. Mid-level practitioners can bill for professional services, but the fees raised may not cover their salary, requiring additional supplementation either from the hospital or the intensivist group. In addition, there are instances when it is not possible for both the intensivist and the mid-level practitioner to bill together. Thus, mid-level practitioner billing can be in competition with intensivist billing. Of note, similar issues have arisen in the provision of anesthesia services and continue to be a point of contention. One is tempted to wish for a less acrimonious process for integrated ICU services.

Tele-ICU

A second approach to extend the reach of the intensivist is the use of telemedicine. Obviously, on-hand procedural expertise will not lend itself easily to telemedicine. However, many other aspects of intensive care delivery involve intellectual input, which may be delivered from a remote location with adequate access to patient information [15, 16]. There is one large commercial tele-ICU service in the USA and the cost of implementation is high. It is likely, with advances in technology, maturation of the field, and expansion of the number of vendors, that the price will drop. The evaluation of tele-ICU efforts has been mixed. For example, two large similarly designed studies from Texas and Massachusetts evaluating the same proprietary tele-ICU product reported opposing results [17, 18]. In Texas, tele-ICU yielded little change in patient outcomes or costs [17]. In contrast, the Massachusetts study found that the use of tele-ICU results in considerable improvement in compliance with a number of quality standards, especially during off-hours, with corresponding improvements in patient outcomes and length of stay [18].

The reasons for the discrepancies are somewhat speculative, and yet are likely key to the future of tele-ICU growth. In Texas, the tele-ICU intensivists were providing care to outlying hospitals where they had no "on the ground" presence, and it appeared that the local clinicians were resistant to having close tele-ICU involvement except during emergencies. In contrast, the same intensivists who provided in-person care during the day administered the Massachusetts tele-ICU intervention. When considering the future role of tele-ICU, it will be important to appreciate that tele-ICU comes in many guises, and is simply one of many tools to help facilitate better quality of care [19]. The tool can no doubt be used wisely or imprudently, and consideration of "how" it is used, specifically in terms of how the care team is constructed, will be crucial to success. In particular, intensivists and tele-ICU product developers will have to address the risk of further depersonalizing an already depersonalized environment. There is an ever-broadening desire to make ICUs more patient and family-friendly. Somehow, the idea of a doctor beamed in by computer monitor has to be an enhancement, not a threat, to that concept.

Nonetheless, it is very tempting to imagine that tele-ICU, perhaps in combination with growth of a mid-level practitioner workforce, or a changed set of job skills for ICU nurses, could help intensivists manage more patients across more hours more efficiently and more safely.

Conclusion

The demand for critical care services has probably always outstripped supply, just as we have always struggled to pay for healthcare and so framing current concerns over healthcare as a "crisis" seems almost to be a misnomer. That said, the scale of healthcare delivery and financing problems is larger now than in the past, and US healthcare delivery is indeed likely to go through revolutionary change in the coming decades. ICU care will be at the "pointy end" of the wedge.

The existing intensivist workforce is neither large enough nor willing to provide in-person 24 h coverage to all critically ill patients. Furthermore, even if we could, there are insufficient funds to pay for such coverage. Solutions to spread the workforce more thinly lie within our grasp, but will not be delivered via a simple top-down government mandate. Professional societies and researchers have a crucial role to play by generating, summarizing and promulgating the evidence on optimal delivery models. These models, especially if financially attractive, will spread most likely via adoption by healthcare providers and networks, especially ACOs, as they organize themselves to respond to payer incentives.

Regardless of the exact nature of the final model, it is likely that intensivists will work in very different ways in the future. First, being "on service" is increasingly likely to feel like a shift. Second, intensivists must embrace partnership with mid-level providers and use of technologies such as tele-ICU to extend their reach. Third, they should also embrace the notion of tiered ICU care, and consideration of who needs ICU care at all. As the intensivist becomes the manager of a larger service, perhaps we should paradoxically be shrinking, not expanding, the workforce. Finally, the training of the new workforce needs to change to provide not only clinical skills but also adequate leadership and managerial skills.

Fundamentally, the looming problems provide very exciting opportunities for US critical care delivery. Until now, we have provided care one patient at a time. There is no national coordinated effort to ensure that all critically ill patients receive the right level of care. Perhaps only now, faced with impending financial disaster, will we start to embrace efficient systems that optimize the available resources in a fashion that best meets the entire population's needs.

References

1. Angus DC, Kelley MA, Schmitz RJ, White A, Popovich Jr J, Committee on Manpower for Pulmonary and Critical Care Societies (COMPACCS). Caring for the critically ill patient. Current and projected workforce requirements for care of the critically ill and patients with pulmonary disease: can we meet the requirements of an aging population? JAMA. 2000;284(21): 2762–70.

2. Gasperino J. The Leapfrog initiative for intensive care unit physician staffing and its impact on intensive care unit performance: a narrative review. Health Policy. 2011;102:223–8.

3. http://www.cbo.gov. Accessed 11 Sept 2011.

4. Leigh JP, Tancred D, Jerant A, Kravitz RL. Annual work hours across physician specialties. Arch Intern Med. 2001;171:1211–3.

5. Wunsch H, Angus DC, Harrison DA, Collange O, Fowler R, Hoste EA, de Keizer NF, Kersten A, Linde-Zwirble WT, Sandiumenge A, Rowan KM. Variation in critical care services across North America and Western Europe. Crit Care Med. 2008;36(10):2787–93. e1-9, Review.

6. Angus DC, Shorr AF, White A, Dremsizov TT, Schmitz RJ, Kelley MA. Critical care delivery in the United States: distribution of services and compliance with Leapfrog recommendations. Crit Care Med. 2006;34:1016–24.

7. Rapoport J, Teres D, Barnett R, Jacobs P, Shustack A, Lemeshow S, Norris C, Hamilton S. A comparison of intensive care unit utilization in Alberta and western Massachusetts. Crit Care Med. 1995;23(8):1336–46.

8. Wunsch H, Linde-Zwirble WT, Harrison DA, Barnato AE, Rowan KM, Angus DC. Use of intensive care services during terminal hospitalizations in England and the United States. Am J Respir Crit Care Med. 2009;180(9):875–80. Epub 2009 Aug 27.

9. http://www.healthcare.gov/law/introduction/index.html. Accessed 11 Sept 2011.

10. Kamerow D. PCORI: odd name, important job, potential trouble. BMJ. 2011;342:d2635.

11. Kahn JM, Goss CH, Heagerty PJ, Kramer AA, O'Brien CR, Rubenfeld GD. Hospital volumen and the outcomes of mechanical ventilation. N Engl J Med. 2006;355:412–50.

12. Kahn JM, Linde-Zwirble WT, Wunsch H, Barnato AE, Iwashyna TJ, Roberts MS, Lave JR, Angus DC. Potential value of regionalized intensive care for mechanically ventilated medical patients. Am J Respir Crit Care Med. 2008;177(3):285–91.

13. Snyder JV, Sirio CA, Angus DC, Hravnak MT, Kobert SN, Sinz EH, Rudy EB. Trial of nurse practitioners in intensive care. New Horiz. 1994;2:296–304.

14. Gershengorn HB, Wunsch H, Wahab R, Leaf D, Brodie D, Li G, Factor P. Impact of nonphysician staffing on outcomes in a medical ICU. Chest. 2011;139(6):1347–53. Epub 2011 Mar 10.

15. Young LB, Chan PS, Lu X, Nallamothu BK, Sasson C, Cram PM. Impact of telemedicine intensive care unit coverage on patient outcomes: a systematic review and meta-analysis. Arch Intern Med. 2011;171(6):498–506. Review.

16. Kahn JM, Hill NS, Lilly CM, Angus DC, Jacobi J, Rubenfeld GD, Rothschild JM, Sales AE, Scales DC, Mathers JA. The research agenda in ICU telemedicine: a statement from the Critical Care Societies Collaborative. Chest. 2011;140(1):230–8.

17. Thomas EJ, Lucke JF, Wueste L, Weavind L, Patel B. Association of telemedicine for remote monitoring of intensive care patients with mortality, complications, and length of stay. JAMA. 2009;302(24):2671–8.

18. Lilly CM, Cody S, Zhao H, Landry K, Baker SP, McIlwaine J, Chandler MW, Irwin RS, University of Massachusetts Memorial Critical Care Operations Group. Hospital mortality, length of stay, and preventable complications among critically ill patients before and after tele-ICU reengineering of critical care processes. JAMA. 2011;305(21):2175–83.

19. Kahn JM. The use and misuse of ICU telemedicine. JAMA. 2011;305(21):2227–8.

Chapter 39
Health Care in the Year 2050

Brian Wowk

> *It is the great glory as well as the great threat of science that everything which is in principle possible can be done if the intention to do it is sufficiently resolute.*
>
> —Sir Peter Medawar

Four decades is a long time. In 1927 Charles Lindbergh flew from New York to Paris in 33 h. 1969 saw the first flight of the supersonic Concorde, which would carry 100 passengers over the Atlantic ten times faster at the edge of space. In 1970 the PDP-11 with 56 kilobytes memory was a popular business computer. In 2011 a pocket smartphone can have a million times more memory.

Sometimes four decades is less significant. Although there have been improvements in safety and economy, the speed of commercial air travel has not increased since 1970. Our computers may be a million times more powerful, but we are not a million times smarter, a million times wealthier, or living a million times longer. Technology may even be breeding a sedentary shorter-lived generation accustomed to having information at their fingertips rather than knowledge in their head. There can be big disconnects between technology and outcomes.

New medical technology is sharply restrained by social factors. Low tolerance for adverse outcomes and associated heavy regulation limit the pace of innovation compared to other fields. What is possible is by no means what will necessarily be done.

Nevertheless, in contemplating health care in 2050 we must be open to the possibility that medicine might become something almost unrecognizable to us today. Historical developments such as vaccines and anesthesia can and have fundamentally changed medicine and public health. Developments of similar magnitude are possible over the next half century. Medicine still operates far from the bounds of what is possible according to known physical law.

B. Wowk, Ph.D. (✉)
21st Century Medicine, Inc., Fontana, CA, USA
e-mail: wowk@wowk.org

D.W. Crippen (ed.), *ICU Resource Allocation in the New Millennium: Will We Say "No"?*,   323
DOI 10.1007/978-1-4614-3866-3_39, © Springer Science+Business Media New York 2013

In attempting to visualize what may come, it is helpful to view health care in three parts: Information, intelligence, and intervention. There is information about health, the process of deciding what to do about it, and what interventions by way of prevention or therapy are available. All three parts will see great change in coming decades.

Information

We are in the midst of an explosion of information. The explosion is driven by both new diagnostic methods and information processing/communication systems. The component density and information storage capacity of computers has been doubling every 2 years since the 1960s (Moore's Law) (see Fig. 39.1). Processing speed has also been increasing exponentially (see Fig. 39.2).

These trends are projected to end circa 2018 as present chip fabrication methods hit physical limits. However, other technologies such as three-dimensional fabrication, molecular electronics, and nanotechnology are waiting in the wings to con-

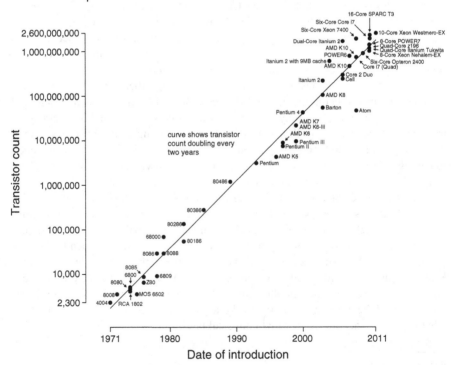

Fig. 39.1 Computer microprocessor transistor counts as a function of year. *Image by Wikipedia user Wgsimon, CC BY-SA 3.0 license*

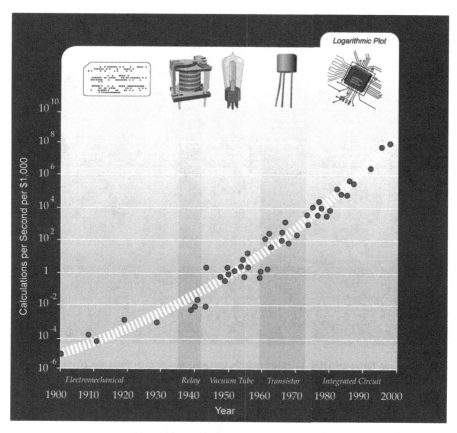

Fig. 39.2 Computer processor speed as a function of year. *Image adapted from Ray Kurzweil and Kurzweil Technologies, Inc., CC BY 1.0 license*

tinue miniaturization trends for perhaps several more decades. The use of small molecules or individual atoms as information storage elements would seem to be the ultimate physical limit for computer miniaturization, a limit to be reached later this century. In the mid-twenty-first century, mere cubic centimeters can be expected to hold 1,000 terabytes of information and parallel processors capable of executing 10^{15} instructions per second. This is the information processing capacity of ten human brains contained in a small electronic device.

This capacity will not go unused. Diagnostic information is undergoing its own exponential increase. The cost of completely sequencing a human genome, approximately one billion US dollars in the 1990s, is now plummeting toward $1,000 and lower [1] (see Fig. 39.3). By 2025 a patient's genome will be part of their medical record, to eventually be joined by their epigenome, proteome, and transcriptome of multiple cell types. The latter parameters of systems biology will be measured with greater frequency and utility as measurement costs continue declining, and understanding of their meaning increases.

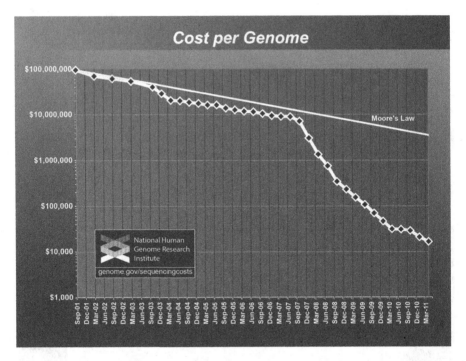

Fig. 39.3 Cost for sequencing an entire human genome. *Image from National Human Genome Research Institute*

"Lab on a chip" miniaturization will decrease the cost and infrastructure required for clinical laboratory tests, moving them to the point of care. Just as photography no longer requires film processing labs, diagnostic tests may no longer require clinical laboratories. Microfluidic laboratories will track numerous biomarkers, analyze circulating cells, and even determine the transcriptome of cells. Cancer will be detected at the earliest stages. Inflammation status will be known in detail, including whether the cause of inflammation is disease, trauma, or pathogens. Real time PCR and other on-chip technologies will permit detection and identification of pathogens on timescales of minutes [2]. By the mid-twenty-first century, these analytical capabilities will be available in very small packages.

Miniaturization of other diagnostic and monitoring technology continues apace. Pill cameras transmitting GI tract images are already a reality. The smallest ultrasound scanner in 2011 is a handheld wand connected to a smartphone [3]. Cardiac monitoring systems with long-term data recording and telemetry will soon be completely unobtrusive. Instrumentation and telemetry of chronically ill patients will contribute to the deluge of information available to physicians. Low cost and unobtrusiveness will lead to greater adoption of personal medical monitoring technologies by the healthy, with or without physician involvement. Some of this technology may be surgically implanted. Many people today elect to have surgical implants for reasons of vanity. In the mid-twenty-first century, implants providing detailed personal health monitoring may be popular (see Fig. 39.4).

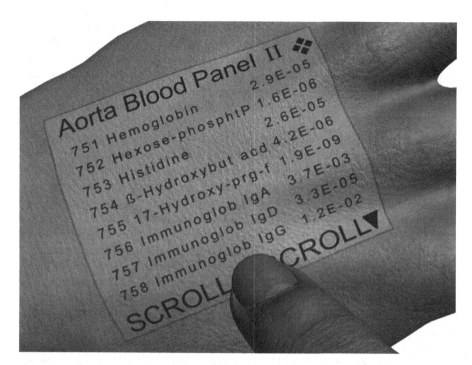

Fig. 39.4 Personal health monitoring in 2050? A dermal thin film display shows readings from implanted sensors. *Concept by Robert Freitas. Artwork by Gina Miller © www.nanogirl.com. Used with permission*

In the fictional television show, "Star Trek," doctors diagnosed by waving a small "tricorder" over a patient. In 2050 real diagnostics may consist of physicians or EMS personnel reading a small device held near a patient, receiving wireless telemetry of vital signs, clinical chemistries, and real-time molecular diagnostics from permanently implanted sensors. Power and substrate for analytical processes might be derived entirely in vivo.

Intelligence

With thousands of measured parameters, and even the patient's entire genome added to the mix, automated processing of diagnostic information will be a vital element of twenty-first century medicine. Computers will digest torrents of information into smaller streams of what is most clinically relevant. Raw measurements will be fed into sophisticated models of cell and organism physiology, drawing upon worldwide biomedical databases to construct the clinical picture of an individual patient.

Clinical decision support systems with computerized physician order entry are already part of medicine. Some of these systems generate automated medication selection and dose recommendations. With increasing complexity of the critical care environment, automation of dosing and administration based on process control feedback is foreseeable. In time-sensitive settings, physicians have already ceded dramatic interventions to automated systems. Implantable cardioverter-defibrillators are an example. The work of the critical care physician of the future may be analogous to that of a modern airline pilot giving direction to automation systems. Most of the flight time of modern airliners is not spent under the direct control of pilots.

Technology will allow the human intelligence of medicine to extend its reach. "eICU" telemedicine systems are already making inroads into critical care. Physicians using the RP-7 robotic telepresence system have run cardiac arrests from home [4]. Information will be also accessible to physicians via their personal computing and communication devices. In 2011 the US FDA approved the first app for image reading and mobile diagnoses by radiologists using the Apple iPhone and iPad [5]. On the patient side, the same personal monitoring technologies that would allow point-of-care "tricorder" readings of vital signs and clinical chemistries could be configured for remote telemetry via a patient's personal communication device. Some remote interventions will also be possible, such as adjustment of implanted therapeutic devices.

It is difficult to predict just how much human intelligence will be replaceable by computers in 2050. Early predictions of the progress of artificial intelligence (AI) made in the 1960s have not come to pass even though the world's most powerful computers now exceed the processing capacity of the human brain. Such capacity will exist on the desktop by 2025. However, human information processing capacity does not necessarily equal human intelligence.

Nowadays AI means expert systems adept at specialized information processing. The original aspiration of AI, the creation of human-like general intelligence in computers, is now called AGI (artificial general intelligence). A few small groups still pursue this objective, with the creation of computer programs capable of self-improvement viewed as an especially important milestone. A general intelligence capable of understanding and improving itself could theoretically lead to the rapid growth of entities with intelligence far greater than the human mind. This hypothetical development has been termed "the singularity." The timing and effects of such a development remain controversial.

In 2011 chat bots are able to pass superficial Turing Tests. Engines able to search the Internet and other deep databases using natural conversational language will be a reality by 2020 if not sooner. A powerful IBM computer named "Watson" beat the best human contestants in the American trivia quiz show, "Jeopardy," in 2011. Watson was designed to be an AI physician and has begun demonstrations in that role [6]. Watson's ability to rapidly digest electronic health records and analyze diagnostic images is expected to make radiology an especially fruitful role [7].

Conservatively, we can predict that by 2050 expert systems will exist that permit patients to discuss medical issues with computers using natural language. For health care professionals, computers with access to deep databases of medical records, journal articles, references works and models of physiology will be able to engage in sophisticated conversation about patients and treatment plans. Medicine will become a partnership between physician, patient, and machine.

NMEs per $B R&D spent (inflation adjusted)

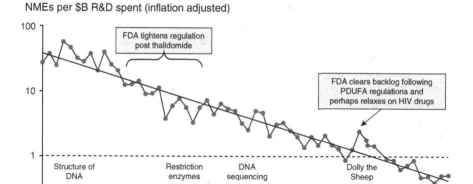

Fig. 39.5 New molecular entities (NMEs) per billion dollars of pharmaceutical R&D (inflation adjusted). *Used with permission from "Life sciences R&D: Changing the innovation equation in India,"* © *2011 courtesy of The Boston Consulting Group*

Intervention

The dramatic advances underway in the information and intelligence of medicine are driven by advances in the unregulated and highly competitive fields of computer software and microelectronics. Intervention is a different story. The number of new drugs (new molecular entities) brought to market per year has been flat since 1940, averaging about 20 a year in the USA [8]. Worse, the productivity of pharmaceutical research has been exponentially *decreasing*. Since the 1962 Kefauver Amendment to the US Food, Drug, and Cosmetic Act, the inflation-adjusted R&D cost to bring a new drug to market has doubled every 7.5 years in a sort-of Moore's Law in reverse [8] (see Fig. 39.5). Entire market sectors are being abandoned by pharmaceutical companies because they cannot afford the contemporary costs of drug development. There are some hopeful signs that the productivity decline may have bottomed out in 2006 [9]. Nevertheless, the present cost and regulatory burden of new drug development makes optimism about treatment progress via the traditional pharmaceutical development pipeline difficult. The days of medical device "hackers" like Walton Lillehei or Willem Kolff inventing new forms of life support in small shops with small budgets are also long gone.

Sociopolitical and business realities aside, the scientific prospects for new disease treatments and cures during the twenty-first century are bright. Detailed understanding of molecular pathways of disease and health will facilitate the development of biologics with greater therapeutic reach than mere enzyme or receptor binding agents. Regenerative medicine will rebuild damaged or defective tissue. Eventually the ability to build and control systems on the molecular scale will profoundly change the nature of medicine itself.

Interventions facilitated by continuing advances in electronics technology are easiest to predict. Surgery will continue to become less invasive as technology permits surgeons to do more work with their hands outside the patient. Microrobotic telepresence will open new frontiers of surgery. Natural orifice transluminal endoscopic surgery will permit some surgeries to be done without ever cutting skin. Brain-computer interface (BCI) technology will lead to prosthetic limb replacements that patients move and feel like their own limbs. Artificial retinas will advance in the twenty-first century as cochlear implants did in the twentieth. Noncortical blindness will be curable.

In the longer term, electronic fixes for sensory or motor deficits are just expensive stop-gap measures. Regenerative medicine, comprising injection of stem cells, transplantation of engineered tissue and organs, and induced regeneration of tissue, organs, and limbs will eventually render prosthetic devices obsolete. A possible exception may be devices that do more than biology can, such as implanted communication/computing devices interfaced directly to the brain. Routine use of such devices is possible by 2050.

Patients with artery disease may be among the early beneficiaries of regenerative medicine. Bone-marrow derived endothelial progenitor cells (EPCs) play a pivotal role in maintenance of vascular endothelium. There is evidence that EPCs prevent and even reverse the damage of atherosclerosis [10]. By the middle of this century, infusion of EPCs derived from rejuvenated pluripotent stem cells may be able to restore a patient's entire vascular endothelium to a youthful state. This one intervention could at once cure heart disease, cerebrovascular disease, peripheral artery disease and prevent at least some forms of dementia.

By the mid-twenty-first century, cancer should be comprehensively curable by biological therapies. It cannot be predicted what specific approaches will be used; however, rare cases of spontaneous remission are a proof-of-concept that malignancies can resolve immunologically. As a matter of physics, any cell that is molecularly distinct from other cells can in principle be identified and destroyed in vivo by technological means, albeit possibly very advanced means. In 2050, nearly a century after Richard Nixon made "the conquest of cancer is a national crusade," it would be wholly remarkable for cancer to remain a major medical problem.

The eventual conquest of cancer, artery disease, and other specific diseases of aging will increasingly expose the aging process itself as a cause of morbidity. Even if one escapes named diseases of aging, the physical and cognitive declines of "healthy" aging are immense and debilitating. It has been said that a pathogen that turned healthy 20-year-olds into healthy 80-year-olds would be thought worse than AIDS [11]. Without rejuvenation of underlying systems, even diseases of aging will just keep recurring like spot fires requiring constant attention. The economic costs would be unbearable. Apart from any normative questions about how long humans should live, if medicine is to avoid therapeutic nihilism it must eventually treat intrinsic biological aging. The economic and human consequences of treating everything but aging will be too severe to ignore.

What of critical care in 2050? Along with more data about what the immune system is doing, better pharmacologic and biologic tools for managing immune function should be available. SIRS may be stoppable in its tracks. Normothermic circulatory arrest of up to 20 min may be survivable without neurological deficit by modulating the postresuscitation inflammatory cascade and other deleterious sequela of reperfusion. Survivability of longer ischemic times will allow more time for placement of cardiac arrest victims on bypass, hypothermic surgical repair of exsanguinated trauma victims, and cerebrovascular interventions.

New life support tools will be available. ECMO and dialysis can presently support cardiopulmonary and renal functions for limited periods of time. In 2050, extracorporeal replacement for all vital organs may be available. Bioartificial life support equipment may consist of integrated cardiopulmonary, renal, hepatic, endocrine, nutritional, and even hematopoietic systems. The role of critical care will increasingly be seen as providing life support for the brain to permit repair or replacement of other organ systems by regenerative medicine as needed. The affordability of such care will strongly depend on the extent to which homeostasis can be automated, as it is in living systems. The distinction between mechanical and biological life support technologies will blur.

Ultimately, the difference between sickness and health and even life and death is a difference in arrangements of atoms and molecules. The final frontier of medicine is therefore detailed control of living systems at the molecular level. While there can be many ways to wield such control, the most powerful enabling technology will be the ability to construct machines with atomic precision. This ability is called molecular nanotechnology. The emerging field of nanomedicine foresees microscopic nanorobotic devices crafted for medical applications [12, 13]. Going beyond mere pharmacologic or biologic signaling of cells, nanorobotic devices could enter cells, even necrotic cells, and perform extensive structural and molecular repairs to restore a healthy state.

In recent years detailed scaling studies have been done of some particular nanorobotic devices. These devices include the respirocyte (an artificial erythrocyte with 200 times the oxygen carrying capacity of red cells) [14], the chromallocyte (a gene therapy vector designed to remove and replace the entire nuclear DNA content of target cells) [15], and the microbivore (an artificial phagocyte) [16]. While the capability to construct such devices lies decades or more in the future, the physical feasibility of anticipated functions can be analyzed today.

The microbivore is illustrative of the therapeutic reach of future nanomedical devices (see Fig. 39.6). The microbivore is an artificial phagocyte $3.4 \times 2.0 \times 2.0 \ \mu m$ in dimension, consisting of 610 billion precisely arranged structural atoms. Programmed to destroy specific pathogens, it would recognize target organisms on contact by species-specific reversible binding, and then ingest them. Inside the microbivore, an ingested pathogen is to be morcellated and enzymatically digested into harmless amino acids, mononucleotides, glycerol, free fatty acids, and simple sugars which are expelled 30 seconds after ingestion. A one terabot (10^{12}) dose of microbivores has been calculated to be able to cleanse the entire blood supply of a patient infected with 100 million CFU/mL bacteria (severe septicemia) in as little as ten minutes [16].

Fig. 39.6 The microbivore, a future nanomedical device for treating sepsis, shown ingesting a bacillus by extensible ciliary action. Perspective in this close-up view makes the device appear larger than red cells, although it is actually smaller. *Used with permission. Image © 2001 Zyvex Corp. and Robert A. Freitas Jr. (http://www.rfreitas.com). Designer Robert Freitas, additional design by Forrest Bishop. All Rights Reserved*

Devices this advanced may not exist as early as 2050, but they are in the direction that medicine is headed should technological progress continue. The future of medicine is the ability to restore and sustain life in a healthy state as we choose to define health on a molecular level. There would be many choices and time to make them.

References

1. New Semiconductors Sequence Human DNA. The State Column, 23 July 2011. (http://www.thestatecolumn.com/health/new-semiconductors-sequence-human-dna).
2. Espy MJ, et al. Real-time PCR in clinical microbiology: applications for routine laboratory testing. Clin Microbiol Rev. 2006;19:165–256.
3. Dolan B. FDA approves Mobisante's smartphone ultrasound. Mobihealthnews, 4 Feb 2011.
4. Harben J. 'The doctors is in' with RP-7 Robotic System. WWW.ARMY.MIL. 28 Sept. 2007.
5. Dilger DE. FDA approves iPad, iPhone radiology app for mobile diagnoses. AppleInsider, 4 Feb. 2011.
6. Murray P. Just months after Jeopardy! Watson wows doctors with medical knowledge. Singularity Hub blog. 6 June 2011.
7. Krishnaraj A. Will Watson replace radiologists? Diagnostic Imaging blog, 24 Feb. 2011.
8. Herper M. The decline of pharamaceutical research, measured in new drugs and dollars. Forbes blog, 27 June 2011.

9. McCormick T. Innovation upturn? New medical entities/Increasing! R&D Returns blog, 29 June 2011.
10. Dong C, Goldschmidt-Clermont PJ. Endothelial progenitor cells: a promising therapeutic alternative for cardiovascular disease. J Interv Cardiol. 2007;20:93–9.
11. Steven B. Harris, MD. Personal communication with geriatrician.
12. Freitas R. Nanomedicine, vol. I: basic capabilities. Landes Bioscience; 1999.
13. Freitas R. Nanomedicine, vol. IIA: biocompatability. Landes Bioscience; 2003.
14. Freitas R. Exploratory design in medical nanotechnology: a mechanical artificial red cell. Artif Cells Blood Substit Immobil Biotechnol. 1998;26:411–30.
15. Freitas R. The ideal gene delivery vector: chromallocytes, cell repair nanorobots for chromosome replacement therapy. J Evolut Technol. 2007;16:1–97.
16. Freitas R. Microbivores: artificial mechanical phagocytes using digest and discharge protocol (2001). http://www.rfreitas.com/Nano/Microbivores.htm.

Afterword

This volume is the third in a series[1] examining consumer demand for ineffective intensive care—medical treatment that serves only to prolong a death spiral but is insisted upon by optimistic surrogates in the face of contrary advice by expert providers of intensive care.

Patients and their families demanding expensive and ill-advised medical care? How can this be, when evidence-based medicine ensures that prognoses are more accurate than at any other time in history?

There are several stock replies to the question of individuals demanding disproportionate health care resources at the expense of the whole population:

1. The problem can always be solved by better communication.
2. This demand doesn't occur often enough to justify the public relations troubles that providers would experience if a protocol limiting such care were instituted.
3. Even when this excessive use of medical resources does occur, the money spent is inconsequential compared to the total expenditures for health care.
4. Physicians do not have an admirable track record of predicting death; there is always the potential for unexpected survival.
5. Health care consumers have a right to use as many resources as they wish, and society must foot the bill no matter the amount because this is the right thing to do in a prosperous society.
6. Saying no is synonymous with "death panels" in terms of hastening the deaths of old, sick people to save money.

The mission of this volume was to address these issues in a multinational forum. Critical care physicians from ten countries reported on their resource allocation before the global financial crash of 2007–2008 and on their plans for resource allocation

---

[1] The previous two books are entitled Crippen D, Kilcullen JK, Kelly DF (eds), Three Patients: International Perspective on Intensive Care at the End of Life (Boston, MA, Kluwer Academic, 2002); and Crippen D (ed), End-of-Life Communication in the ICU: A Global Perspective (New York, Springer, 2008).

D.W. Crippen (ed.), *ICU Resource Allocation in the New Millennium: Will We Say "No"?*,   335
DOI 10.1007/978-1-4614-3866-3, © Springer Science+Business Media New York 2013

in the following decade. Among these ten countries was the world's largest consumer of medical services, in both private practice and academic medicine: the United States.

Data on use of medical resources in the non-American countries represented in this book are no surprise. These countries endeavor to provide health care for all or most of their populations using limited resources. In so doing, they must prioritize health care expenses, just as other national budget items are prioritized. Some enforceable form of rationing is used to fund universal health care for their citizens. These multinational data also suggest that the funds for all national services will diminish in the future, and collective belts will have to be tightened to maintain health care coverage.

The United States, on the other hand, indemnifies only a portion of its population, and at a high cost that increases yearly. The United States stands apart in another way as well: it is unwilling or unable to say no to inappropriate use of funds for the most expensive care rendered in medical facilities—intensive care.

The expert analysts considering these issues for this volume mostly arrived at conclusions that have been expressed for years. Nearly every contributor to this book agreed that there is a problem with inappropriate demands for ill-advised medical services, especially at the end of life. The proposed remedies run the gamut from better communication to various means of creative arm-twisting, but all the authors stopped short of advocacy of saying no as an enforceable policy.

Dr. Sprung suggested that most recalcitrant surrogates will come around in time if worn down by the usual process, but he stopped short of advocacy of saying no as a matter of enforceable policy. He wrote: "Doctor's actions not honoring patient or family requests especially if based more on society's needs than those of the individual patient will undermine trust in the medical profession." Drs. Rie and Chalfin described a health care system as it would be in a perfect world but stopped short of describing objective means to regulate it. Drs. Kuiper and Hollenberg suggested that it is appropriate to have a floor and a ceiling for medical care services, but they stopped short of advocacy of saying no as an effective means of achieving that goal.

All the contributors to this volume agreed that some regulatory means are necessary to maintain the integrity of a health care system, but none expressed a desire to look unreasonable patients and their families in the eye and say no. Several authors related that they do say no on occasion but always after the surrogates finally agree to stopping ineffective intensive care. Saying no in the face of adamant refusal by surrogates is another matter entirely, and few if any physicians will back down in such situations. Many physicians are concerned about threats of legal action by surrogates. Mr. Ross suggested that such lawsuits are unlikely but hastened to add that there is no solid legal precedent protecting providers from legal threats.

The hypothetical Fair and Equitable Health Care Act (FEHCA) was created for this volume to allow experts in the field of medical resource provision to evaluate and criticize a system that prioritizes health care coverage by setting enforceable limits. This plan was formulated to offer the most benefit to the most consumers and is sensitive not to their desires or demands but to needs that are amenable to indemnification.

Dr. Kilcullen suggested that the time has come to consider plans like the FEHCA as cost-effective methods of maximizing care and avoiding bankruptcy, and that such plans will have to be accepted, regardless of public expectations. Drs. Rie and Kofke proposed that the plan does not go far enough in its provisions, will not be accepted by the public, and will be watered down by the political process.

Dr. Whetstine, ethicist, suggested that saying no does not violate traditional ethical canons but would be a significant practical problem for a society that has become accustomed to entitlement. Expected customer service may trump resource allocation.

These criticisms are remarkably similar to criticisms of the Simpson–Bowles plan[2] to spread the pain of cuts in the US budget. Everyone agreed cuts needed to be made, no one agreed on how to accomplish this, and so the economy continues its headlong rush to disaster.

The potential for financial collapse looms large for the rest of the world as well. Drs. Kuiper and Hollenberg suggested that the global village will adapt to the availability of fewer resources by accepting that universal health care does not mean coverage of all care desired by the consumer. Longer queues and limitations on the use of some resources because of cost are inevitable.

The resource pie has a finite volume and can be cut a finite number of ways. Fairness dictates that everyone has access to the pie, but the size of the slices cannot be dictated by consumer desire. Saying no will not only be effective for cost control, but also it will be mandatory. Drs. Kuiper and Hollenberg stopped short, however, of detailing precisely how decisions will be made about who will be covered for what, arguing that such issues depend on local considerations.

Traditionally in the United States, escalating health care costs have been accommodated through increased charges on the national credit card, avoidance of payment for medical services after they have been performed, and decreases in the pool of consumers by means of pricing care beyond their ability to pay. All these rationing devices became technically obsolete when the American Congress passed the Patient Protection and Affordable Care Act in March 2010.[3]

Among other provisions, this legislation ends the private insurance industry's propensity to discriminate against sicker patients, and will indemnify 96 % of Americans in a portable manner. This means a lot more resources will be needed to support the expected additional 31 million medical consumers in an already overheated provision system.

Dr. Angus suggested that under such a system, American physicians will use technologically advanced command centers to direct cost-efficient bedside-care managers in the form of nurse practitioners. This approach will solve part of the problem, but the issue of limits and saying no continues to be avoided, and will not be resolved quietly.

---

[2] http://www.fiscalcommission.gov/sites/fiscalcommission.gov/files/documents/TheMomentofTruth12_1_2010.pdf.

[3] http://www.gpo.gov/fdsys/pkg/BILLS-111hr3590enr/pdf/BILLS-111hr3590enr.pdf.

The current refusal by the United States Congress to cut spending and raise taxes will mean a static or decreasing amount of funds for rising health care expenses. Back in the day, the supply of clinical resources was limited by the ability of consumers to pay for it. Among other things, this meant the need for health care was provided by "teaching hospitals," the likes of Bellevue, Cook Country, Grady, and Charity hospitals dedicated to serving the public at large cheaply and training house staff in the process. There were large 30-bed wards, no creature comforts, TV sets, physical or occupational therapy, and only rudimentary rehabilitation.

Patients were given the benefit of an augmented healing process for treatable diseases commensurate with cost constraints. Technology was sparse. Survival of the fittest ruled. If a patient became septic, or developed respiratory failure, they died. There were few stragglers to land in ICUs for months of expensive care to produce a relatively few survivors. Then with the advent of Medicare and Medicaid in 1965, money became available to finance the medical–industrial revolution, and survivors of formerly fatal diseases dramatically increased, as did the cost to get them there.

Now as we enter the probable inability to finance the inverse pyramid of cost, the reverse scenario is likely to occur. Expensive programs serving the fewest patients will become extremely difficult to fund, elective resources that expensively promote patient comfort will dry up. The attitude that health care providers must necessarily be well rested and comfortable may as well. It isn't out of the question that we may return to 30 bed wards and patients' families carrying out half the nursing load.

Private insurance companies compelled to suspend rating risks for illnesses will stop writing insurance, forcing indemnification onto the government. The tendency of the government to reimburse only a portion of providers' expenses will result in more hospitals going out of business and more providers refusing to deal with government-insured patients. This downward spiral can be slowed by cost-efficient measures that sidestep the difficult issue of saying no, but such measures will only prolong the problem of too much demand and not enough resources.

Our inability or unwillingness to say no is an immovable object, and our inability to fund unlimited demand is an irresistible force. There is a strong likelihood that American health care will burst in totally unpredicted ways as the irresistible force meets the immovable object. We will all be observers of this process in our lifetime.

Index